Mobile Crane Operations

Level One

Trainee Guide

Pearson

Boston Columbus Indianapolis New York San Francisco Amsterdam
Cape Town Dubai London Madrid Milan Munich Paris Montreal Toronto Delhi
Mexico City Sao Paulo Sydney Hong Kong Seoul Singapore Taipei Tokyo

NCCER

President: Don Whyte
Vice President: Steve Greene
Chief Operations Officer: Katrina Kersch
Mobile Crane Operations Curriculum Project Manager:
  Chris Wilson
Senior Development Manager: Mark Thomas

Senior Production Manager: Tim Davis
Quality Assurance Coordinator: Karyn Payne
Desktop Publishing Coordinator: James McKay
Permissions Specialists: Kelly Sadler
Production Specialist: Kelly Sadler
Editors: Graham Hack, Debie Hicks

Writing and development services provided by Topaz Publications, Liverpool, NY

Lead Writer/Project Manager: Troy Staton
Desktop Publisher: Joanne Hart
Art Director: Alison Richmond

Permissions Editor: Andrea LaBarge
Writers: Troy Staton, Terry Egolf

Pearson

Director of Alliance/Partnership Management: Andrew Taylor
Editorial Assistant: Collin Lamothe
Program Manager: Alexandrina B. Wolf
Assistant Content Producer: Alma Dabral
Digital Content Producer: Jose Carchi
Director of Marketing: Leigh Ann Simms

Senior Marketing Manager: Brian Hoehl
Composition: NCCER
Printer/Binder: LSC Communications
Cover Printer: LSC Communications
Text Fonts: Palatino and Univers

Credits and acknowledgments for content borrowed from other sources and reproduced, with permission, in this textbook appear at the end of each module.

ISBN-13:   978-0-13-473825-3
ISBN-10:   0-13-473825-X

Preface

To the Trainee

Mobile cranes are found on almost every construction job site and in every facet of the construction industry. Residential and commercial construction companies, industrial and petrochemical plants, shipbuilders, and highway and bridge construction companies all use mobile cranes to do the heavy lifting.

Mobile crane operators use cranes to move heavy loads around a job site with precision. They are required to inspect and make adjustments to the crane. They use gauges and monitors to understand the stresses on the machinery and to maintain safety during a lift. Of course, safety is of paramount importance when using such large machines to move heavy loads. Training as a mobile crane operator, therefore, is extremely important to the safety of those who will work around a lifting site.

The need for mobile crane operators will continue to increase in all industries as experienced operators retire. Those who meet the qualifications under OSHA 29 *CFR* 1926, Subpart CC, *Cranes and Derricks in Construction* will continue to be in high demand in this field that the US Department of Labor expects to experience growth over the next several years.

Finally, this edition of *Mobile Crane Operations* also incorporates NCCER's *Basic Rigger* and *Signal Person* training. Trainees who complete the *Core Curriculum* and *Mobile Crane Operations Level One* will also now earn a *Basic Rigger* and *Signal Person* credential by successfully completing all of the modules.

New with *Mobile Crane Operations*

NCCER is pleased to release *Mobile Crane Operations* in full color, with new photographs and figures. This version is now presented in NCCER's improved instructional systems design, in which the sections of each module are directly tied to learning objectives.

The training in *Mobile Crane Operations* is more thorough with this edition. The "Crane Communications" module offers NCCER's complete *Signal Person* training module, which includes all OSHA hand signals. The safety information has been greatly expanded with the inclusion of "Crane Safety and Emergency Procedures". The "Rigging Practices" module has been greatly expanded to include additional training on rigging hardware, hoisting equipment, and jacks. Finally, each module has additional study questions to help you progress through the training and onward towards your mobile crane operator certification.

We wish you success as you progress through this training program. If you have any comments on how NCCER might improve upon this textbook, please complete the User Update form located at the back of each module and send it to us. We will always consider and respond to input from our customers.

We invite you to visit the NCCER website at **www.nccer.org** for information on the latest product releases and training, as well as online versions of the *Cornerstone* magazine and Pearson's NCCER product catalog.

Your feedback is welcome. You may email your comments to **curriculum@nccer.org** or send general comments and inquiries to **info@nccer.org**.

NCCER Standardized Curricula

NCCER is a not-for-profit 501(c)(3) education foundation established in 1996 by the world's largest and most progressive construction companies and national construction associations. It was founded to address the severe workforce shortage facing the industry and to develop a standardized training process and curricula. Today, NCCER is supported by hundreds of leading construction and maintenance companies, manufacturers, and national associations. The NCCER Standardized Curricula was developed by NCCER in partnership with Pearson, the world's largest educational publisher.

Some features of the NCCER Standardized Curricula are as follows:

- An industry-proven record of success
- Curricula developed by the industry, for the industry
- National standardization providing portability of learned job skills and educational credits
- Compliance with the Office of Apprenticeship requirements for related classroom training (*CFR 29:29*)
- Well-illustrated, up-to-date, and practical information

NCCER also maintains the NCCER Registry, which provides transcripts, certificates, and wallet cards to individuals who have successfully completed a level of training within a craft in NCCER's Curricula. *Training programs must be delivered by an NCCER Accredited Training Sponsor in order to receive these credentials.*

Special Features

In an effort to provide a comprehensive and user-friendly training resource, this curriculum showcases several informative features. Whether you are a visual or hands-on learner, these features are intended to enhance your knowledge of the construction industry as you progress in your training. Some of the features you may find in the curriculum are explained below.

Introduction

This introductory page, found at the beginning of each module, lists the module Objectives, Performance Tasks, and Trade Terms. The Objectives list the knowledge you will acquire after successfully completing the module. The Performance Tasks give you an opportunity to apply your knowledge to real-world tasks. The Trade Terms are industry-specific vocabulary that you will learn as you study this module.

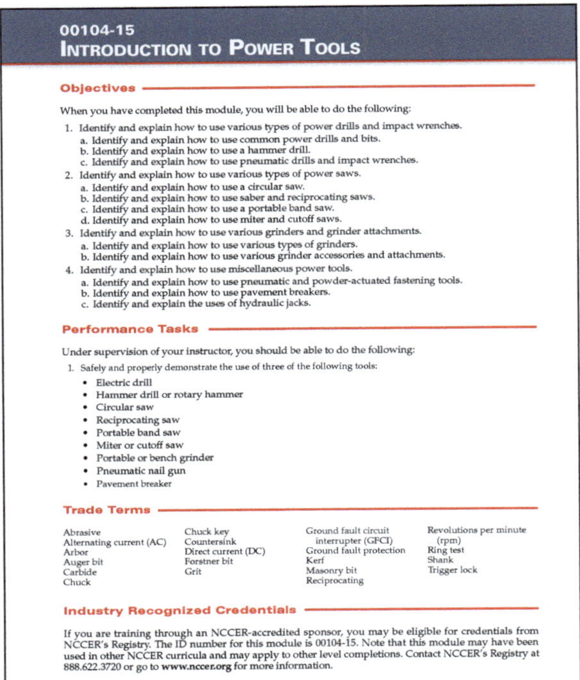

Figures and Tables

Photographs, drawings, diagrams, and tables are used throughout each module to illustrate important concepts and provide clarity for complex instructions. Text references to figures and tables are emphasized with *italic* type.

Notes, Cautions, and Warnings

Safety features are set off from the main text in highlighted boxes and categorized according to the potential danger involved. Notes simply provide additional information. Cautions flag a hazardous issue that could cause damage to materials or equipment. Warnings stress a potentially dangerous situation that could result in injury or death to workers.

Trade Features

Trade features present technical tips and professional practices based on real-life scenarios similar to those you might encounter on the job site.

Bowline Trivia

Some people use this saying to help them remember how to tie a bowline: "The rabbit comes out of his hole, around a tree, and back into the hole."

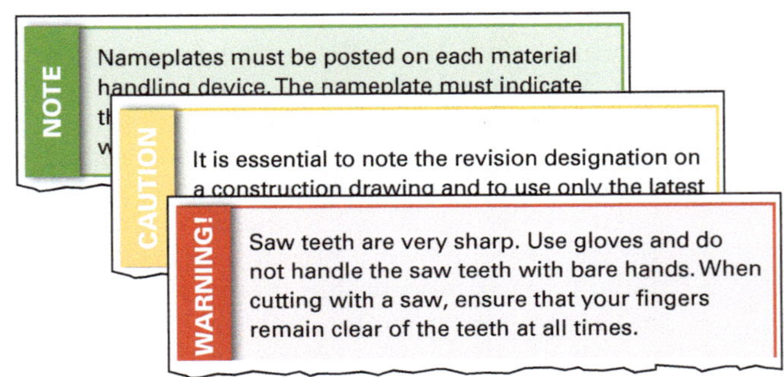

Case History

Case History features emphasize the importance of safety by citing examples of the costly (and often devastating) consequences of ignoring best practices or OSHA regulations.

Case History

Requesting an Outage

An electrical contractor requested an outage when asked to install two bolt-in, 240V breakers in panels in a data processing room. It was denied due to the 24/7 worldwide information processing hosted by the facility. The contractor agreed to proceed only if the client would sign a letter agreeing not to hold them responsible if an event occurred that damaged computers or resulted in loss of data. No member of upper management would accept liability for this possibility, and the outage was scheduled.

The Bottom Line: If you can communicate the liability associated with an electrical event, you can influence management's decision to work energized.

Going Green

Going Green features present steps being taken within the construction industry to protect the environment and save energy, emphasizing choices that can be made on the job to preserve the health of the planet.

GOING GREEN

Reducing Your Carbon Footprint

Many companies are taking part in the paperless movement. They reduce their environmental impact by reducing the amount of paper they use. Using email helps to reduce the amount of paper used,

Did You Know

Did You Know features introduce historical tidbits or interesting and sometimes surprising facts about the trade.

Did You Know?

Safety First

Safety training is required for all activities. Never operate tools, machinery, or equipment without prior training. Always refer to the manufacturer's instructions.

Step-by-Step Instructions

Step-by-step instructions are used throughout to guide you through technical procedures and tasks from start to finish. These steps show you how to perform a task safely and efficiently.

Perform the following steps to erect this system area scaffold:

Step 1 Gather and inspect all scaffold equipment for the scaffold arrangement.

Step 2 Place appropriate mudsills in their approximate locations.

Step 3 Attach the screw jacks to the mudsills.

Trade Terms

Each module presents a list of Trade Terms that are discussed within the text and defined in the Glossary at the end of the module. These terms are presented in the text with **bold, blue** type upon their first occurrence. To make searches for key information easier, a comprehensive Glossary of Trade Terms from all modules is located at the back of this book.

During a rigging operation, the **load** being lifted or moved must be connected to the apparatus, such as a crane, that will provide the power for movement. The connector—the link between the load and the apparatus—is often a sling made of synthetic, chain, or **wire rope** materials. This section focuses on three types of slings:

Section Review

Each section of the module wraps up with a list of Additional Resources for further study and Section Review questions designed to test your knowledge of the Objectives for that section.

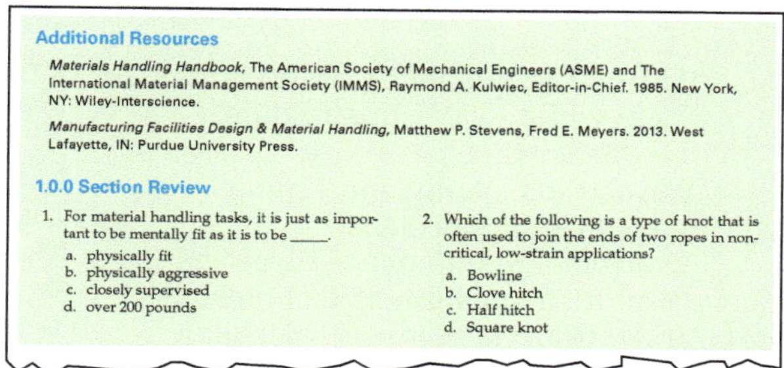

Review Questions

The end-of-module Review Questions can be used to measure and reinforce your knowledge of the module's content.

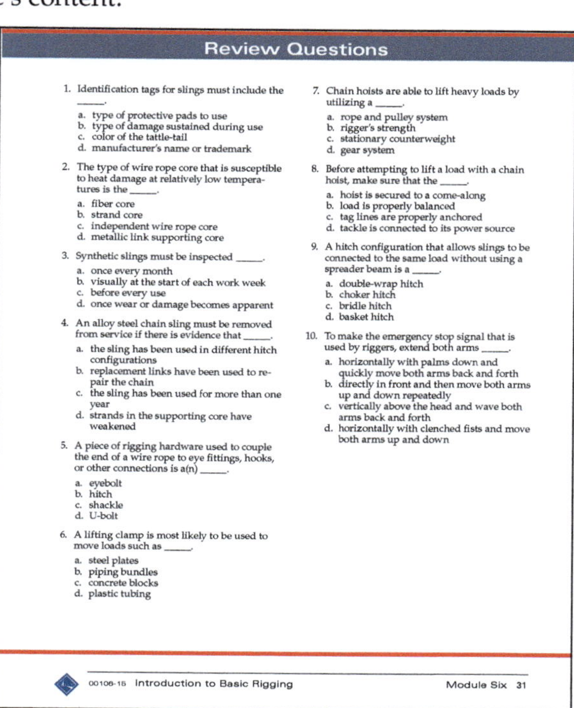

NCCER Standardized Curricula

NCCER's training programs comprise more than 80 construction, maintenance, pipeline, and utility areas and include skills assessments, safety training, and management education.

Boilermaking
Cabinetmaking
Carpentry
Concrete Finishing
Construction Craft Laborer
Construction Technology
Core Curriculum: Introductory
 Craft Skills
Drywall
Electrical
Electronic Systems Technician
Heating, Ventilating, and Air
 Conditioning
Heavy Equipment Operations
Heavy Highway Construction
Hydroblasting
Industrial Coating and Lining
 Application Specialist
Industrial Maintenance Electrical
 and Instrumentation Technician
Industrial Maintenance Mechanic
Instrumentation
Ironworking
Manufactured Construction
 Technology
Masonry
Mechanical Insulating
Millwright
Mobile Crane Operations
Painting
Painting, Industrial
Pipefitting
Pipelayer
Plumbing
Reinforcing Ironwork
Rigging
Scaffolding
Sheet Metal
Signal Person
Site Layout
Sprinkler Fitting
Tower Crane Operator
Welding

Maritime

Maritime Industry Fundamentals
Maritime Pipefitting
Maritime Structural Fitter

Green/Sustainable Construction

Building Auditor
Fundamentals of Weatherization
Introduction to Weatherization
Sustainable Construction
 Supervisor
Weatherization Crew Chief
Weatherization Technician
Your Role in the Green
 Environment

Energy

Alternative Energy
Introduction to the Power Industry
Introduction to Solar Photovoltaics
Power Generation Maintenance
 Electrician
Power Generation I&C
 Maintenance Technician
Power Generation Maintenance
 Mechanic
Power Line Worker
Power Line Worker: Distribution
Power Line Worker: Substation
Power Line Worker: Transmission
Solar Photovoltaic Systems Installer
Wind Energy
Wind Turbine Maintenance
 Technician

Pipeline

Abnormal Operating Conditions,
 Control Center
Abnormal Operating Conditions,
 Field and Gas
Corrosion Control
Electrical and Instrumentation
Field and Control Center
 Operations
Introduction to the Pipeline
 Industry
Maintenance
Mechanical

Safety

Field Safety
Safety Orientation
Safety Technology

Supplemental Titles

Applied Construction Math
Tools for Success

Management

Construction Workforce
 Development Professional
Fundamentals of Crew Leadership
Mentoring for Craft Professionals
Project Management
Project Supervision

Spanish Titles

Acabado de concreto: nivel uno
 (*Concrete Finishing Level One*)
Aislamiento: nivel uno
 (*Insulating Level One*)
Albañilería: nivel uno
 (*Masonry Level One*)
Andamios (*Scaffolding*)
Carpintería: Formas para
 carpintería, nivel tres
 (*Carpentry: Carpentry Forms, Level
 Three*)
Currículo básico: habilidades
 introductorias del oficio
 (*Core Curriculum: Introductory Craft
 Skills*)
Electricidad: nivel uno
 (*Electrical Level One*)
Herrería: nivel uno
 (*Ironworking Level One*)
Herrería de refuerzo: nivel uno
 (*Reinforcing Ironwork Level One*)
Instalación de rociadores: nivel uno
 (*Sprinkler Fitting Level One*)
Instalación de tuberías: nivel uno
 (*Pipefitting Level One*)
Instrumentación: nivel uno, nivel
 dos, nivel tres, nivel cuatro
 (*Instrumentation Levels One through
 Four*)
Orientación de seguridad
 (*Safety Orientation*)
Paneles de yeso: nivel uno
 (*Drywall Level One*)
Seguridad de campo
 (*Field Safety*)

Acknowledgments

This curriculum was revised as a result of the farsightedness and leadership of the following sponsors:

ABC Pelican Chapter
Bay Ltd.
Bo-Mac Contractors, Ltd.
Cowboyscranes.com
Exelon Generation
Jacobs Field Services

KBR, Inc.
Kelley Construction
Mammoet USA
North American Crane Bureau
Orion Marine Group
Southland Safety

This curriculum would not exist were it not for the dedication and unselfish energy of those volunteers who served on the Authoring Team. A sincere thanks is extended to the following:

Ed Burke
Robert Capelli
Anthony Johnson
Richard Laird
Don McDonald

Timothy Prakop
Larry "Cowboy" Proemsey
Joseph Watts
Harold Williamson

A sincere thanks is also extended to the dedication and assistance provided by the following technical advisors:

Grant Chutz

Keith Denham

Frank Jones

NCCER Partners

American Council for Construction Education
American Fire Sprinkler Association
Associated Builders and Contractors, Inc.
Associated General Contractors of America
Association for Career and Technical Education
Association for Skilled and Technical Sciences
Construction Industry Institute
Construction Users Roundtable
Design Build Institute of America
GSSC – Gulf States Shipbuilders Consortium
ISN
Manufacturing Institute
Mason Contractors Association of America
Merit Contractors Association of Canada
NACE International
National Association of Women in Construction
National Insulation Association
National Technical Honor Society
National Utility Contractors Association
NAWIC Education Foundation
North American Crane Bureau
North American Technician Excellence
Pearson

Prov
SkillsUSA®
Steel Erectors Association of America
U.S. Army Corps of Engineers
University of Florida, M. E. Rinker Sr., School of
  Construction Management
Women Construction Owners & Executives,
  USA

Contents

This course map shows all of the modules in *Mobile Crane Operations Level One*. The suggested training order begins at the bottom and proceeds up. Skill levels increase as you advance on the course map. The local Training Program Sponsor may adjust the training order.

Orientation to the Trade

OVERVIEW

Mobile cranes provide services on construction jobsites that cannot be provided by any other type of equipment. Any number of hazards can be created when a heavy load is lifted into the air by mechanical means. As a result, the mobile crane operator assumes a great deal of responsibility and must be competent. This module introduces the mobile crane operator craft as well as the basics of craft apprenticeships.

Module 21101

Objectives

When you have completed this module, you will be able to do the following:

1. Describe the mobile crane trade and related career opportunities.
   a. Describe how mobile cranes are used and identify applicable standards.
   b. Identify mobile crane-related career opportunities and operator responsibilities.
   c. Describe qualified and certified crane operators.
2. Describe NCCER standardized training, apprenticeships, and the desirable characteristics of a craft professional.
   a. Explain how NCCER training programs are structured.
   b. Describe apprenticeship programs.
   c. Describe personal characteristics common to successful craft professionals.

Performance Tasks

This is a knowledge-based module; there are no Performance Tasks.

Trade Terms

Accreditation
American National Standards Institute (ANSI)
American Society of Mechanical Engineers (ASME)
Apprenticeship
Certification
Lift director
Load rating chart
Master Trainer
National consensus standard

North American Crane Bureau, Inc. (NACB)
Occupational Safety and Health Administration (OSHA)
Practical examiner
Qualified person
Rigger
Site supervisor
Tilt-up construction
Training credentials

Industry Recognized Credentials

If you are training through an NCCER-accredited sponsor, you may be eligible for credentials from NCCER's Registry. The ID number for this module is 21101. Note that this module may have been used in other NCCER curricula and may apply to other level completions. Contact NCCER's Registry at 888.622.3720 or go to **www.nccer.org** for more information.

Contents

Figures

1.0.0 THE MOBILE CRANE OPERATOR TRADE

Objective

Describe the mobile crane trade and related career opportunities.

a. Describe how mobile cranes are used and identify applicable standards.
b. Identify mobile crane-related career opportunities and operator responsibilities.
c. Describe qualified and certified crane operators.

Trade Terms

Accreditation: In the context of crane-operator certification, the act of granting credit or recognition to an organization or institution that maintains suitable standards of testing for the certification process.

American National Standards Institute (ANSI): An organization that oversees the creation, promulgation, and use of thousands of norms and guidelines that directly impact businesses. ANSI also provides third-party accreditation to products and personnel certification programs, recognizing the competence of organizations to carry out these activities in accordance with requirements defined in international standards.

American Society of Mechanical Engineers (ASME): ASME promotes the art, science, and practice of multidisciplinary engineering and allied sciences around the globe through continuing education, training and professional development, codes and standards, research, conferences and publications, government relations, and other forms of outreach. The organization is therefore an engineering society as well as a standards-developing organization.

Certification: Refers to a candidate's successful completion of a written test of knowledge as well as a practical examination. Certification testing must be based on the equipment capacity and type that will be operated, and can be administered by an accredited third-party organization (such as NCCER) or through an audited testing program operated by the employer.

Lift director: As defined in ASME Standard B30.5, the lift director directly oversees the work being performed by a crane and the associated rigging and signaling crew. The lift director is responsible for addressing all hazards related to the lift as well as those brought forward by the crane operator, signal persons, riggers, or other members of the crew.

Load rating chart: One or more charts associated with each individual crane model that specify the crane's capacity at various distances and angles, and in various configurations. Per *ASME Standard B30.5*, Section 5.1.1.3, load charts should include a "full and complete range of manufacturer's crane load ratings at all stated operating radii, boom angles, work areas, boom lengths and configurations, jib lengths and angles (or offset), as well as alternate ratings for the use and nonuse of optional equipment on the crane, such as outriggers and extra counterweights, which affect ratings."

Master Trainer: An individual certified in accordance with NCCER's Master Trainer Instructor Certification Training Program (MTICTP) and authorized to train NCCER Craft Instructors, Performance Evaluators, and Proctors.

National consensus standard: As defined in 29 *CFR* 1910.2(g), "any standard or modification thereof which (1) has been adopted and promulgated by a nationally recognized standards-producing organization under procedures whereby it can be determined by the Secretary of Labor or by the Assistant Secretary of Labor that persons interested and affected by the scope or provisions of the standard have reached substantial agreement on its adoption, (2) was formulated in a manner which afforded an opportunity for diverse views to be considered, and (3) has been designated as such a standard by the Secretary or the Assistant Secretary, after consultation with other appropriate Federal agencies."

North American Crane Bureau, Inc. (NACB): NACB's responsibility is to provide the most comprehensive services available in crane training and inspection services, while maintaining current state-of-the-art safety information applicable to crane operation. NACB is committed to providing the industry with a standard of excellence unparalleled in the field of safety. A division of the NACB, NACB Interactive Educational Systems, Inc., provides and distributes crane training simulators. NCCER and NACB work as partners in support of NCCER's Mobile Crane Operator Certification Program.

The mobile crane may be the most powerful piece of equipment in the construction industry. With maximum mast heights that can approach 1,000 feet (305 meters) and equally massive lift capacities, mobile cranes are essential in building and maintaining bridges, highways, buildings, pipelines, ships, and towers. Mobile crane operation requires physical coordination, stamina, focus, and concentration. A mobile crane operator must also be able to read a load rating chart and comprehend various forms of nonverbal communication. An understanding of physical concepts, such as balance and leverage, is also essential.

Operating a mobile crane can be exciting and challenging. It can also be hazardous and stressful, as a tremendous amount of power and responsibility lie in the hands of the operator (*Figure 1*). The key to being a successful crane operator is to be well-trained and well-practiced. All crane operators, regardless of the size or capacity of the machine under their control, should aspire to understand and use the best practices of the industry.

1.1.0 What Mobile Cranes Do

A mobile crane is a machine designed to raise, move, and lower heavy weights. In many cases, a load must be raised and placed in a different place. There are three basic types of cranes: mobile cranes, tower cranes, and overhead cranes. Tower cranes can be permanent or temporary structures but are stationary and typically integrated into the structure they are being used to build. Overhead cranes move along fixed lengths of track and include bridge cranes and gantry cranes. Mobile cranes are usually mounted on trucks or crawlers, but they may also be mounted on railcars, boats, or barges (*Figure 2*). The main distinguishing feature of mobile cranes is that they are equipped with a means of propulsion and/or travel. Cranes do not need to be designed for highway travel to be considered mobile. Many mobile cranes cannot be driven on the open road.

In bridge and highway construction, mobile cranes are used to lift large loads into place and to drive pilings. *Figure 3* shows a mobile crane being used to drive pilings for an overpass using a diesel pile hammer. Mobile cranes are also used in building construction, particularly for tilt-up construction. In tilt-up construction, concrete wall panels are formed at the site and then lifted into place using a crane (*Figure 4*).

Mobile cranes are also used to lift heavy objects, such as heating, ventilating, and air conditioning (HVAC) units onto the tops of buildings. The energy industry uses mobile cranes to build and maintain power plants, transmission towers, pipelines, and oil rigs. The list of applications for mobile cranes is virtually endless.

1.1.1 Relevant Standards

Mobile crane operations are primarily governed by detailed standards from two different organizations. The Occupational Safety and Health Administration (OSHA) publishes safety standards for the industry that are of great importance in the United States. The most relevant standards are 29 *CFR 1910.180, Crawler Locomotive and Truck Cranes*, and 29 *CFR 1926.1400*, Subpart CC,

Figure 1 The Manitowoc 31000 crawler crane shown on the right has a nominal capacity of 2,300 tons (2,086 metric tons).

Figure 2 Crawler crane on a barge.

Figure 3 A mobile crane being used to drive pilings.

Cranes and Derricks in Construction. The 29 *CFR* 1910 standard is related to general industry use, while the 29 *CFR* 1926 standard governs the construction industry. OSHA standards are enforceable by law and must be considered mandatory.

The **American Society of Mechanical Engineers (ASME)** was created to "serve diverse global communities by advancing, disseminating and applying engineering knowledge for improving the quality of life." Over the years, ASME has developed many standards that provide guidance to a variety of industries. *ASME Standard B30.5, Mobile and Locomotive Cranes* is a **national consensus standard** directly related to mobile cranes. However, ASME is not a function of national government.

ASME Standard B30.5 is divided into four chapters:

- *Chapter 5-0, Scope, Definitions, Personnel Competence, and References* – The topics clearly identified in the chapter title are included here.
- *Chapter 5-1, Construction and Characteristics* – This section describes how cranes should be built and assembled.
- *Chapter 5-2, Testing, Inspection, and Maintenance* – This section describes how to inspect and maintain a crane to help ensure safe and optimal performance.
- *Chapter 5-3, Operation* – This section explains how a crane is to be set up and operated.

Figure 4 Lifting a tilt-up panel.

Note that there are other ASME standards in the B30 series that are indirectly related to mobile cranes. For example, *ASME Standard B30.9, Slings,* covers the construction and usage of slings used for rigging.

Mobile crane operators must meet certain physical, academic, and performance standards as outlined in the OSHA and ASME standards. These standards are designed to contribute to the safe operation of all mobile cranes and ensure that the operator is capable of operating the crane in every situation.

These standards and the guidance they provide will be discussed in greater detail as this training progresses.

1.2.0 Opportunities and Responsibilities

Mobile crane operators have many different career opportunities available to them. Cranes are an integral part of construction in every conceivable environment. The construction of large wind turbines, for example, requires a large crane, several other cranes to assist, and sophisticated rigging plans (*Figure 5*). These projects take place in remote areas. While operating a mobile crane places a great amount of responsibility on the operator, the craft offers exciting and rewarding career paths.

Given the complexity of operating a crane, prospective operators or operators-in-training spend a significant amount of their time on the job assisting with the setup, operation, and transportation of a mobile crane without actually operating the crane. These relevant tasks are just as important to a successful lift as the lift itself. Once a person is certified and experienced, many additional opportunities become available. These include acting as a trainer, classroom instructor, supervisor, or consultant. Some of these opportunities require additional training and certification, and may even seem out of reach at this point in your training. However, they are absolutely achievable with hard work and dedicated service to the career field.

The mobile crane operator has many responsibilities on the jobsite. Mobile cranes are very sophisticated machines and the projects in which cranes are used have become more complex. This means that crane operators must be fully aware of their surroundings, their own physical condition, and any emotional or mental stress that could distract them from the task at hand.

Safety must always be the primary concern of the mobile crane operator. To prevent accidents, the operator must be able to recognize potential hazards and be both willing and able to stop the operation until the hazard is neutralized. According to 29 *CFR* 1926.1418, the crane operator must always have the authority to stop and refuse to handle loads until a qualified person has determined a safe condition exists. Each operator must be willing to exercise this authority when conditions demand it.

Large Telescoping Cranes

One of the heavyweights in the mobile crane industry is the Liebherr LTM 11200-9.1. This mobile monster has a maximum lifting capacity of 2,646,000 pounds (1,200,205 kg). The telescopic boom extends 328 feet (100 m), but accessory lattice booms can be added to reach a maximum hoist height of 617 feet (188 m). With nine axles and a travel weight of 230,000 pounds (104,000 kg), this model is not likely to park neatly in your driveway. Add the maximum 445,000 pounds (202,000 kg) of counterweight to that, and it becomes a truly massive and impressive piece of mechanical engineering.

In the past, crane operators were assigned the majority of responsibility for the overall safety of a lift operation. Standards now place greater responsibility on the **site supervisor** and **lift director** for the safety of the lift and site. However, this does not relieve the operator of the responsibility to communicate consistently with supervision and be aware of any safety or operational concerns, and to share any concerns of his/her own. *ASME Standard B30.5* defines five positions related to crane operation as follows:

Figure 5 Cranes erecting a wind turbine.

- *Crane owner* – Has custodial control of a crane through a lease or ownership
- *Crane user* – Arranges for the crane's presence on a jobsite and controls its use there
- *Site supervisor* – Exercises supervisory control over the jobsite on which a crane is being used and over the work being performed on the site
- *Lift director* – Directly oversees the work being performed by a crane and the rigging crew
- *Crane operator* – Directly controls the functions of the crane

The ASME standard outlines the specific responsibilities of each individual listed here. Their responsibilities are explored in greater detail as this training progresses.

Qualified riggers and signal persons are also important components of lift teams. A **rigger** is trained to prepare and attach loads to the crane to ensure the security of the load as well as the safety of the property and personnel in the general area. Signal persons use both verbal and nonverbal techniques to communicate with one or more crane operators and direct their actions.

Crane operators should shoulder some of the responsibility for the work area around their cranes. Before operating the crane, an operator should inspect the area and be fully aware of local hazards. Hazards may include pedestrian or vehicular traffic, power lines, and unsafe or unstable ground conditions. The operator should ensure that the area around the crane is properly barricaded and that people, equipment, and materials are kept out of danger. The barricades shown in *Figure 6* are designed to keep nearby workers out of the operating radius of the crane. The operator must stay alert for potential hazards during operations such as overhead power lines,

Figure 6 A mobile crane with proper barricades around it.

Figure 7 Two cranes working in tandem.

pedestrians, and other equipment operating in the area. *Figure 7* shows two cranes working on a jobsite with several potential hazards in the area. The crane operator should examine the load rigging before beginning any lift. The operator also needs to make sure that the load and lift characteristics fall within the capabilities of the crane.

The operator's responsibilities all revolve around safety. Operators must be confident in their ability to operate the crane safely, ensure that the jobsite is prepared and secured to support safe operation, and ensure that the load is rigged properly and within the capacity specifications of the crane.

Experienced crane operators have additional career opportunities they can pursue, if desired. These include the following:

- Mobile crane inspector
- Crane operation trainer
- Crane operation supervisor
- Crane maintenance and repair specialist
- Lift planning consultant
- Lift director
- Crane safety consultant or safety manager
- Crane accident investigator
- Signal person
- Rigger

Some of these positions require years of experience and expertise in the operation and performance of mobile cranes. They may also require further training, education, and certification.

1.3.0 Becoming a Mobile Crane Operator

Becoming a proficient mobile crane operator involves study, a lot of practice, and jobsite experience. Structured training programs, such as this one, will assist those who want to become operators in attaining the necessary skills and knowledge. It is very important that prospective operators also work with cranes in other ways, such as assisting with their setup and transportation. While learning how to operate and control a mobile crane, a trainee might also serve as a rigger or signal person, with the proper training. Setup work includes creating a clear path from the road to the work site, leveling the ground or installing pads so that the crane will remain level and stable during operation, and setting up barricades to keep pedestrians and vehicles clear of hazards.

Trainees may also operate equipment other than cranes while they learn the trade. These include machines that use principles similar to those that apply to crane operation, but they are much less complicated to learn and use safely. These might include skid-steer loaders, forklifts, backhoes, and aerial lifts. A trainee will initially practice crane operations under close supervision in a safe and isolated environment to gain experience. When both the trainee and supervisor or instructor feel ready, and all necessary testing has been completed successfully, the trainee becomes a crane operator. New operators generally perform low-risk operations on construction sites under close supervision to gain additional knowledge and confidence. More complicated lifts are generally assigned to those with experience and proven skill.

Operating a crane requires the operator to recognize and quickly respond to events in the work environment. This requires good vision, hearing, coordination, and reflexes. The physical qualifications for mobile crane operators are detailed in *ASME Standard B30.5*. Note that 29 *CFR* 1926, Subpart CC does not specify physical characteristics for a crane operator candidate. The following

Crane Simulators

Crane simulators, like the one shown here, are often used to train new operators, assess the skills of experienced operators, and test potential new employees. The simulators duplicate the cab environment and controls. The simulation software causes the screen images to respond to the controls as the real machine would. Trainees using the simulator learn to perform tasks the safest and most efficient way. The use of a simulator can result in significant savings in terms of crane ownership or rental, cost of fuel, and reduced wear and tear on cranes. Simulators are available for many types of heavy equipment in addition to cranes.

list summarizes the physical characteristics described in the ASME standard:

- Vision quality of at least 20/30 in one eye and 20/50 in the other, with or without corrective lenses.
- The ability to distinguish colors, regardless of position, if color differentiation is required.
- Adequate hearing to meet operational demands, with or without a hearing aid.
- Sufficient strength, endurance, agility, coordination, and speed of reaction to meet operation demands.
- Normal depth perception, field of vision, reaction time, manual dexterity, coordination, and no tendencies toward dizziness, seizures, or similar undesirable characteristics.
- A negative result in substance abuse testing conducted by a recognized laboratory service.
- No evidence of physical defects or emotional instability that might create a hazard for the operator or others. Any physical issue that an examiner feels could interfere with an operator's ability to perform can be cause for disqualification.

Operators also have specific academic and performance guidelines detailed in *ASME Standard B30.5* as well as in the current 29 *CFR 1926.1400,* Subpart CC and 1910.180. In addition to passing a physical examination, the following list summarizes the academic and performance guidelines found in the ASME standard:

- Satisfactory completion of a written examination covering operational characteristics, controls, and emergency response skills. Questions are appropriate for the crane type in which certification is sought.

- The ability to read, write, comprehend, and use arithmetic and load rating charts from the crane manufacturer's operating instructions in the language in which it is provided.
- A passing grade on a combination written and verbal test related to load rating chart usage based on various crane configurations.
- Successful completion of an operational test demonstrating proficiency in handling the specific type of crane for which qualification is sought. This not only includes common crane operations such as lifting, but also prestart and post-operation inspections, securing procedures, and crane travel.
- A demonstrated understanding of all applicable *ASME B30* standards, as well as federal, state, and local requirements.

Qualified operators may be required to requalify periodically, at the discretion of the employer. Physical exams are required every three years, and more often if deemed necessary by the employing organization.

1.3.1 Qualification and Certification

It is important to first understand the difference between *qualified* and *certified* crane operators. The terms are frequently misunderstood, and for good reason. As discussed previously, there are a number of physical qualifications for crane operators found in the ASME standards. The qualifications for a crane operator, as prescribed in *ASME Standard B30.5*, include the successful completion of a written test and a practical test of skill. One who meets the listed qualifications and passes the both the knowledge and skill tests is considered by ASME to be a qualified crane operator. The

Simple and Effective Crane Barricades

Mobile cranes, once in position for a lift, must be barricaded to help prevent workers from getting too close to the crane itself. Doing so represents a major hazard, such as being caught between the crane and another piece of equipment. The body of the crane may also rotate++-into a distracted worker, causing serious injury and possibly death. OSHA requires a proper barricade.

Barricades are often field-prepared using caution tape, rope, and similar materials to identify the boundary. The tape or rope must be supported at various intervals around the crane. Since cranes operate on many different types of surfaces, supporting the boundary material is often a challenge. Cranes typically do not have storage areas that can be devoted to various poles, stakes, and drums that can accommodate any ground surface.

But what if the barricade could be mounted on the crane itself? One experienced crane operator asked himself this question, which led to a new solution. The product line from Barricade Buddy, LLC represents an innovative approach to maintaining OSHA compliance when securing the area around the swing radius of a mobile crane. Barricade Buddy barricades have already been designed to fit many crane types, and more are in the works. The barricades provide a highly visible, lightweight, noncorrosive, nonconductive, and easily assembled boundary that can move with the crane when necessary. The best innovations often come from craft professionals themselves. More information about the Barricade Buddy system is available at **www.barricadebuddy.com**.

term *certification* does not even appear in *ASME Standard B30.5-2014.*

However, although the ASME standard is a national consensus standard, it is not enforceable by law. 29 *CFR* 1926.1427, a legally enforceable body of work, speaks directly to the subject of qualification and certification. This standard associates certification only with testing that is administered by a third-party organization, such as the certification program offered by NCCER. A mobile crane operator certification that is obtained through NCCER (or another accredited third-party organization) is portable, meaning that it should be accepted by a wide variety of employers and organizations.

OSHA also allows three other options for testing: testing through an audited employer's program; testing by the US military; and testing by other government agencies. However, these three options lead to what OSHA considers to be a qualified operator. The significant difference is that any qualification received through these options is not portable. For example, if a crane operator is qualified through testing by an employer, it is no longer valid when the operator changes employers. The operator must either seek certification through a third-party organization, or be requalified by the new employer.

When 29 *CFR* 1926, Subpart CC was revised and released in 2010, crane operator qualification or certification became a requirement. The revised standard originally required that all crane operators involved in construction be certified by November, 2014. Many crane operators across the country sought certification during this three-year period. However, in September, 2014, the deadline to comply with the requirement was further extended to November 10, 2017. Some states have had certification requirements for many years, and both new and experienced crane operators have continued to seek qualification or certification during this extended period.

29 *CFR* 1926.1427 requires all third-party, crane operator certification programs to seek accreditation by an appropriate agency. The Mobile Crane Operator Certification Program offered by NCCER has been accredited by the American National Standards Institute (ANSI). This indicates that the materials and process developed by NCCER meet the high standards for an effective

and recognized source of certification and certification materials. Anyone who earns certification through these programs will receive an NCCER gold wallet card with their photo (*Figure 8*). In addition to the ANSI accreditation, NCCER's Mobile Crane Operator Certification Program is recognized by the following organizations:

OSHA

- Association of General Contractors (AGC) of America
- Associated Builders and Contractors, Inc. (ABC)

1.3.2 NCCER's Certification Program

NCCER and the North American Crane Bureau, Inc. (NACB) have built upon the substantial strengths of their organizations to develop certification tools for the crane industry. All written assessments (knowledge tests) and practical examinations (skill tests) of NCCER's Mobile Crane Operator Certification Program adhere to strict professional guidelines endorsed by the American Psychological Association. Certification testing meets or exceeds current *ASME Standard B30.5* and 29 *CFR* 1926, Subpart CC requirements. NACB utilizes a professional staff to train, test, and certify practical examiner candidates. There are specific qualification requirements for this role, and individuals must attend a one-day training program conducted by an NCCER-certified Master Trainer after those initial qualifications are accepted.

NCCER currently offers certification for four distinct yet related construction industry crafts: mobile crane operators, tower crane operators, riggers, and signal persons. In all four crafts, candidates must pass both a written assessment and a practical exam to receive certification.

Figure 8 Gold wallet card for successful certification program candidates.

Per the OSHA requirements, a crane operator's certification is valid for five years. Certified operators must take the written assessment and practical exam again every five years to retain their certification.

For certification, mobile cranes are divided into four broad categories, with a written assessment associated with each one. These crane categories are the following:

- Industrial/All-Purpose
- Lattice Boom
- Telescopic Boom
- Boom Truck

Under these four categories, there are four different practical exams that narrow down the specific certification types available. Refer to *Figure 9*. The successful completion of the written assessment and practical exam combinations shown provide a variety of equipment-specific certifications. Crane operators can choose the appropriate path for their situation, and may elect to seek certification in all categories to maximize their flexibility and value.

Figure 9 Paths to NCCER Mobile Crane Operator Certification.

Additional Resources

ASME Standard B30.5, Mobile and Locomotive Cranes. Current Edition. New York, NY: American Society of Mechanical Engineers.

29 *CFR* 1926.1400, *Cranes and Derricks in Construction.* **www.ecfr.gov**

Mobile Crane Safety Manual. 2014. Milwaukee, WI: Association of Equipment Manufacturers.

North American Crane Bureau, Inc. **www.cranesafe.com**

1.0.0 Section Review

1. Which of the following ASME standards applies directly to mobile cranes?

   a. *ASME Standard A3.1*
   b. *ASME Standard A9.4*
   c. *ASME Standard B15.2*
   d. *ASME Standard B30.5*

2. The primary concern of every mobile crane operator must be ______.

   a. physical fitness
   b. efficiency
   c. speed
   d. safety

3. Per OSHA standards, a mobile crane operator's certification is valid for ______.

   a. three years
   b. five years
   c. seven years
   d. ten years

2.0.0 NCCER STANDARDIZED TRAINING AND APPRENTICESHIPS

Objective

Describe NCCER standardized training, apprenticeships, and the desirable characteristics of a craft professional.

a. Explain how NCCER training programs are structured.
b. Describe apprenticeship programs.
c. Describe personal characteristics common to successful craft professionals.

Trade Terms

Apprenticeship: As defined by the US Department of Labor, a combination of on-the-job training and related technical instruction in which workers learn the practical and theoretical aspects of a highly skilled occupation. Apprenticeship programs can be sponsored by individual employers, joint employer and labor groups, and/or employer associations.

Training credentials: Recognition provided by documents and records such as transcripts, certificates, and wallet cards that provide proof of a trainee's successful completion of specific training curricula through an NCCER-accredited organization.

There are several ways to gain entry into the operator's seat of a mobile crane, and multiple ways to advance your career in the construction trades. Apprenticeship is one of the most potentially rewarding paths to success, since it provides a recognized and well-organized path to proficiency.

2.1.0 NCCER Training Structure

NCCER is an independent, not-for-profit educational foundation founded and funded by the construction industry to address the training needs of the industry. NCCER offers industry-driven training and education programs that provide training credentials recognized nationwide to deserving craft professionals.

The US Department of Labor (DOL), Office of Apprenticeship sets the minimum standards for apprenticeship programs across the country. These programs rely on mandatory classroom instruction (referred to by the DOL as *related technical instruction*) and on-the-job learning (OJL). To qualify as an apprenticeship, training programs must include at least 144 hours of related technical instruction along with 2,000 hours of OJL per year. NCCER's mobile crane operator's apprenticeship program consists of three years of apprenticeship training by DOL standards. A year of NCCER curricula instruction is commonly referred to as a *level*. The NCCER training program for mobile crane operators, therefore, has three levels. This training module is part of Level One.

The *Mobile Crane Operations Level One* Trainee Guide, as well as the training modules that comprise the *Core Curriculum*, provide at least 144 hours of instruction. The other two levels of this NCCER curriculum contain a variety of mobile crane topics for instruction, and each level contains at least 144 hours of instruction.

NCCER uses the DOL apprenticeship standards as a foundation on which to build comprehensive curricula that provide trainees with in-depth classroom and OJL experiences. Standardized NCCER curricula provide trainees with industry-driven training and education, using a competency-based learning approach. This means that trainees must not only demonstrate that they possess the knowledge needed for their chosen craft, but must also demonstrate the skills needed to safely perform related hands-on tasks.

When the instructor has documented that a trainee has acquired the necessary knowledge and skills from a given training module, that information is sent to NCCER and kept in the NCCER Registry System. The Registry System provides a means of verifying the NCCER training completed by a craft professional as he/she moves to new positions within a company, from company to company, or even from state to state. See the *Appendix* for examples of the credentials issued by NCCER.

Did You Know?

Active Apprenticeships

According to the DOL, there were 448,000 trainees enrolled in registered craft apprenticeship programs in 2015. More than 52,500 apprentices graduated from a registered program during the same time period. George Washington (surveyor), Benjamin Franklin (printer), and Paul Revere (silversmith) all served as apprentices in a craft.

Whether you enroll in an NCCER training program or another apprenticeship program, make sure you work for an employer or sponsor who supports a nationally standardized training program that includes credentials to clearly document your skill development.

2.2.0 Apprenticeship Training

Apprenticeship training goes back thousands of years. Its basic principles have not changed over time. First, it is a means for a person entering a craft to learn from those who have mastered the craft. Second, it includes learning by doing; real skills versus theory. Some theory is presented in the classroom, however, it is presented in a way that helps the trainee understand the purpose behind the skill that is to be learned.

2.2.1 Apprenticeship Standards

All apprenticeship standards prescribe certain work-related or OJL activities. This OJL is broken down into specific tasks in which the apprentice receives hands-on training. In addition, a specified number of hours is required in each task. The total number of OJL hours for an apprenticeship program traditionally ranges from 4,000 to 8,000. In a typical apprenticeship program, the required OJL may be acquired in increments of 2,000 hours per year. In a competency-based program, it may be possible to shorten this time by testing out of specific tasks through a series of performance evaluations.

The apprentice must log all work hours and provide it to the apprenticeship committee so that accurate records can be maintained. After each 1,000 hours of related work (six months for a full-time worker), an apprentice may receive a pay increase if so prescribed by the apprenticeship standards. Incremental pay increases are a part of many apprenticeship programs.

For those entering an apprenticeship program, a high school education or training through a career and technical education (CTE) school is desirable. Courses in various shop processes, mechanical or architectural drawing, and general mathematics are always helpful to support a career as a craft professional. Manual dexterity, good physical conditioning, and quick reflexes are important. The ability to solve problems quickly and accurately and to work closely with others is essential. An apprentice must also consider safety as the highest priority.

The prospective apprentice must submit certain information to the apprenticeship committee. This information generally includes the following:

- Aptitude test (General Aptitude Test Battery [GATB] Form Test) results, usually administered by the local Employment Security Commission
- Proof of educational background (high school transcripts should be sent to the committee)
- Letters of reference from past employers and friends
- Proof of age
- A copy of Form DD214, Certificate of Release or Discharge from Active Duty, if the candidate is a veteran
- A record of technical training received that relates to the construction industry and/or a record of any pre-apprenticeship training

Some companies have physical activity requirements that must also be met by apprentices. These requirements can vary from company to company.

An apprentice must agree to do the following:

- Wear proper safety equipment on the job
- Purchase and maintain tools of the trade as needed and required by the contractor
- Submit a monthly on-the-job training report to the committee
- Report to the committee if a change in employment status occurs
- Attend classroom-related instruction and adhere to all classroom regulations such as attendance requirements

2.2.2 Youth Apprenticeship Programs

The DOL apprenticeship guidelines also allows students to begin their apprentice training while still in high school. In some crafts, a student entering a program in 11th grade may complete as much as two years of an NCCER standardized four-year curriculum before high school graduation. In addition, the program, in cooperation with local craft-worker employers, allows students to work in the trade and earn money while still in school. Upon graduation, the student can enter the industry at a higher level and with more pay than someone just beginning the training. These programs provide students with an opportunity to prepare for a career as a craft professional and enter the workforce as a productive employee at an earlier age. Note that a student in a youth apprentice program must graduate from high school in order to be eligible for entry into a contractor's registered apprenticeship program.

Student progress is recognized through official transcripts, and a student can enter the next year of the program whenever and wherever it is offered. They may also have the option of applying the credits at a two-year or four-year college that offers degree or certification programs in the construction trades.

2.3.0 Characteristics of Successful Craft Professionals

In order to be successful, a craft professional must have the skills to use current trade materials, tools, and equipment to safely finish the task in a minimum period of time. A crane operator must be adept at adjusting to different situations. Crane operators must train continuously to remain knowledgeable about technical advancements in crane operation and control. A professional never takes chances with regard to personal safety or the safety of others.

The word *professionalism* is a broad term that describes the desired overall behavior and attitude expected in the workplace. Professionalism is too often absent from the construction site and the various trades. Most people would argue that professionalism must start at the top in order to be successful. It is true that management demonstration of professionalism is important to success in the workplace, but it is more important that individuals recognize their own responsibility to act professionally.

Professionalism includes honesty, productivity, safety, civility, cooperation, teamwork, clear and concise communication, and being on time and prepared for work. It can be demonstrated in a variety of ways during every day spent in the workplace. Most important is that you do not accept the unprofessional behavior of others too casually. This is not to say that you shun the unprofessional worker; instead, demonstrate professional behavior and serve as an example to effect change.

Professionalism is a benefit both to the employer and the employee. It is a personal responsibility. The construction industry is what each individual chooses to make of it. If you choose professionalism, the industry image will follow. The sections that follow provide additional guidance in various aspects of professionalism.

2.3.1 Honesty

Honesty and personal integrity are important traits of the successful professional. Professionals pride themselves in performing a job well, and in being punctual and dependable. Each job is completed in a professional way, never by cutting corners or reducing materials. A valued professional maintains work attitudes and ethics that protect property such as tools and materials belonging to employers, customers, and other trades from damage or theft at the shop or jobsite.

Honesty and success go hand-in-hand. It is not simply a choice between good and bad, but a choice between success and failure. Dishonesty will always catch up with you. Whether you are stealing materials, tools, or equipment from the jobsite or simply lying about your work, it will not take long for your employer to find out. Perhaps you can find another employer, but this option will eventually disappear. A craftworker's reputation tends to have a life of its own, and each person must protect it through desirable behavior and performance.

Honesty means more than just avoiding criminal acts. It means giving a fair day's work for a fair day's pay. It also means that you accept responsibility for any errors you might make. It means carrying out your side of a bargain. Employers place a high value on an employee who prioritizes honesty.

2.3.2 Loyalty

Employees expect their employers to look out for their best interests, to provide them with steady employment, and to promote the deserving to better positions as openings occur. Employers feel that they, too, have a right to expect their employees to be loyal to them—to keep the company's interests in mind, to speak well of them to others, to keep negative information within the workplace, and to keep all matters that pertain to the business absolutely confidential. Both employers and employees should keep in mind that loyalty is not something to be demanded; rather, it is something to be earned and returned.

2.3.3 Willingness to Learn

Every employer and jobsite has a certain way of doing things. Employers expect their workers to be willing to learn their ways. Adapting to change and being willing to learn new methods and procedures as quickly as possible are key. Sometimes changes in safety regulations or the purchase of new equipment makes it necessary for even experienced employees to learn new methods and operations. Employees often resent having to accept improvements due to the retraining involved, especially when it comes as a surprise. However, employers will no doubt expect employees to put forth the necessary effort. Methods must be kept up to date in order to meet competition and progress as a business. It is this profit that enables the owner to continue in business and provide jobs for the employees.

2.3.4 Willingness to Take Responsibility

Most employers expect their employees to see what needs to be done, then go ahead and do it. It is very tiresome to have to ask again and again for a task to be done. It should be obvious that, having been asked once, an employee should assume responsibility for it from then on. A craft professional's primary responsibilities should be clearly stated by the employer, providing guidance and clarity to the worker. Take those responsibilities seriously.

Craft professionals also need to be engaged enough in their role to know when it is time to assume responsibility for something without direction. In addition, a craft professional should always report issues or problems that affect safety or the product in any meaningful way, and take responsibility for any personal errors.

2.3.5 Willingness to Cooperate

To cooperate means to work together. In the modern business world, cooperation is the key to getting things done. Learn to work as a member of a team with your employer, supervisor, and fellow craft professionals towards the common goal of getting the work done efficiently, safely, and on time (*Figure 10*). Look for ways to contribute that support the team's goals.

People can work together well only if there is some understanding about what work is to be done, when and how it will be done, and who will do it. Rules and regulations are a necessity in any work situation and should be supported by all employees.

2.3.6 Present and On Time

Tardiness means being late for work, while absenteeism means being off the job for one reason or another. Consistent tardiness and frequent absences may be an indication of poor personal habits, unprofessional ethics, and/or a lack of commitment to your craft.

Your working life is significantly regulated by the clock. Each worker is required to be at work at a specified time, along with everyone else. Failure to get to work on time results in confusion, lost time, and resentment on the part of team members who do come on time. In addition, it may lead to penalties, including dismissal. It is your

Figure 10 Learn to work as a team toward a common goal.

obligation to be at work at the time indicated. In many cases, a missing team member cripples the team and stalls the project. The company can experience a significant loss from only a few minutes of lost time for an entire crew.

Consider the matter from the point of view of a supervisor. Supervisors cannot keep track of workers who arrive at random times. It is not fair to others for the tardiness of one to be ignored. Acting in the best interests of the team and the company, a supervisor must address a worker's tardiness. Employers resort to docking pay, demotion, and even dismissal in an effort to control tardiness and absenteeism. No employer likes to impose discipline or restrictions on employees. However, to be fair, an employer is sometimes forced to discipline those who fail to meet the standard.

It is sometimes necessary to take time off from work. No one can be expected to work when sick or when there is a serious personal matter at hand. However, it is possible to get into the habit of letting unimportant and unnecessary matters get in the way of the job. This results in lost production and hardship on those who try to carry on the work with less help. Do not allow trivial reasons to interfere with your work. Get a sufficient amount of rest before going to work the next day. If you are ill, use the time at home to recover as quickly as possible. Each of us would expect no less from someone hired to do an important job. If it is necessary to stay home, phone the supervisor early in the morning so planning for your absence can begin.

A practiced morning routine helps prevent being in a rush and breathless, late arrivals. In fact, arriving a little early shows your interest and enthusiasm for the work, which is appreciated by employers. Tardiness is one of those habits that can easily stand in the way of promotion. A worker that cannot manage his/her own time is not often chosen for a position of leadership.

The most frequent causes of absenteeism are illness, death in the family, accidents, personal business, and dissatisfaction with the job. Some of these causes are legitimate and unavoidable, while others can be controlled somewhat. One can usually plan to carry on most personal business affairs after working hours. If dissatisfaction with the job is a problem, make the necessary plans for change. However, do so in a professional manner, giving your employer as much effort through the last day on the job as was provided on the first day.

2.0.0 Section Review

1. In order to qualify as an apprenticeship program, a training program must include at least _____.

   a. 2,000 hours of technical instruction per year
   b. 2,000 hours of technical instruction in the entire program
   c. 144 hours of technical instruction per year
   d. 144 hours of technical instruction in the entire program

2. The responsibility to document the working hours of an apprentice and provide it to the apprenticeship committee belongs to the _____.

   a. committee itself
   b. classroom instructor
   c. employer
   d. apprentice

3. Loyalty is an important quality _____.

   a. only for employees
   b. only for employers
   c. for both employees and employers
   d. for crane operators without a certification

SUMMARY

Cranes are impressive and powerful machines. The power and size of cranes means that they must be operated with great care and attention. To work as a mobile crane operator, you must be skilled and knowledgeable, as well as mentally and physically fit. Knowing and following all applicable standards will help you to work safely and efficiently. Career opportunities for trained and certified mobile crane operators are numerous and varied.

A training program that combines competency-based, hands-on training with classroom instruction has proven to be the most effective means for a person to learn and advance in the mobile crane industry. Developing the job skills is only part of the solution; it is just as important to learn good work habits, convey a positive, cooperative attitude to those around you, and practice good safety habits every day.

1. The mast heights of mobile cranes can approach ______.

    a. 400 feet (122 meters)
    b. 700 feet (213 meters)
    c. 1,000 feet (305 meters)
    d. 1,800 feet (549 meters)

2. Tilt-up building construction involves ______.

    a. site-built wooden stud walls
    b. site-poured concrete wall panels
    c. modular concrete floors
    d. corrugated sheet metal walls

3. Which of the following statements is *true*?

    a. ASME is a major manufacturer of cranes and related equipment.
    b. ASME standards have nothing to do with the qualifications of mobile crane operators.
    c. Only a single OSHA standard is relevant to mobile cranes.
    d. 29 *CFR* 1926.1400, Subpart CC is devoted to cranes and derricks in construction.

4. As a mobile crane operator, if you see a significant safety hazard, you should ______.

    a. simply avoid it
    b. tell a co-worker
    c. call OSHA
    d. safely stop the lift process

5. Which of the following represents the ASME requirements related to eyesight for a mobile crane operator?

    a. 20/30 vision in both eyes, without correction
    b. 20/30 vision in both eyes, with or without correction
    c. 20/30 vision in one eye and 20/50 in the other, with or without correction
    d. 20/50 vision in both eyes, with or without correction

6. Candidates for mobile crane operator certification must pass ______.

    a. only a practical exam
    b. only a written assessment
    c. both a written test and a practical skills test
    d. a written assessment, practical exam, and examiner interview

7. Which of the following is *not* one of the four broad categories of mobile cranes for which NCCER certification is provided?

    a. All-Terrain
    b. Telescopic Boom
    c. Industrial/All-Purpose
    d. Boom Truck

8. The total on-the-job learning hours needed for a trainee to complete NCCER's apprenticeship program is ______.

    a. 2,000 hours
    b. 4,000 hours
    c. 6,000 hours
    d. 8,000 hours

9. Which organization determines the minimum standards for an apprenticeship program?

    a. DOL
    b. OSHA
    c. ASME
    d. NACB

10. If one of your co-workers complains about your company, your professionalism would best be demonstrated by ______.

    a. contributing your own complaints in support
    b. agreeing with the person to avoid conflict
    c. suggesting that the person leave the jobsite immediately
    d. finding some good things to say about the company

Fill in the blank with the correct term that you learned from your study of this module.

1. An individual certified by NCCER to train NCCER Craft Instructors, Performance Evaluators, and Proctors is referred to as a(n) __________.

2. An organization that is a division of the US Department of Labor is the __________.

3. If an organization is recognized for maintaining suitable testing standards for the mobile crane operator certification process, it receives __________.

4. When concrete wall sections are formed at a construction site and then put in place to form a wall, the process is referred to as __________.

5. A(n) __________ is responsible for supervisory control of the larger work area where a crane may be used, and must be knowledgeable about any site conditions that could affect a crane lift.

6. A person defined by OSHA as having successfully demonstrated the ability to solve problems related to a specific topic, and possessing recognized educational documentation or the appropriate experience to do so competently, is called a(n) __________.

7. Information related to the capacity of each specific crane model under various conditions and configurations is collected and placed in a(n) __________.

8. The __________ is actively engaged in the accreditation of products and personnel certification programs by determining their conformance to prescribed standards.

9. The individual that directly oversees the work being performed by a crane and the associated rigging crew is called the __________.

10. Trainees that successfully complete specific craft training programs through an accredited training organization receive __________ to recognize their achievement.

11. The US DOL defines a(n) __________ as a combination of on-the-job training and related technical instruction in which workers learn the practical and theoretical aspects of a highly skilled occupation.

12. An organization that is both an engineering society as well as a standards-developing organization is the __________.

13. The __________ has a responsibility to provide the most comprehensive services available in crane training and inspection services, while maintaining current state-of-the-art safety information applicable to crane operation.

14. An individual certified to administer practical skill tests for the certification of mobile crane operators is call a(n) __________.

15. A mobile crane operator candidate that successfully completes both a written and practical test based on the equipment capacity and type to be operated is awarded __________.

16. A __________ specializes in the preparation and attachment of loads to cranes and similar equipment, and is also trained in the preparation and movement of loads without the use of cranes and similar equipment.

17. A __________ has been adopted and promulgated by a nationally recognized organization under procedures whereby it can be determined by the Secretary of Labor or by the Assistant Secretary of Labor that persons interested and affected by its scope or provisions have reached substantial agreement on its adoption.

Trade Terms

Accreditation
American National Standards Institute (ANSI)
American Society of Mechanical Engineers (ASME)
Apprenticeship
Certification

Lift director
Load rating chart
Master Trainer
National consensus standard
North American Crane Bureau, Inc. (NACB)
Occupational Safety and Health

Administration (OSHA)
Practical examiner
Qualified person
Rigger
Site supervisor
Tilt-up construction
Training credentials

Joey Watts

CEO/Partner, Southland Safety, LLC

Inspired by his beginnings in the oilfields of East Texas, Joey Watts launched into a productive and successful career as CEO and partner in a company providing valuable safety and rescue services, training deployment and development, and certifications in the mobile crane industry. He credits his success to listening to the valuable insights shared by his mentors and the master craft professionals that came before him.

Please give a brief synopsis of your construction career and your current position.

At Southland Safety, we specialize in on-site safety, training, and related certifications and serve as industry consultants. I also serve as a Mobile Crane Practical Examiner and a Rigging Examiner.

How did you get started in the construction industry?

My career began with hard work in the oilfields of East Texas, and I continually progressed in knowledge and skills until becoming CEO and partner of Southland Safety.

Who or what inspired you to enter the construction industry?

I have been most inspired by the many people that I have been fortunate enough to work with in my career. I've enjoyed them immensely.

Why do you think credentials are important in the construction industry?

I have worked with and learned from mentors both young and old. But in my opinion, having industry-recognized credentials allows an individual to measure their own craft improvement and be recognized for their progressive achievements.

What do you enjoy most about your career?

I really enjoy the people I get to work with. Construction workers are some of the most colorful and intelligent people you will ever meet!

Would you recommend construction as a career to others? Why?

Yes. There are many opportunities for life improvement and advancing your career in the construction industry.

What advice would you give to someone who is new to the construction industry?

Be open to learning new things and always focus on improvement in everything you do. Never close your ears, and keep listening and learning. Always humble yourself before the master craft professionals that precede you and learn as much as you can from them, while you can.

How do you define craftsmanship?

Having pride of craftsmanship in a job well done is achieved by applying your knowledge, skills, and ability to produce a superior product or service.

Trade Terms Introduced in This Module

Accreditation: In the context of crane-operator certification, the act of granting credit or recognition to an organization or institution that maintains suitable standards of testing for the certification process.

American National Standards Institute (ANSI): An organization that oversees the creation, promulgation, and use of thousands of norms and guidelines that directly impact businesses. ANSI also provides third-party accreditation to products and personnel certification programs, recognizing the competence of organizations to carry out these activities in accordance with requirements defined in international standards.

American Society of Mechanical Engineers (ASME): ASME promotes the art, science, and practice of multidisciplinary engineering and allied sciences around the globe through continuing education, training and professional development, codes and standards, research, conferences and publications, government relations, and other forms of outreach. The organization is therefore an engineering society as well as a standards-developing organization.

Apprenticeship: As defined by the US Department of Labor, a combination of on-the-job training and related technical instruction in which workers learn the practical and theoretical aspects of a highly skilled occupation. Apprenticeship programs can be sponsored by individual employers, joint employer and labor groups, and/or employer associations.

Certification: Refers to a candidate's successful completion of a written test of knowledge as well as a practical examination. Certification testing must be based on the equipment capacity and type that will be operated, and can be administered by an accredited third-party organization (such as NCCER) or through an audited testing program operated by the employer.

Lift director: As defined in *ASME Standard B30.5*, the lift director directly oversees the work being performed by a crane and the associated rigging and signaling crew. The lift director is responsible for addressing all hazards related to the lift as well as those brought forward by the crane operator, signal persons, riggers, or other members of the crew.

Load rating chart: One or more charts associated with each individual crane model that specify the crane's capacity at various distances and angles, and in various configurations. Per *ASME Standard B30.5*, Section 5.1.1.3, load charts should include a "full and complete range of manufacturer's crane load ratings at all stated operating radii, boom angles, work areas, boom lengths and configurations, jib lengths and angles (or offset), as well as alternate ratings for the use and nonuse of optional equipment on the crane, such as outriggers and extra counterweights, which affect ratings."

Master Trainer: An individual certified in accordance with NCCER's Master Trainer Instructor Certification Training Program (MT-ICTP) and authorized to train NCCER Craft Instructors, Performance Evaluators, and Proctors.

National consensus standard: As defined in 29 *CFR* 1910.2(g), "any standard or modification thereof which (1) has been adopted and promulgated by a nationally recognized standards-producing organization under procedures whereby it can be determined by the Secretary of Labor or by the Assistant Secretary of Labor that persons interested and affected by the scope or provisions of the standard have reached substantial agreement on its adoption, (2) was formulated in a manner which afforded an opportunity for diverse views to be considered, and (3) has been designated as such a standard by the Secretary or the Assistant Secretary, after consultation with other appropriate Federal agencies."

North American Crane Bureau, Inc. (NACB): NACB's responsibility is to provide the most comprehensive services available in crane training and inspection services, while maintaining current state-of-the-art safety information applicable to crane operation. NACB is committed to providing the industry with a standard of excellence unparalleled in the field of safety. A division of the NACB, NACB Interactive Educational Systems, Inc., provides and distributes crane training simulators. NCCER and NACB work as partners in support of NCCER's Mobile Crane Operator Certification Program.

Occupational Safety and Health Administration (OSHA): A division of the US Department of Labor that is responsible for establishing and enforcing safety and health standards for all places of employment in the United States.

Practical examiner: An NCCER-certified individual with the appropriate qualifications and training to administer practical examinations for NCCER certification programs.

Qualified person: As defined in 29 *CFR* 1926.1401, a person who, by possession of a recognized degree, certificate, or professional standing, or who by extensive knowledge, training, or experience, has successfully demonstrated the ability to solve problems relating to the subject matter, the work, or the project.

Rigger: An individual who specializes in the preparation and movement of often large and heavy objects. Riggers attach all types of loads to cranes and other equipment using cables, slings, hooks, and other specialized hardware, and also specialize in techniques used to move and position objects without the use of cranes and similar heavy equipment.

Site supervisor: As defined in *ASME Standard B30.5*, a site supervisor exercises supervisory control over the work site on which a crane is being used and over the work that is being performed on that site. Responsibilities include ensuring that a qualified person is designated as a lift director and being fully aware of any site conditions that may affect the lift, such as buried utilities.

Tilt-up construction: A building technique involving concrete walls. The walls are formed and poured at the construction site, creating large concrete panels. Each panel is then raised into position to form the exterior walls as the panels are joined.

Training credentials: Recognition provided by documents and records such as transcripts, certificates, and wallet cards that provide proof of a trainee's successful completion of specific training curricula through an NCCER-accredited organization.

SAMPLES OF NCCER TRAINING CREDENTIALS

Additional Resources

This module is intended as a thorough resource for task training. The following reference works are suggested for further study.

ASME Standard B30.5, Mobile and Locomotive Cranes. Current edition. New York, NY: American Society of Mechanical Engineers.
29 *CFR* 1926.1400, *Cranes and Derricks in Construction*. **www.ecfr.gov**
Mobile Crane Safety Manual. 2014. Milwaukee, WI: Association of Equipment Manufacturers.
North American Crane Bureau, Inc. **www.cranesafe.com**
US Department of Labor, Bureau of Labor Statistics. **www.bls.gov**

Figure Credits

The Manitowoc Company, Inc., Module opener, Figures 1, 7, SA04
Carolina Bridge Co., Figures 2, 3
Liebherr USA, Co., SA01
Robert I. Carr, Ph.D., P.E., Figure 5
Courtesy of DOE/NREL, Figure 7
Barricade Buddy, LLC, Figure 6, SA02
© CoolKengzz/Shutterstock.com, Figure 10
NACB Interactive Educational Systems, Inc. (NACB-IES), SA03

Answer	Section Reference	Objective
Section One		
1. d	1.1.1	1a
2. d	1.2.0	1b
3. b	1.3.2	1c
Section Two		
1. c	2.1.0	2a
2. d	2.2.1	2b
3. c	2.3.2	2c

NCCER CURRICULA — USER UPDATE

NCCER makes every effort to keep its textbooks up-to-date and free of technical errors. We appreciate your help in this process. If you find an error, a typographical mistake, or an inaccuracy in NCCER's curricula, please fill out this form (or a photocopy), or complete the online form at **www.nccer.org/olf**. Be sure to include the exact module ID number, page number, a detailed description, and your recommended correction. Your input will be brought to the attention of the Authoring Team. Thank you for your assistance.

Instructors – If you have an idea for improving this textbook, or have found that additional materials were necessary to teach this module effectively, please let us know so that we may present your suggestions to the Authoring Team.

NCCER Product Development and Revision

13614 Progress Blvd., Alachua, FL 32615

Email: curriculum@nccer.org
Online: www.nccer.org/olf

❏ Trainee Guide ❏ Lesson Plans ❏ Exam ❏ PowerPoints Other _______________

Craft / Level: _______________________________________ Copyright Date: __________

Module ID Number / Title: ____

Section Number(s): ____

Description: ____

____

____

____

____

Recommended Correction: ____

____

____

____

____

Your Name: ____

Address: ____

____

Email: ____ Phone: _______________

Basic Principles of Cranes

OVERVIEW

Mobile cranes range from small, simple models to large, mechanically and electrically complex pieces of equipment. In addition, cranes that appear similar often differ in their operation and configuration. A fundamental understanding of crane types and how they move from place to place is important to riggers and signal persons as well as to crane operators. Cranes also have different types of lifting booms, each of which has its own strengths and weaknesses. This module presents specific features of various cranes and booms, and introduces the basic principles of lifting and leverage.

Module 21102

Basic Principles of Cranes

Objectives

When you have completed this module, you will be able to do the following:

1. Identify and describe various types of cranes and crane components.
   a. Identify and describe mobile cranes based on their means of travel.
   b. Identify and describe various types of crane booms.
   c. Identify and describe common crane attachments and accessories.
   d. Describe common crane instrumentation and safety devices.
   e. Identify and describe various crane reeving patterns.
2. Identify factors related to lifting capacity and explain their significance.
   a. Explain the significance of ground conditions and a level surface.
   b. Describe the bearing surface and explain how to determine the required blocking.
   c. Define and describe the significance of the center of gravity and the quadrants of operation.
   d. Describe the significance of boom length, angle, operating radius, and elevation.
   e. Explain how to use a load chart and understand the basic concepts of critical lifts.

Performance Tasks

Under the supervision of your instructor, you should be able to do the following:

1. Verify the boom length of a telescopic- and/or lattice-boom crane using manufacturer's data or a measuring tape.
2. Measure the operating radius of a telescopic- and/or lattice-boom crane using a measuring tape.
3. Calculate the amount of blocking needed for the outrigger of a specific crane.
4. Verify that a crane is level.

Trade Terms

Anti-two-blocking device	Duty cycle	Jib backstay	Parts of line
Backfill	Dynamic loads	Jib forestay	Pendants
Base mounting	Effective weight	Jib mast	Quadrant of operation
Base section	Floats	Lattice boom	Reach
Block and tackle	Grapples	Leads	Reeving
Blocking	Gross capacity	Leverage	Ring gear drive
Boom torque	Hardpan	Load moment	Sheave
Carbody	Headache ball	Lowboy	Swallow
Carrier	Hoist drum	Luffing	Tipping fulcrum
Center of gravity (CG)	Hoist reeving	Luffing jib	Upperworks
Check valve	Hydraulic motors	Net capacity	Wheelbase
Counterweights	Idlers	Non-ferrous	Whip line
Crane mat	Impact loads	Open-throat boom	
Crawler frames	Interpolation	Operating radius	
Critical lift	Jib	Outriggers	

Industry Recognized Credentials

If you are training through an NCCER-accredited sponsor, you may be eligible for credentials from NCCER's Registry. The ID number for this module is 21102. Note that this module may have been used in other NCCER curricula and may apply to other level completions. Contact NCCER's Registry at 888.622.3720 or go to **www.nccer.org** for more information.

Contents

1.0.0 INTRODUCTION TO MOBILE CRANES

Objective

Identify and describe various types of cranes and crane components.

a. Identify and describe mobile cranes based on their means of travel.
b. Identify and describe various types of crane booms.
c. Identify and describe common crane attachments and accessories.
d. Describe common crane instrumentation and safety devices.
e. Identify and describe various crane reeving patterns.

Trade Terms

Anti-two-blocking device: Two-blocking refers to a condition in which the lower load block or hook assembly comes in contact with the boom tip, boom tip sheave assembly or any other component above it as it is being raised. If this occurs, continuing to apply lifting power to the cable can result is serious equipment damage and/or failure of the hoist line. An anti-two-blocking device, therefore, prevents this condition from occurring.

Base mounting: A crawler crane assembly consisting primarily of the carbody, ring gear drive, crawler frames, and tracks.

Base section: The lowest portion of a telescopic boom that houses the other telescopic sections but does not extend.

Block and tackle: A system of two or more pulleys, which form a block, with a rope or cable threaded between them, reducing the force needed to lift or pull heavy loads.

Blocking: Wood or a similar material used under outrigger floats to support and distribute loads to the ground. Also referred to as *cribbing*.

Boom torque: A twisting force applied to the crane boom, typically resulting from imbalanced reeving of the boom tip sheave assembly ropes.

Carbody: The part of a crawler-crane base mounting that carries the rotating upperworks.

Carrier: The base of a wheeled crane that provides crane movement and supports the upperworks.

Check valve: A valve designed to allow flow in one direction but closes as necessary to prevent flow reversal.

Counterweights: Weights added to the crane, usually on the end opposite the boom, to help counter the weight of the load and improve stability.

Crane mat: A portable platform, typically made of large wooden timbers bolted together, used to support and spread the weight of a crane over a larger ground area.

Crawler frames: Crane assemblies comprised of the crawler tracks, track idlers, and track power sources of a crawler crane. Also called *tread members* or *track assemblies*.

Duty cycle: An expression of equipment use over time. In the case of mobile cranes, an 8-, 16-, or 24-hour rating expressed as a percentage.

Grapples: Devices used to pick up bulk items, containers, rocks, trees and tree limbs, etc. Grapples typically have several jaws that operate like fingers to pick up material, using mechanical or hydraulic power.

Headache ball: A heavy round weight often attached to a load line to provide sufficient weight to allow the load line to unspool from the drum when there is no live load. Larger versions of headache balls are used to swing into structures to demolish them.

Hoist drum: A drum is a cylindrical component around which a rope is wound. The hoist drum is used to wind or unwind the rope for hoisting or lowering the load; the part of a crane that spools and unspools the lifting line.

Hoist reeving: The reeving pattern applied to the hoist sheaves. Single- or multiple-line hoist reeving is used for whip, boom, and jib lines.

Telescopic boom: A crane boom that extends and retracts in sections that slide in and out, powered by hydraulic pressure.

Hydraulic motors: Motors powered by hydraulic pressure provided by an external pump. Hydraulic motors are often used to power the tracks of crawler cranes, instead of complex drive systems connected directly to the diesel engine.

Idlers: Pulleys, wheels, or rollers that do not transmit power, but guide or place tension on a belt or crawler-crane track.

Jibs: Extensions attached to the boom point to provide added boom length for reaching and lifting loads. Jibs may be in line with the boom, offset to another angle, or adjustable to a variety of angles. A jib is sometimes referred to as a *fly*.

Jib backstay: A piece of standing rigging that is routed from the jib mast back to the main boom to help support the jib.

Jib forestay: A piece of standing rigging that is routed from the far tip of the jib back to the jib mast, holding the tip of the jib up.

Jib mast: A structure mounted on the main boom that provides a fixed distance for the point of connection of the jib forestay and jib backstay. Also referred to as a *jib strut*.

Lattice boom: A boom constructed of steel angles or tubing to create a relatively lightweight but strong, rigid structure.

Leads: Steel structures that provide support for a pile hammer and help to align and position the hammer with the pile to be driven. The hammer can travel up or down in the leads as necessary.

Load moment: The force applied to the crane by the load; the leverage of the load, opposing the leverage of the crane. The load moment is calculated by multiplying the gross load weight by the horizontal distance from the tipping fulcrum to the center of gravity of the suspended load. The load moment is usually reported to the operator as a percentage of the crane's capacity at the present set of conditions. As those conditions change, such as the boom angle, the load moment changes as well.

Lowboy: A trailer with a low frame for transporting very tall or heavy loads. A typical lowboy has two drops in deck height: one right after the gooseneck connecting it to the tractor, and one right before the wheels. This allows the trailer deck to be extremely low compared with common trailers.

Luffing: Changing a boom angle by varying the length of the suspension ropes.

Luffing jib: A jib mounted on the end of a boom that can be positioned at different angles relative to the main boom.

Non-ferrous: Having no iron. Ferrous metals, such as steel, contain iron and are magnetic as a result.

Open-throat boom: A lattice boom with an opening in the boom structure near the far end, allowing the hoist lines to drop through the boom rather than over the end of the boom.

Outriggers: Extendable or fixed members attached to a crane base that rest on ground supports at the outer end to stabilize and support the crane.

Parts of line: When a line is reeved more than once, the resulting number of lines that are supporting the load block.

Pendants: Ropes or strands of a specified length with fixed end connections, used to support a lattice boom or boom components. According to 29 *CFR* 1926.1401, a pendant may also consist of a solid bar.

Reach: The combined operating height and radius of a boom, or the combination of boom and jib.

Reeving: A method often used to multiply the pulling or lifting capability by using wire rope routed through multiple pulleys or sheaves a number of times.

Ring gear drive: Sometimes referred to as the swing circle. An assembly that provides the point of attachment and pivot point for the upperworks of a crane. The ring gear is typically driven by hydraulic pressure, allowing the upperworks to rotate on a set of bearings that reduce friction and transfer the weight of the upperworks (and any load) to the carbody.

Sheaves: Wheels that have a groove for a belt, rope, or cable to run in. The terms *sheave* and *pulley* are often used interchangeably.

Swallow: The space between the sheave and the frame of a block, through which the rope is passed.

Upperworks: A term that refers to the assembly of components above the ring gear drive; the rotating collection of components on top of the base mounting or carrier; may also be referred to as the *house*, or as the *superstructure* as defined in 29 *CFR* 1926.1401.

Wheelbase: The distance between the front and rear axles of a vehicle.

Whip line: A secondary hoisting rope usually of lower capacity than that provided by the main hoisting system. When a whip line exists, it is typically out at the tip of a jib, while the main hoist line is closer to the crane and operated from the tip of the main boom.

This section describes the capabilities and limitations of various crane types used in construction and related applications. Jobsite environmental conditions and how these conditions determine the loads that can be safely lifted are also discussed. Even cranes of the same basic type differ greatly from one manufacturer to another, so operators must always be prepared to familiarize themselves with the information supplied by the manufacturer for the specific crane in use.

Crane selection for a particular task depends upon the technical requirements of crane capacity, reach, site clearances, and site conditions. Economic factors may also dictate the type of crane to be used. The most significant differences between mobile cranes revolve around two distinct characteristics:

- How they travel from one place to another
- The type of boom equipped

1.1.0 Approaches to Travel

As the name implies, mobile cranes are those that have a means of propulsion. A crane does not need to be fast, nor does it need to be drivable on the highway to be considered mobile. Although many mobile cranes can be driven on the highway, the mobility implied by their name is more about their ability to move around in general. The vast majority of mobile cranes are powered by diesel engines due to the amount of torque they develop and their durability. Diesel engines are especially suited for mobile cranes, just as they are for large trucks. A variety of crane types, based on their means of movement, are presented in the sections that follow.

1.1.1 Crawler Cranes

Crawler cranes are available in a variety of configurations. *Figure 1* shows a typical crawler crane with a lattice boom. Crawler cranes are also available with a telescopic boom. Crawler cranes are so named because they move slowly on tracks that provide superior traction and stability. The track assemblies are called crawler frames or *tread members*. The hoist drum and ring gear drive in this model are powered by hydraulic motors that provide very precise control. The unit shown has a diesel engine driving hydraulic pumps that in turn drive the hydraulic motors. The hoist drum manages the wire rope used for lifting, and the ring gear drive rotates and positions the upper portion of the crane.

The motors that power the tracks, placing the crane in motion, are also hydraulic (*Figure 2*). A number of hydraulic pumps often exist in a single crane. The counterweights can be lowered all the way to the ground using hydraulic actuators. The operator cab shown here can tilt up to 20 degrees to provide better visibility of the boom and/or load. The tracks of very large crawler cranes may be powered by a drive shaft from the engine instead.

The lifting and rotating assembly, called the upperworks, is mounted on top of the crawler assembly. The complete assembly beneath the

(A) CRAWLER CRANE

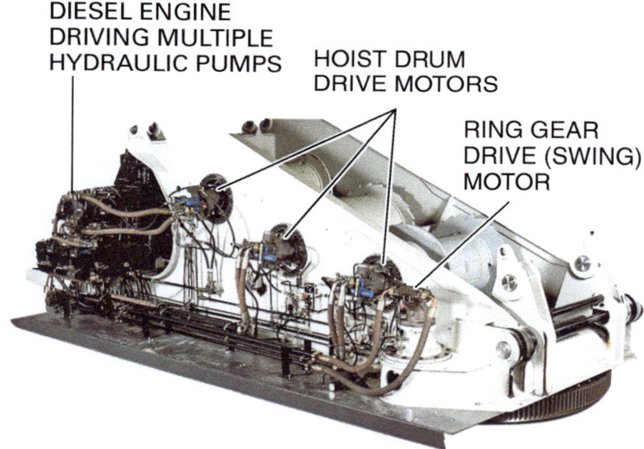

**(B) HYDRAULIC HOIST DRUM AND
RING GEAR POWER UNIT**

Figure 1 Typical crawler crane.

upperworks is referred to as the base mounting. The base mounting of the crawler crane transmits loads imposed on the upperworks down to the ground. To accomplish this effectively, the base needs to be extremely stiff. The base is stiffened by using complex castings or heavy weldments. The base mounting, shown in *Figure 3*, consists primarily of the carbody, ring gear drive or swing circle, crawler frames, tracks, and the propelling mechanism. Note that the base of a crawler crane should not be called a carrier. (Only wheeled crane bases are called *carriers*.)

Figure 2 Hydraulic drive motor.

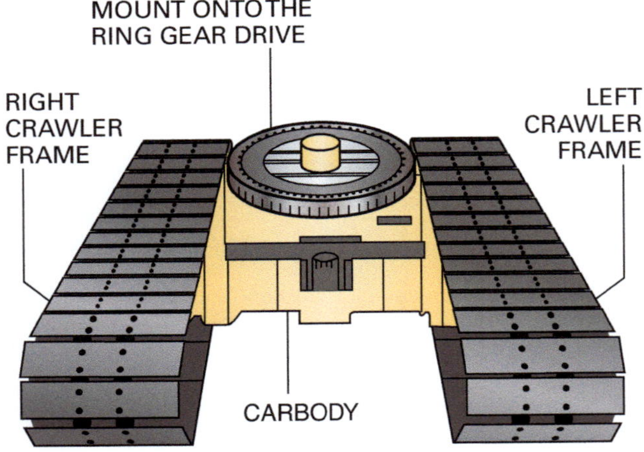

Figure 3 Crawler base mounting.

A carbody is the central or main portion of the base, with the ring gear drive fixed on top. The ring gear drive is the point where the upperworks is joined to the base mounting. The ring gear drive transmits the loads imposed on the upperworks to the base, and must also serve as a low-resistance surface for the swinging motion. Crawler frames are mounted on each side of the carbody. The crawler frames hold the crawler track motors, idlers, and crawler tracks (*Figure 4*).

Most crawler cranes are powered by an engine located in the upperworks. Older cranes use a mechanical system of drive shafts to couple power from the engine to the crawler tracks. Newer machines have hydraulic motors mounted directly to the crawler frames to propel the unit. Instead of complex drive arrangements, hydraulic tubing from the pump(s) is routed to the crawler motors. However, some larger machines have separate engines mounted on the base mounting

Figure 4 Crawler track and track idlers.

to provide power to the crawler tracks via mechanical drive shafts.

The width between the crawler tracks affects the stability of the crawler crane. The greater the width between the tracks, the greater the lifting capacity and stability. However, transporting cranes with widely spaced tracks can be difficult. This has led to the development of crawlers with extendable/retractable crawler frames. The cranes are transported while the tracks are retracted, making the crane as narrow as possible. When they arrive at the site, the tracks are extended to their full width to achieve increased stability. Most crawler frames are designed to extend or retract without assistance, while others require jacks, blocking, and other equipment to get the job done. Others may be equipped with removable crawler frames to further reduce the crane width and allow easier transportation by trailer (*Figure 5*). Lowboy trailers and railcars are commonly used to transport these cranes. Crawler cranes are definitely not designed to travel on the open road under their own power.

When comparing crawler cranes to other cranes with equal lifting capacity, crawler cranes generally offer lower rental rates than other cranes, but their transit and erection costs are higher. If the nature of the work and the ground conditions make it necessary to operate a machine from a crane mat and/or from a number of different positions, siting costs must also be considered. The cost of putting a truck crane in place is usually far less than that of placing and leveling a supportive mat for a crawler. However, when power and endurance are critical characteristics, a crawler crane is the best choice.

NCCER – *Mobile Crane Operations*

Figure 5 Crawler crane disassembled and loaded for transport.

1.1.2 Wheeled Truck Cranes

Wheeled truck cranes are available in configurations similar to crawler cranes; they are available with both lattice and telescopic booms. A truck crane carrier is analogous to a crawler crane's base mounting, and should not be confused with an ordinary commercial truck chassis. Truck crane carriers are designed and manufactured with a completely different priority than commercial trucks. Some of the larger units may have nine axles or more.

Truck crane carriers may be designed for moving around the job site only, traveling on the highway and smooth ground, or both. To accommodate these extremes of travel in a single crane, transmissions with more than 30 forward gears, as well as special creeping gears, are available. The creeping gears are used for very slow movement at the job site.

Highway speeds can range from 35 to 70 mph (48 to 113 kph). Job site ramps up to approximately 40 percent grade can generally be climbed by a wheeled truck crane with no load. Brakes are designed to hold position on similar grades. *Figure 6* and *Figure 7* show typical truck cranes. Note that some large truck cranes must have their counterweights removed before they are driven on the highway due to handling and highway-loading concerns.

Some wheeled truck cranes that are designed to be superior in off-road situations. Known as rough-terrain cranes (*Figure 8*), they are often found on construction sites due to their versatility, maneuverability, and ability to handle the

Figure 6 Telescopic-boom truck crane with rotating controls.

Figure 7 Lattice-boom truck crane with rotating controls.

Figure 8 Rough-terrain crane with rotating controls.

terrain. Rough-terrain cranes are usually dual-axle carriers with telescopic booms and a relatively short wheelbase. They typically have larger tires and more versatile steering capabilities than common wheeled truck cranes. Steering options include both two-wheel and all-wheel steering capabilities. Two-wheel and all-wheel drive capabilities are also available. Having only two axles and a short wheelbase allows the crane to turn sharply and take full advantage of the four-wheel steering capability.

Many rough-terrain cranes have a single cab. The crane is both driven and operated from the same cab. The cab may swing with the boom (a crane with rotating controls, referred to as a *swing cab*) or remain fixed (a crane with fixed or stationary controls, referred to as a *fixed cab*). Rough-terrain cranes generally travel at less than highway speeds (about 30 mph, or 48 kph) due to their gear ratios, and their short wheelbase delivers a poor highway ride. As a result, they are usually transported to the job site on low-boy trailers. Heavier lifts are made over the front of the crane, as that is its position of greatest lift capacity.

There are two small crane types that might be considered a subset of rough-terrain cranes. Pick-and-carry cranes are often based on rough-terrain designs. A true pick-and-carry crane, also called a *cherry picker*, does not have outriggers, as it is designed to lift a load and travel (at low speeds) with it suspended. However, they are generally designed for highway travel from site to site. Another small crane type is the carry-deck crane, which is considered an American version of the pick-and-carry crane. The carry-deck crane has a small deck that can accommodate a load as it is moved from one place to another. While some are suitable for highway driving, others are not. They can be designed with rough-terrain

characteristics, or be designed for shipping/receiving, fabrication shop, or similar duties where rough terrain characteristics are not important.

There is one additional wheeled truck crane type that fills the void between standard models that excel on the highway and those that excel in rough territory—the all-terrain crane (*Figure 9*). All-terrain cranes can be considered a hybrid between standard wheeled truck cranes and rough-terrain cranes. This type of crane is quite capable of highway travel. At the same time, it is also designed to traverse reasonably rough terrain, within limits.

All-terrain cranes typically have more axles and are available in greater capacities than rough-terrain models. For example, the largest rough-terrain crane presently offered by one company has a nominal capacity of 135 tons (122 metric tons); the largest all-terrain model offered by the same company is rated at 450 tons (408 metric tons). All-terrain cranes are not generally as maneuverable in tight quarters as rough-terrain models because their length and additional axles limit maneuverability. They also may not be equipped with the same steering options that are available on rough-terrain models. However, they do offer performance on the highway as well as on a rough job site, filling the needs of many users. All-terrain cranes generally set-up quick and have long boom capabilities with additional jibs for added reach. These features make them a favorite with crane rental services.

As a general rule, wheeled truck cranes do not offer the stability of a crawler crane. As a result, the vast majority of them are equipped with outriggers. In spite of its ability to maneuver in challenging terrain, no mobile crane is designed to be used on unstable ground or when it is not level. Outriggers are designed to provide stability and a means to level the crane. Lift capacity and safety are both increased as a result. The outriggers should remove the weight of the crane and the lifted load from the wheels and tires altogether. *Figure 10* shows the two most common types of hydraulic outriggers. Cantilever outriggers are used on smaller cranes, since their reach is limited; larger cranes require outriggers of greater strength and able to provide a wider stance. Boom trucks and other small cranes may have outriggers that must be deployed and positioned manually.

1.1.3 Railcar-Mounted Cranes

Railcar-mounted cranes (*Figure 11*) can also have either a telescopic or a lattice boom. The two types of booms have different weight capacities and lift heights. Like lattice-boom crawler cranes, the lattice boom railcar-mounted crane has greater

NCCER – *Mobile Crane Operations*

Figure 9 All-terrain crane.

weight capacity and height potential than a telescopic-boom version. However, the telescopic boom offers greater versatility and flexibility, as well as quick setup, when that is an important consideration. Self-propelled railcar-mounted cranes are used primarily for railroad track and bridge repair, railyard maintenance, loading and unloading rail freight, and to clear train derailments.

1.2.0 Crane Booms

Cranes not only differ by means of travel; they also differ in the types of booms they employ. In addition, there are special accessories that can be attached to, or suspended under, the crane boom.

1.2.1 Telescopic Booms

Telescopic-boom cranes are among the most widely used types of mobile cranes. This is partially because of their ease of setup and teardown. Hydraulic power for a wide variety of applications expanded dramatically in the 1950s, making the telescopic boom possible. The simplest telescopic-boom crane is the boom truck (*Figure 12*). Boom trucks are commonly used to transport and deliver or pick up materials. The boom is operated from outside the cab, in a standing position. Note that the deck area of the crane can be used to move materials from one place to another, and then be unloaded by the crane.

Telescopic booms are sectional, as shown in *Figure 13*. The crane shown here has a total of four boom sections, one of which serves as the

housing for the other three. A headache ball is often attached to the load line to provide sufficient weight to allow the load line to unspool from the drum when there is no live load. Telescopic booms allow the operator to change the length of the boom at any time by extending or retracting the boom sections, even while the crane is loaded. Hydraulic actuators, often hidden from view, extend and retract the individual sections. The base section—the lower portion of the boom that does not extend—is raised up and down by one or more hydraulic actuators. Operators must be very careful when extending the boom of a crane, as the crane's capacity changes dynamically as the boom length changes. This increases the possibility of the crane tipping over.

1.2.2 Lattice Booms

Lattice booms can be much longer than most telescopic booms. However, lattice-boom cranes are more time-consuming to prepare for a lift, and to prepare for travel. The site preparation and erection of lattice-boom cranes may take anywhere from a day for a smaller crane to possibly weeks for very large cranes. The length of setup time depends on the length of the boom and any attachments to be used.

Newer lattice-boom cranes are powered by computer-controlled systems. Older models are very mechanical in their control, using complex linkage. Computer-controlled models have redundant fail-safe devices to assure operational safety. These devices consist of many components, including: automatic braking systems,

(A) CROSSBEAM OUTRIGGERS

(B) CANTILEVER OUTRIGGERS

Figure 10 Hydraulic outriggers.

load-moment measuring devices, mechanical-system monitoring devices, and function-lockout systems that stop operation when a system fault occurs. Such controls provide for better safety, control, and accuracy than older friction-operated models.

As noted earlier, lattice booms are available on both crawler and wheeled truck cranes. Lattice-boom crawler cranes are specifically designed for heavy-duty service. The reliability and versatility of the lattice-boom crawler crane makes it a widely applied crane design. Besides lifting heavy loads to great heights, these cranes are also ideal for applications with a high **duty cycle**. The duty cycle refers to the consistency and repetitive nature of the work over a period of time. *Figure 14* shows an example of a lattice-boom crawler crane. The load block shown connects and moves up and down with the load, riding on the crane **reeving**.

Lattice-boom truck cranes provide the mobility of a truck crane with the extreme lifting capacity of a lattice-boom crane. Depending on the size of the crane and its gross vehicle weight, some components such as boom sections, counterweights, and outriggers usually have to be removed before highway travel to meet local, state, and federal weight and physical size restrictions. *Figure 15* shows a typical lattice-boom truck crane. Many parts of the crane and boom are identified as well. The identified parts of the lattice-boom assembly are the same for crawler cranes.

There are a number of possible configurations for lattice booms, as shown in *Figure 16*. Lattice booms that are sectional (very common) begin with a boom base—the section that attaches to the upperworks of the crane. Additional sections can be added to increase boom length. A lattice boom is supported by a network of wire ropes or lines called **pendants**. The boom **hoist reeving** and boom pendants are used to raise and lower the boom. If the crane is rigged with a jib, it is supported by a **jib forestay** and a **jib backstay**. Note that a jib may also be referred to by some in the industry as a *fly*. A **luffing jib** is raised and lowered with jib hoist reeving and luffing jib pendants, as shown in *Figure 15*. The process of adjusting the jib angle is called **luffing**. Both fixed

Inventor of the Hydraulic Crane

Sir William Armstrong (1810–1900) is generally credited with development of the first hydraulic crane. A scientist and inventor, he was responsible for the first home in England to be powered by hydroelectricity. He is also considered the father of modern artillery.

Armstrong was inspired to create a hydraulic crane (in this case, the powering fluid was water) as part of a water-piping project to move water from reservoirs to distant homes in Newcastle. The water pressure was more than needed in areas of lower elevation. He proposed to use the excess water pressure to power a crane for unloading ships more effectively. The experiment proved extremely successful and more cranes were added to further increase the efficiency of the harbor.

NCCER – *Mobile Crane Operations*

Figure 11 Railcar-mounted crane working in the yard.

CRANE CONTROLS

Figure 12 Telescopic-boom truck with stand-up control.

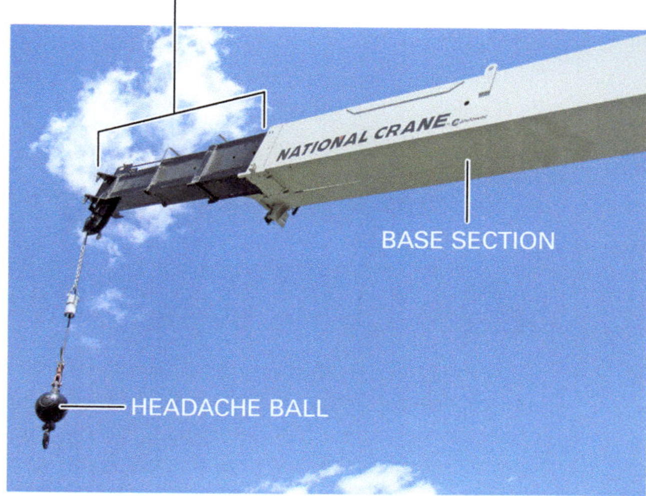

Figure 13 Sections of a telescopic boom.

and luffing jibs require a **jib mast**, or jib strut, to support the forestay/backstay or the jib hoist reeving. The load hoist line and its reeving are used to actually lift the load. Note that some of the captions shown in *Figure 16* refer to an **open-throat boom**. This refers to an opening in the boom structure near the tip. The boom tip is open on the bottom side and is designed to provide clearance for multi-part reevings at a high boom angle when the main **sheaves** are positioned along the centerline of the boom. When the main sheaves are offset below the center line of the boom, the need for an open-throat boom is eliminated. Many new lattice booms manufactured at present are built with the main sheaves offset to eliminate the open-throat design.

Figure 14 Lattice-boom crawler crane.

1.2.3 Jibs

The boom is sometimes extended by adding a jib (*Figure 17*). A jib can be added to either a telescopic boom or to a lattice boom. The jib may be fixed in position to the boom, or it may be a luffing jib. A luffing jib allows the angle between the main boom and jib to be adjusted. An adjustable jib angle provides more flexibility in the lifting arrangement, allowing the crane to reach areas that may not be accessible otherwise. Standards require all jibs to have a mechanical stop that will prevent the jib from being positioned more than 5 degrees above the center line of the main boom.

The lifting capacity of a jib declines as the jib moves closer to a horizontal position. *Figure 18* demonstrates the reduction of capacity at different jib angles.

1.3.0 Crane Attachments

Various attachments have been developed to enable a crane to perform specific tasks beyond simple lifting. These attachments allow cranes to be used for tasks such as excavation, concrete pouring, scrap or debris removal, and loose-material lifting. Refer to *ASME Standard B30.20* for below-the-hook lifting device requirements.

1.3.1 Specialty Buckets

Cranes are often used to pick up and move materials such as dirt, concrete, or even water. Specialty buckets designed for these tasks include clamshell buckets, concrete buckets, and drag buckets.

Clamshell buckets (*Figure 19*) are hinged at the top center, allowing the bottom of the bucket to open and close like a clamshell. Digging or dredging clamshell buckets are built with counter weights, levers, and teeth that enable the bucket to open up and dig sharply into the ground as it is lowered. As it is lifted, the clamshell closes and

Duty-Cycle Work

You may hear the terms *duty-cycle crane* and *lift crane* used at times. A duty-cycle crane is generally considered to be a crane designed for long-term, repetitive work that may not approach the maximum lift capacity of a crane. Lift cranes need to be lightweight and maneuverable enough to get to where the load is. A duty-cycle crane generally works in a single area, perhaps for years at a time. Duty-cycle cranes generally have larger engines and dissipate heat more efficiently. Duty-cycle cranes also need faster hoist/fall speeds and hydraulic systems with individual pumps dedicated to a task. Lift cranes often use one or more hydraulic pumps that provide pressure to a manifold that supplies all of the cranes systems with pressure as needed. Many duty-cycle cranes are fixed in position, but mobile cranes are also needed for duty-cycle applications. The dredging activity shown here is a good example of duty-cycle work for a mobile crane.

Figure Credit: Liebherr USA, Co.

NCCER – *Mobile Crane Operations*

Figure 15 Lattice-boom truck crane.

removes a bucket load of material. Rehandling clamshell buckets operate on the same principle as digging clamshell buckets but, instead of digging, are used to move free-flowing stockpile materials such as coal, fertilizer, or wood chips. Clamshell buckets for cranes are available with single, two-, and three-rope operating configurations to match the control capabilities of the crane.

Concrete buckets (*Figure 20*) are used to lift loads of cement up to a pour site. Concrete buckets may lift as much as 8 cubic yards (6 cubic meters) to workers well above ground level. They may be used to lift and dispense cement under the complete control of the crane operator, or the bucket may be controlled by another party on the site using a remote control.

Drag buckets are used in dredging, mining, or material-handling operations. They can also handle large loads of 8 cubic yards (6 cubic meters) or

more. Drag buckets are equipped with teeth and are pulled across a surface by an auxiliary winch on a crane. When full, they are lifted by the crane, and the contents are typically dispensed into a truck or railcar, or onto a barge.

Self-dumping bins are also popular for moving loose or dry, granular materials. The type shown in *Figure 21* is not controlled or tipped through a rope or cable. The crane operator merely sets the bin down on a firm surface and then moves the boom slightly to allow a latch to open on the bin lifting arm. As the bin is then lifted with the latch open, the contents are dumped in a controlled manner. The latch can be reset by again placing the bin on a firm surface and moving the boom slightly. This is a simple design that allows a crane operator to function more independently and efficiently.

Figure 16 shows various lattice-boom configurations on crawler cranes, labeled:

Figure 16 Examples of lattice-boom configurations on crawler cranes.

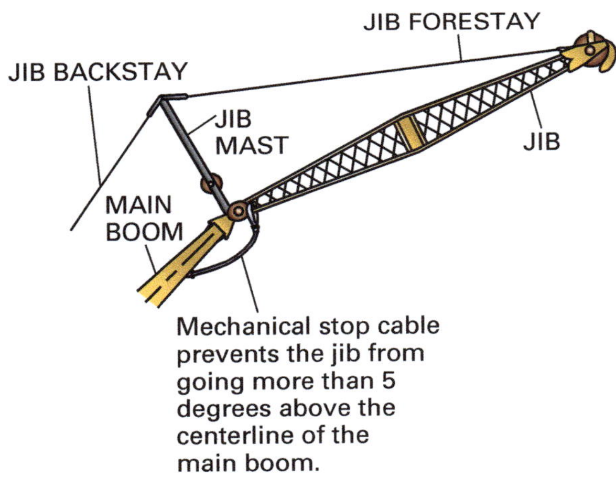

Figure 17 Typical jib.

1.3.2 Grapples and Magnets

Grapples and magnets are used to pick up and move materials that are irregular in size and shape such as scrap metal or tree branches. Grapples (*Figure 22*) usually have from two to six tines that interlock, acting much like fingers to grab material. The style of grapple shown here is sometimes referred to as an orange-peel grapple. Grapples can be specifically designed for scrap metal, rocks, logs, or brush.

Magnets are used to pick up magnetic metal objects. *Figure 23* shows an electromagnet designed for crane use. The crane operator is usually in control of the power supply to the electromagnet. The magnet is energized to pick up a load of metal, and de-energized to drop the load. Remember that magnets do not work on **nonferrous** materials such as aluminum, brass, and copper.

1.3.3 Special Lifting Devices

Cranes can be fitted with clamps and similar devices designed to easily lift and move specific and unique loads. Clamps can be attached to a crane to move stacks of bricks and blocks, concrete highway barriers, pipe, and many other objects. Product manufacturers will also build custom clamps and hooks for specific tasks. *Figure 24* shows several types of specialty accessories. Small clamping devices are also available, designed to clamp onto and lift sheet goods such as steel plate and sheet piles.

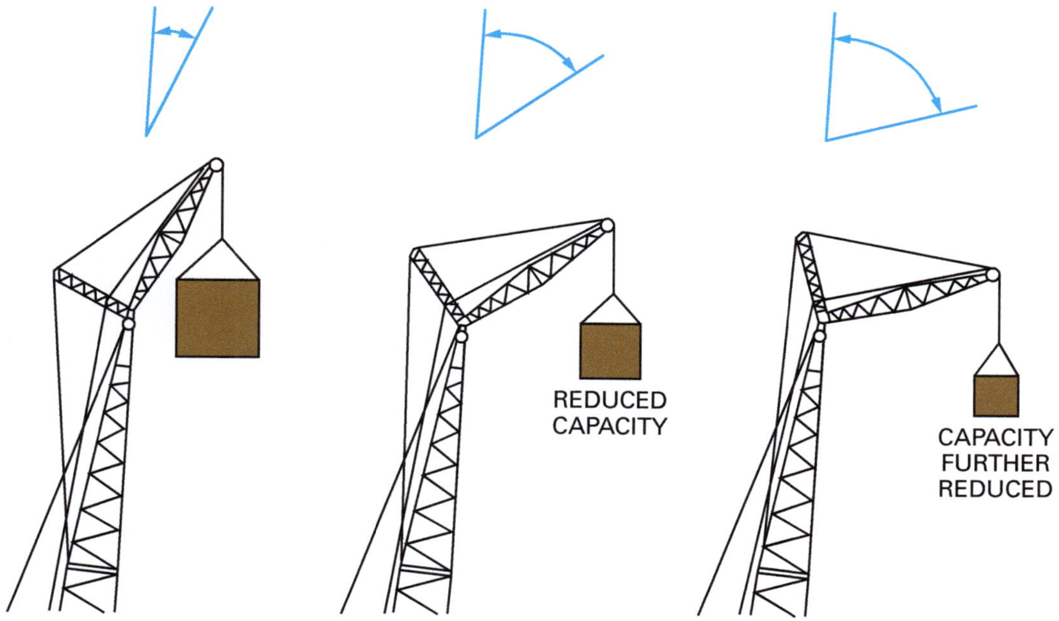

Figure 18 Reduction of capacity as the jib is lowered.

The Spiderman of Cranes

Sometimes you need lifting capability in places where a crane cannot be driven, such as inside a building or many miles into the desert. This mini-crawler, with tracks tucked underneath the body, can get to the spot. Once in place, the spider-like legs are extended, lifting the rig off of its tracks and creating a stable lifting platform. Manufactured under the name of SpyderCrane®, these cranes are extremely versatile. Glaziers installing glass in tall buildings, for example, use special suction-cup accessories that allow the crane to pick up large sheets of glass and position them in the wall opening for installation from inside the building. There are also used for remote military and civil operations, where they can be flown in and dropped. There is no cab; they are remotely controlled.

Figure Credit: Smiley Lifting Solutions - www.spydercrane.com

Figure 19 Clamshell bucket.

Figure 20 Concrete bucket.

Figure 21 Self-dumping bin.

Figure 22 Grapple.

1.3.4 *Personnel Platforms*

There are occasions when cranes are needed to lift personnel. 29 *CFR* 1926.1431 provides guidance in this area. The standard specifies that lifting by crane is only allowed if any other means to do so represents a greater hazard, or if there is no alternative. It also provides guidance in crane operation, the construction of the personnel platform, and other relevant areas. *ASME Standard B30.23, Personnel Lifting Systems,* is also dedicated to this topic.

Figure 25 shows a personnel platform that meets the OSHA and ASME requirements for a personnel platform. Platforms are available in a number of shapes and sizes. Some have no roof, while others have a fixed or removable roof section. Those with roof sections provide a greater margin of safety.

1.3.5 *Pile Drivers*

Cranes, especially lattice-boom crawler cranes, are an essential part of heavy pile-driving operations. Pile driving is heavy, repetitive work, imposing a great deal of wear and tear on the equipment. Lattice-boom crawler cranes are uniquely suited for the task.

There are two primary types of pile drivers typically managed by cranes: pile hammers and vibratory pile drivers. Pile hammers, either diesel-powered or hydraulic, represent the traditional pile-driving equipment class. They generally require a set of leads that support and position the hammer and pile, allowing the hammer to move freely up and down as the pile is driven. *Figure 26* shows a diesel-powered pile hammer and a pile in a set of swinging leads. Pile hammers like this

NCCER – *Mobile Crane Operations*

Figure 23 Crane electromagnet.

are almost exclusively handled by lattice boom crawler cranes.

Vibratory pile drivers use high-frequency vibration and their own weight to drive piles. They are best suited for driving sheet piles, but they can also drive concrete and wooden piles in favorable soil conditions, as shown in *Figure 27*. The vibration is created by an external hydraulic power source. Although lattice-boom crawler cranes are often used, vibratory units can be used with telescopic-boom cranes as well. A limited amount of vibration and stress is transmitted to the crane through the ropes.

1.3.6 Counterweights

The tipping force created by the weight of the boom and the lifted load and must be countered or the crane will tip over. Large counterweights are mounted to the backs of crane upperworks to balance the loads, increasing lift capacity. *Figure 28* shows a heavy-lift crane fitted with counterweights at the back of the upperworks to directly counterbalance the expected load. The weights are modular, so that weight can be added or removed incrementally. Counterweights can also be added to the crawler frames, as shown, to provide additional counterbalance for lifting over the side.

1.4.0 Safety Devices and Operational Aids

Prior to any mobile crane movement or operation, the operator should note the location of all the operating gauges and check their current readings. Instruments and gauges should be located in clear view of the operator, by design. The operator must always be aware of the status of the crane and keep a constant eye on the controls as well as the load. The first indication of a crane system failure is likely to come to an operator's attention through the instrumentation. *Table 1* describes the most basic instruments and gauges related to a crane's power plant. These instruments are similar to those provided in today's cars.

29 *CFR* 1926.1415 and 1926.1416 list the safety devices and operational aids that are required on mobile cranes. Of course, many more such devices are often applied to cranes; the OSHA standards outline the minimum requirements. The list of safety devices required by 29 *CFR* 1926.1415 includes the following items:

- A crane level indicator, either built into the equipment or readily available on it
- Boom stops, except for derricks and telescopic booms
- Jib stops, except for derricks
- A brake-locking feature on cranes equipped with foot-pedal brakes
- An integral holding device or hydraulic check valve for hydraulic outrigger and stabilizer jacks
- Rail clamps and stops on railcar-mounted cranes
- A working horn

The standard requires that all the devices above be in proper working order. If one or more of these devices fail to operate properly during a lift, the operation must be brought to a safe halt.

1.4.1 Operational Aids

29 *CFR* 1926.1416 outlines the operational aids required for mobile cranes. Like safety devices, the specified aids must be in proper working order; their failure during a lift requires the operator to bring the operation to a safe halt. However, the standard does allow specific periods of time for repairs to be made as long as specific alternative measures are taken to replace the function of the aid. For example, if a boom hoist limiting device has failed, the standard lists three acceptable alternative practices to be used while repairs are planned and executed.

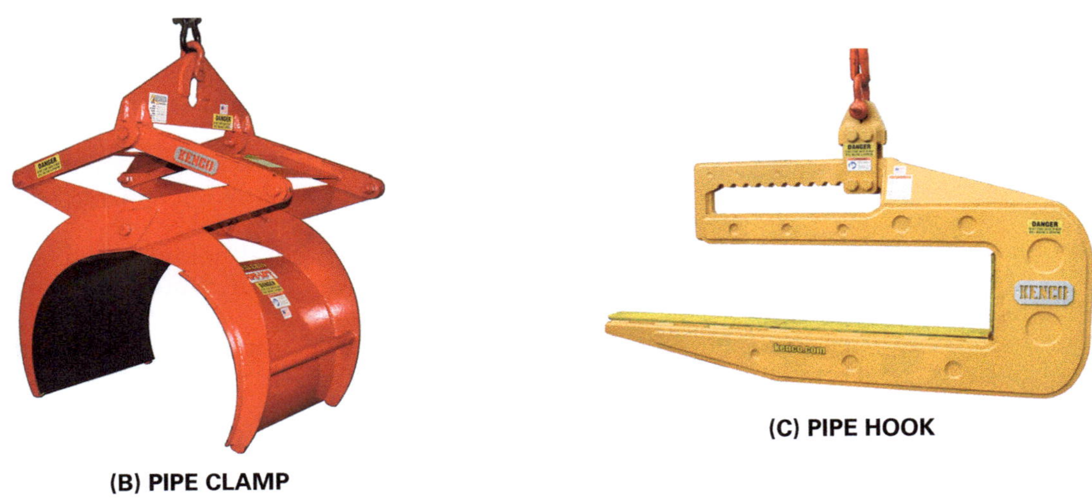

Figure 24 Special lifting devices.

Diesel Fuel

Diesel engines have long been preferred for heavy, hard-working motor vehicles of all types. Diesel fuel however, produced significant amounts of pollutants for many years. The sulfur content of diesel fuel has been one of the primary concerns. Nearly 100 percent of the sulfur in fuel is emitted in the form of sulfur oxides, primarily sulfur dioxide (SO_2). SO_2 causes respiratory problems and reacts with other airborne chemicals to form sulfate particles that collect in the lungs.

The US Environmental Protection Agency (EPA) has been working to reduce the pollutants that originate from diesel fuel through the issuance of quality standards. Although the fuel has certainly risen in cost, the sulfur content of diesel fuel has been reduced to a tiny fraction of what it once was. Today's diesel fuel is referred to as ultra-low sulfur diesel, or ULSD, and contains less than 0.5 percent of the sulfur that it contained in the 1980s. The emission standards for diesel engines have also been steadily improved to reduce their impact on the environment.

Figure 25 Personnel lifting platform.

Figure 26 Pile hammer in leads.

OSHA has placed the required list of operational aids into two designated categories: Category I and Category II. Category I devices must be repaired within seven days; Category II devices must be repaired within 30 days. The list of Category I operational aids includes the following:

- Boom hoist limiting device
- Luffing jib limiting device
- **Anti-two-blocking device**

Category II operational aids include the following:

- Boom angle or radius indicator
- Jib angle indicator (if a luffing jib is installed)
- Boom length indicator (for cranes with a telescopic boom)
- Load-weighing device
- Outrigger/stabilizer position monitor (only required for recently manufactured equipment)
- Hoist drum rotation indicator (for hoist drums not visible to the operator; only required for recently manufactured equipment)

There are a number of exceptions included in the standards, such as those related to the age of the crane. For additional details on these exceptions, consult the listed OSHA standards (found in the *Code of Federal Regulations* [*CFR*]) directly.

Newer cranes may be equipped with sophisticated monitoring panels, providing the operator with a great deal of information and control in real time. Such panels and systems fall into the broad category of human-machine interfaces (HMIs). HMIs also include joysticks and similar electronic controls. Crane manufacturers often have specific names for their HMI systems, such as the Electronic Crane Operating System (ECOS) from Manitowoc and the LICCON System from Liebherr.

Graphical displays and color are extremely useful features that keep the operator constantly informed of the crane's capacity and **load moment**. Load charts are built into the interface, and information about the lift can be programmed into the system during the planning stages.

Touch-screen control is also becoming more common. *Figure 29* is an example of one HMI panel and some of the functions they typically provide. Remember however, that every crane brand and model is different, and new features are consistently being introduced. Documenting all of the control options available in today's cranes is virtually impossible. Each crane operator must study and understand each individual crane's features and controls.

Figure 27 Vibratory pile driver at work.

Figure 28 A lattice-boom crawler crane with
counterweights.

Table 1 Common Crane Instruments and Gauges

WATER TEMPERATURE GAUGE	This gauge displays the current engine coolant temperature of water-cooled engines. It monitors the coolant through a sensor unit located in the engine block. This gauge is not present on air-cooled engines.
ENGINE OIL PRESSURE GAUGE	This gauge lets the operator know the status of the crane engine oil pressure.
VOLTMETER (BATTERY VOLTAGE)	Prior to starting the crane but with the key in the ON position, this gauge displays the crane battery's state of charge. With the crane running, this gauge shows the voltage output of the alternator.
FUEL GAUGE	With the ignition in either the ON or ACC position, this gauge indicates how much fuel is in the crane engine fuel tank.
TACHOMETER	This instrument indicates the crane engine's rotations per minute (rpm).

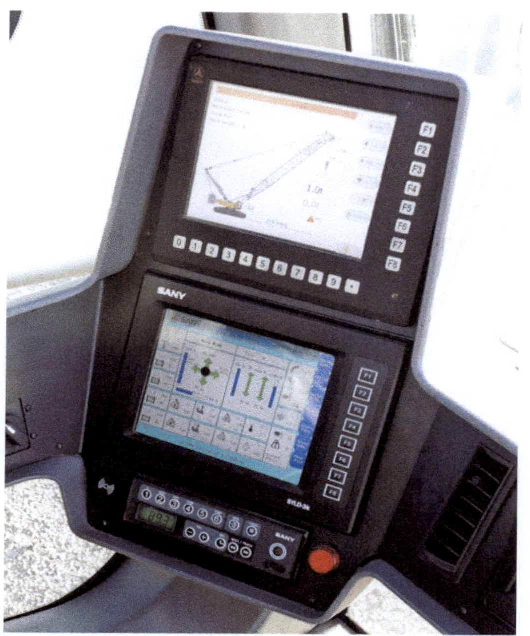

Figure 29 Crane operation monitoring panel.

« HMI PANEL FOR CRANE OPERATION MONITORING – COMMON FUNCTIONS:

- Graphic representation of machine configuration
- Graphical step-by-step machine set-up
- Boom length & angle
- Jib length & angle
- Load on hook
- Rated load
- Load radius
- Tip height
- Anti-two block warning & function limiters
- Operating mode
- Audio/visual warning when the load on hook is within a preset percentage of the crane's rated load
- Audio/visual warning and limits functions when the load on hook is at a preset percentage of the crane's rated load
- Operator settable alarms

Note that all warning plates, both inside and outside the cab, should be kept readable and in place. They should be considered an integral part of the equipment, and should not be defaced or removed.

1.5.0 Crane Reeving

Reeving for cranes refers to the routing of ropes or cables through the swallow and around the sheaves of a block. Mobile cranes use the same single- and multiple-part reeving patterns as manually powered block and tackle arrangements use to multiply power.

When a single hoist line is reeved for a lift, the line must be run on the center sheave, or on the sheave next to the center when an even number of sheaves exists. Reeving the line on a sheave at or near one side of the boom causes boom torque that could damage the boom. The twisting of the boom also causes the hoist line to rub against the side of the sheave, causing excessive wear to both the sheave and the line. *Figure 30* demonstrates the effect of boom torque. Single-line reeving

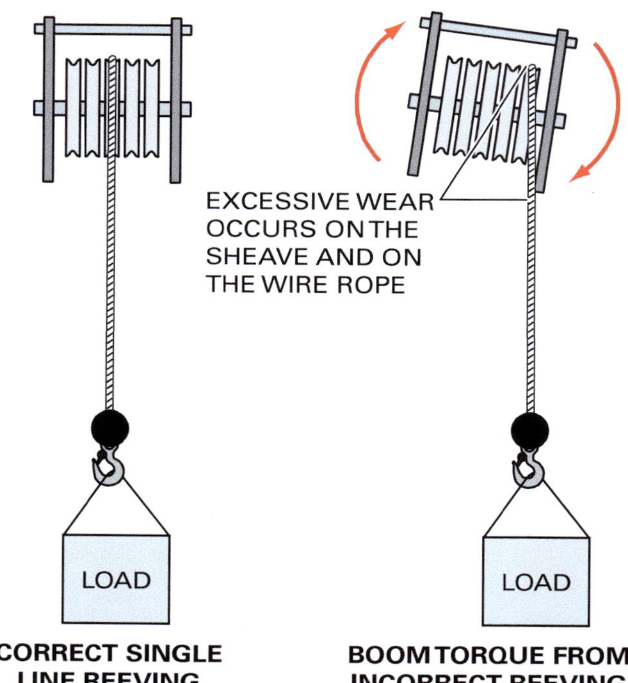

Figure 30 Effect of incorrect single-line reeving.

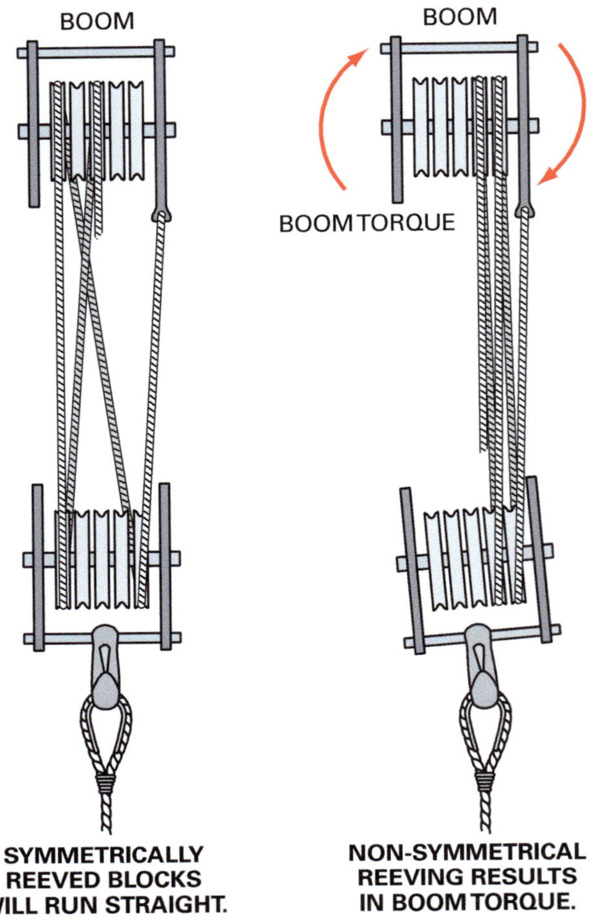

Figure 31 Balanced and unbalanced reeving.

may be used on smaller cranes or boom trucks, and for the whip line on larger cranes.

It is also important to reeve the boom and load block symmetrically for multiple-line lifts. If the load block is not reeved symmetrically and balanced, it will not run straight and will cause the same type of wear associated with boom torque. *Figure 31* shows balanced and unbalanced reeving for multiple-line lifts.

When a line is reeved more than once, the resulting number of lines that are supporting the load block are referred to as parts of line. *Figure 32* shows the correct reeving patterns for three- to five-part reeving. *Figure 33* shows the correct reeving patterns for six- to eight-part reeving. Very large cranes may require even more parts. Using these patterns will result in balanced reeving and smooth-running blocks with minimal boom torque. Notice that the hoist line terminates at the boom for even-numbered parts of line, but terminates at the load block for odd-numbered parts of line. Additional information on wire rope and the reeving process is provided in NCCER Module 21204, "Wire Rope" from *Mobile Crane Operations Level Two*.

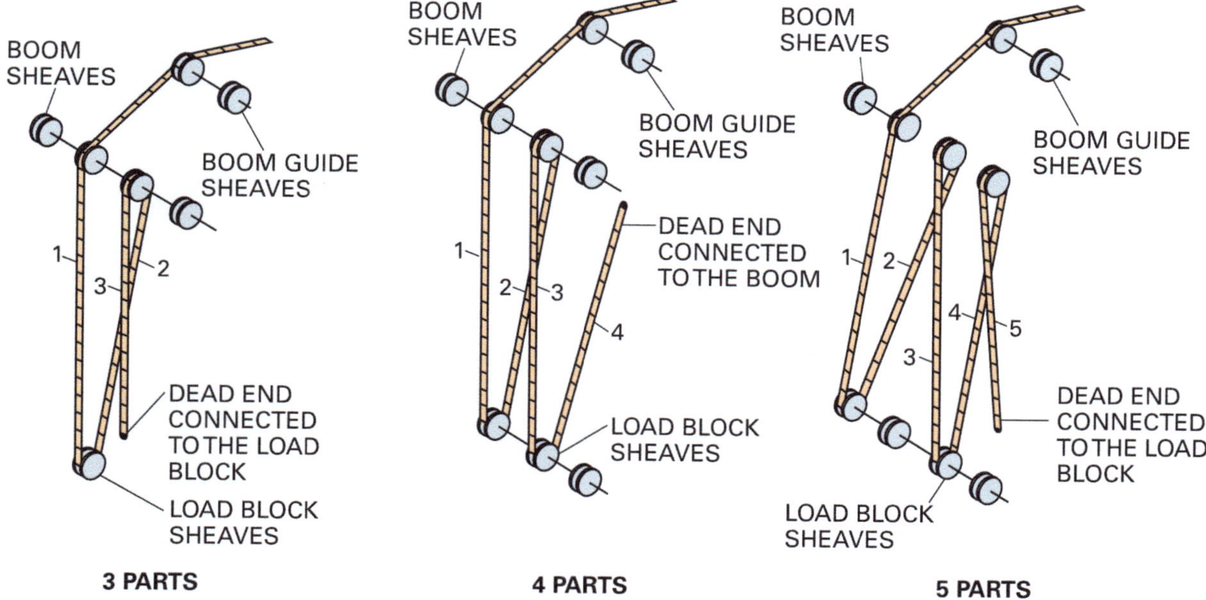

Figure 32 Three-, four-, and five-part reeving.

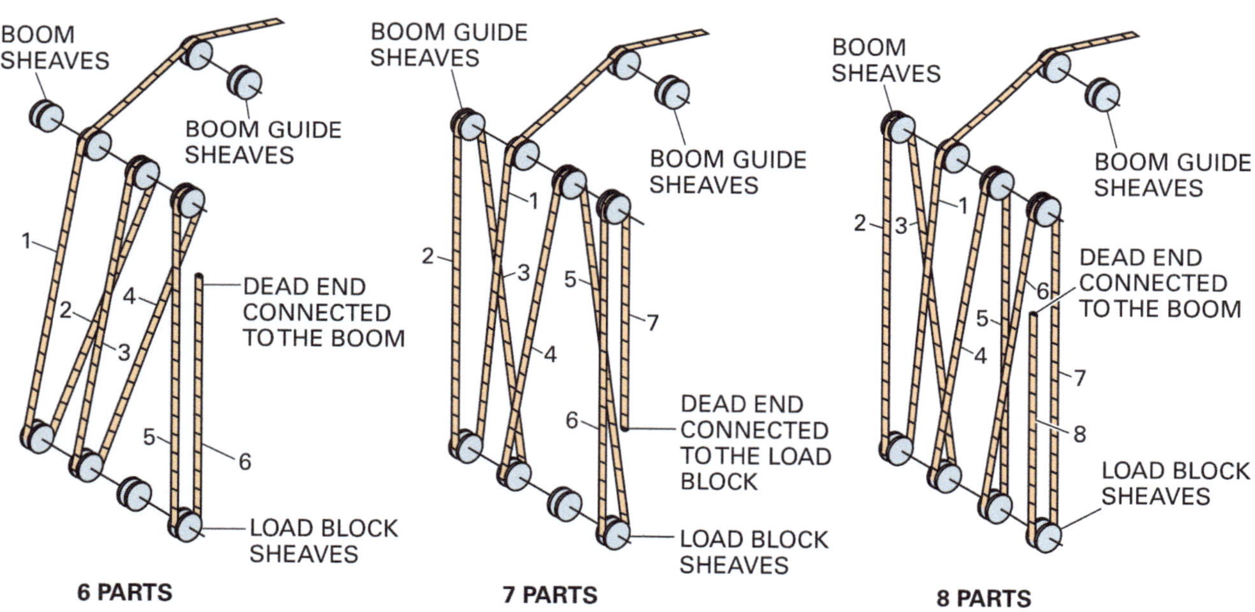

Figure 33 Six-, seven-, and eight-part reeving.

Additional Resources

ASME Standard B30.5, Mobile and Locomotive Cranes. Current edition. New York, NY: American Society of Mechanical Engineers.

29 CFR 1926.1400, **www.ecfr.gov**

Mobile Crane Safety Manual (AEM MC-1407). 2014. Milwaukee, WI: Association of Equipment Manufacturers.

Cranes: Design, Practice, and Maintenance, Ing J. Verschoof. Second Edition. 2002. Hoboken, NJ: John Wiley and Sons, Inc.

IPT's Crane and Rigging Handbook, Ronald G. Garby. Current Edition. Spruce Grove, Alberta, Canada: IPT Publishing and Training Ltd.

North American Crane Bureau, Inc. website offers resources for products and training, **www.cranesafe.com**

1.0.0   Section Review

1. A crane with rotating controls has _____.
   a. an operator's cab that swings with the boom
   b. an operator's cab that remains stationary as the crane rotates
   c. a control console operated from a standing position
   d. controls mounted directly on the boom

2. If a jib can be adjusted to change its angle in relation to the main boom, it is called a(n) _____.
   a. lattice jib
   b. hydraulic jib
   c. luffing jib
   d. open-throat jib

3. A rehandling clamshell bucket is designed to _____.
   a. lift large volumes of water or oil
   b. dig into and lift hard soil
   c. pick up tree stumps and boulders
   d. pick up free-flowing material

4. Which of the following is a safety device required by OSHA standards?
   a. A brake-locking feature on cranes with foot-pedal braking
   b. A boom angle indicator
   c. A jib angle indicator
   d. An electronic crane operating system (ECOS)

5. When a single hoist line is reeved for a lift and there are an even number of sheeves, the hoist line must be run on the sheave _____.
   a. nearest the crane
   b. nearest the boom
   c. next to the center
   d. furthest from the load

2.0.0 BASIC LIFTING CONCEPTS

Objective

Identify factors related to lifting capacity and explain their significance.

a. Explain the significance of ground conditions and a level surface.
b. Describe the bearing surface and explain how to determine the required blocking.
c. Define and describe the significance of the center of gravity and the quadrants of operation.
d. Describe the significance of boom length, angle, operating radius, and elevation.
e. Explain how to use a load chart and understand the basic concepts of critical lifts.

Performance Tasks

1. Verify the boom length of a telescopic- and/or lattice-boom crane using manufacturer's data or a measuring tape.
2. Measure the operating radius of a telescopic- and/or lattice-boom crane using a measuring tape.
3. Calculate the amount of blocking needed for the outrigger of a specific crane.
4. Verify that a crane is level.

Trade Terms

Backfill: Soil and rock used to level an area or fill voids, such as the perimeter of building foundations or trenches. Areas with fresh backfill may not be stable enough to support a crane.

Center of gravity (CG): The point at which the entire weight of an object is considered to be concentrated, such that supporting the object at this specific point would result in its remaining balanced in position.

Critical lift: Defined in *ASME Standard B30.5* as a hoisting or lifting operation that has been determined to present an increased level of risk beyond normal lifting activities. For example, increased risk may relate to personnel injury, damage to property, interruption of plant production, delays in schedule, release of hazards to the environment, or other significant factors.

Dynamic loads: A load on a structure (in this case, a crane) that is not constant, but consistently changing as the result of one or more changes in various factors. Also referred to as shock loading, significant dynamic loads can be applied to a crane through abrupt motions and lifting a load from its support too quickly.

Effective weight: The weight of an accessory such as a boom extension or jib that reflects the effect of its weight on the lift, usually based on its position, rather than its actual weight. For example, a jib folded and stored on the main boom will have different effective weight than when it is installed on the main boom tip.

Floats: The portion of outriggers that touches the ground; the feet of the outriggers.

Gross capacity: The total amount a crane can safely lift under a given set of conditions. The gross capacity includes but is not limited to the load block, ropes, and rigging as well the primary load.

Hardpan: A hard, compacted layer of subsoil, usually with a major clay component.

Impact load: The dynamic effect on a stationary or mobile body as imparted by the forcible contact of another moving body or the sudden stop of a fall.

Interpolation: The process of estimating or calculating unknown values between two known values.

Leverage: The mechanical advantage in power gained by using a lever.

Net capacity: The weight of the item(s) that can be lifted by the crane; the gross capacity of a crane minus all noted capacity deductions.

Operating radius: The distance from the center of the boom's mounting point (usually the ring gear drive) to the center of gravity of the load.

Quadrant of operation: The direction of the boom relative to the base mounting or carrier body.

Tipping fulcrum: The point of crane contact with the ground where it would pivot if it were to tip over; the fulcrum of the leverage applied by the load. Depending on the attitude and type of crane, the tipping fulcrum may be the edge of one crawler assembly, one or more outriggers, or similar locations.

The lifting capacities of mobile cranes are affected by a wide range of conditions, from the type of soil beneath them to the height of the boom tip. These conditions include but are not limited to the following:

- Condition of the ground
- Crane being positioned within 1 degree of level
- Type and size of the bearing surfaces
- Type of crane (wheeled versus crawler)
- Type of boom
- The crane's capacity and actual configuration
- The center of gravity (CG) as the lift progresses
- Quadrant of operation
- Boom length, boom angle, and the resulting operating radius
- Swing out, side loading, and dynamic loads

Each of these factors will be explored further as you progress through the section.

2.1.0 Ground Conditions and Leveling

Ground conditions are a major factor in the ability of any crane to perform a lifting task safely. Even a small crane with a light load can overturn if the ground beneath gives way. In addition, the crane must be in a level position to ensure stability. It is important to note that load charts are only valid if the crane is level. Load charts assume a level condition.

2.1.1 Ground Conditions

Ground conditions around a job site can vary widely. Some areas may have been graded down to a hardpan, while excavations or depressions may have been filled but not compacted. Mobile cranes may encounter many different conditions just moving from one side of the site to the other. The ground beneath the crane must be able to support all the loads that are placed on it. This includes the weight of the crane, the load, and all rigging. In addition, there are impact loads that occur and dynamic loads that result from swinging, hoisting, lowering, and traveling. *Table 2* shows the weight-bearing capacities of various soil types.

Table 2 Bearing Capacities of Various Soil Types

SOIL	CAPACITY	
Cemented sand & gravel	135 PSI	931 kPa
Sand & gravel compact	110 PSI	758 kPa
Sand, coarse to medium compact	60 PSI	414 kPa
Sand, fine to silty compact	54 PSI	372 kPa
Clay compact	54 PSI	372 kPa
Silt compact	40 PSI	276 kPa

The backfill adjacent to new buildings is often not thoroughly compacted and cannot provide proper support for cranes. Other dangerous areas include the edges of excavations. These areas may give way beneath the crane as it approaches to place piping or tanks into the excavation. This is also dangerous for any workers who may be in the trench. Existing sewer and water lines pose hazards as well. Excavations or grading may remove enough covering soil so that the weight or vibration of a crane traveling or working above the line could cause it to collapse. *Figure 34* shows examples of each of these ground condition hazards.

For working near excavations, any crane component bearing weight (such as an outrigger) must be positioned at least 1.5 times the depth of the excavation away from the edge. For example, if a trench is 4 feet (1.2 meters) deep, the outrigger floats should be no closer than 6 feet (1.8 meters) from the edge. For crawler cranes, the leading edge of the track should also be positioned at least this far away. Different soil conditions may require that this distance be increased.

> **WARNING!**
>
> The clearance distance calculation for excavations provided here, although common, is not sufficient for every situation. Ensure that all factors related to the required clearance have been considered before positioning a crane or any other heavy equipment near an excavation. The unexpected failure of an excavation wall while a crane is positioned nearby can lead to fatalities in addition to serious personal injuries and/or property damage.

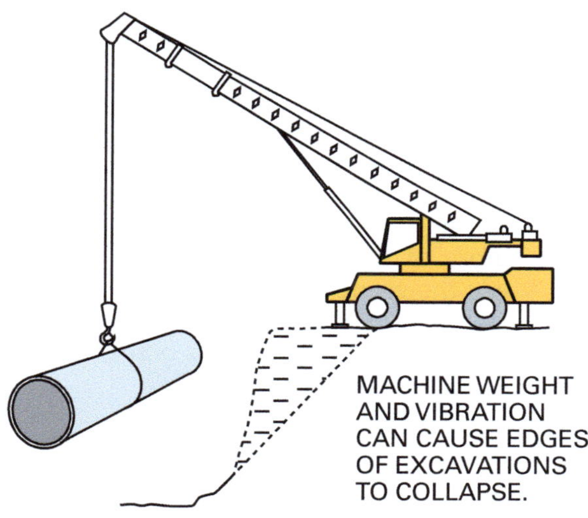

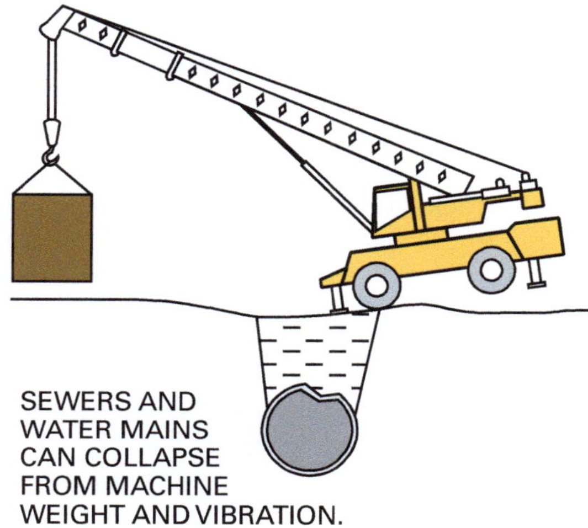

Figure 34 Examples of ground condition hazards.

2.1.2 Leveling

Most load chart ratings require the crane to be perfectly level or level within a one-percent grade (0.57 degrees). This applies to all cranes including pick-and-carry models. When it is impossible to level a given crane, the manufacturer can sometimes provide a load chart that specifies how far off level the crane may be and the effect on load capacity. For example, a crane on a barge will be accompanied by a barge chart that considers the position of the barge in the water.

When a crane is not perfectly level, the following three problems can arise:

- The boom can be side loaded.
- The boom angle changes when the crane swings, which in turn changes the radius.
- The crane becomes more difficult to operate.

Table 3 demonstrates how capacity is reduced due to boom side-loading when the crane is not perfectly level.

An example would be found on the load chart for a crane with a 200-foot boom at an angle of 70 degrees that is out of level by just 1 degree (1.75 percent). Swinging the boom from the high side to the low side changes the operating radius by 6.5 feet (2 meters). With the crane out of level, the boom or counterweight will also try to swing downhill, complicating control.

The operator must know what the crane manufacturer considers a level condition, and then adjust the outriggers or level the supporting surface to meet those requirements.

The small target levels often provided by the crane manufacturers may not be accurate enough to confidently level the crane and meet the requirements of the load chart. On the other hand, some cranes are equipped with accurate electronic levels that can be read directly on the crane

Table 3 Crane Capacity Decrease Due to an Off-Level Condition

CRANE CAPACITY DECREASE DUE TO OFF-LEVEL CONDITION			
Boom length or Radius	1.75% or 1 degree	3.5% or 2 degrees	5.25% or 3 degrees
Short boom, minimum radius	10%	20%	30%
Short boom, maximum radius	8%	15%	20%
Long boom, minimum radius	30%	41%	50%
Long boom, maximum radius	5%	10%	15%

instrument panel. When the equipped level is unreliable or the accuracy needs to be checked, the following two methods can be used to level the crane accurately:

- Place a carpenter's level in two different positions, 90 degrees apart, on a sturdy part of the crane, such as on top of the ring gear. The longer the level, the more accurate the reading. The bubble of the level should be between the lines in both positions.
- Boom up to the highest boom angle possible for the crane configuration. At two positions, 90 degrees apart, look straight ahead at the boom and determine if the ball is hanging straight in-line with the boom (plumb). If it is, the crane is reasonably level. In this case, the headache ball or load block is being used as a plumb bob.

2.2.0 Bearing Surface and Ground Pressure

The bearing surface refers to the points where the crane makes ground contact during a lift. It is a shortened form of the term loadbearing surface. The weight of both the crane and the load is concentrated on the bearing surfaces.

Crane bases are different and each is designed to place weight on various types of bearing surfaces. The three primary types of bearing surfaces used by cranes are the following:

- Outriggers
- Rubber (sitting directly on the tires, like a pick-and-carry crane)
- Crawler tracks

For a crane to be considered on its outriggers, the outriggers must be fully extended and the crane tires must be relieved of all weight. Some crane manufacturers supply capacity ratings based on partial outrigger extension, but that represents an exception, not the rule. If the outriggers are not fully extended, the tires are bearing weight, or both, the capacity chart for the crane on outriggers will be invalid.

The term *on rubber* refers to a wheeled truck crane being operated without any outriggers extended. The crane is being operated with its wheels and tires as the sole method of support and stability. Pick-and-carry cranes operate this way in most cases. Lifting capacity is significantly reduced as a result, but maximum lifting capacity is often not a priority for them.

Crawler cranes typically have capacity ratings for both crawlers extended and crawlers retracted when that feature is available. Extending

the tracks provides a broader base and therefore greater capacity and stability.

The crane base in use has a great deal to do with ground pressure. For crawler cranes, the weight is distributed across the total surface area of the tracks in contact with the ground. When on rubber, the weight is distributed to the numerous small patches of area where the tires contact the ground. When outriggers are in use, the weight is being distributed across the total area of the floats in contact with the ground.

When the crane is sitting idle, the weight is distributed among the various bearing surfaces. The weight distribution is not likely to be uniform, but relatively so. Some bearing surfaces, such as those at the opposite end from the boom near the counterweights, may be carrying extra weight that will be balanced by the load. Once lifting begins, the weight distribution changes. Ground pressure increases most at the bearing surfaces nearest the load. The ground pressure at an outrigger can instantly become much greater than that under a track on similar lifts, due to the smaller bearing surface offered by the outrigger. *Figure 35* shows the relative ground pressures of track- and outrigger-supported cranes with the load supported over different areas of the crane. The darker red areas represent greater pressure. If the crane is not properly supported, one or more of these bearing surfaces can sink, allowing the crane to tilt away from level. As it moves away from a level condition, the lift capacity is dropping and stability is being compromised.

2.2.1 *Mats and Blocking*

To compensate for ground conditions and the changing ground pressures, crawler cranes are often supported on crane mats that are made of steel or wood. Wooden mats can be made from 8- to 12-inch (20- to 30-cm) square timbers that have been bolted together at uniform intervals. These mats, fabricated in manageable sections, are then combined to make a mat that extends at least 2 feet (0.6 meters) beyond the ends and edges of the crane tracks. Once the crane is positioned, timber blocking is secured to the mat at the ends and sides of the tracks. This helps prevent the crane from sliding easily in any direction. *Figure 36* shows a typical mat arrangement for a crawler crane.

For wheeled cranes, the load on the outrigger floats must also be spread across the ground surface. The material used to transfer and spread loads is called *blocking* or *cribbing*. Crane manufacturers must make outrigger floats small and light enough so that one person can move them, since

NCCER – *Mobile Crane Operations*

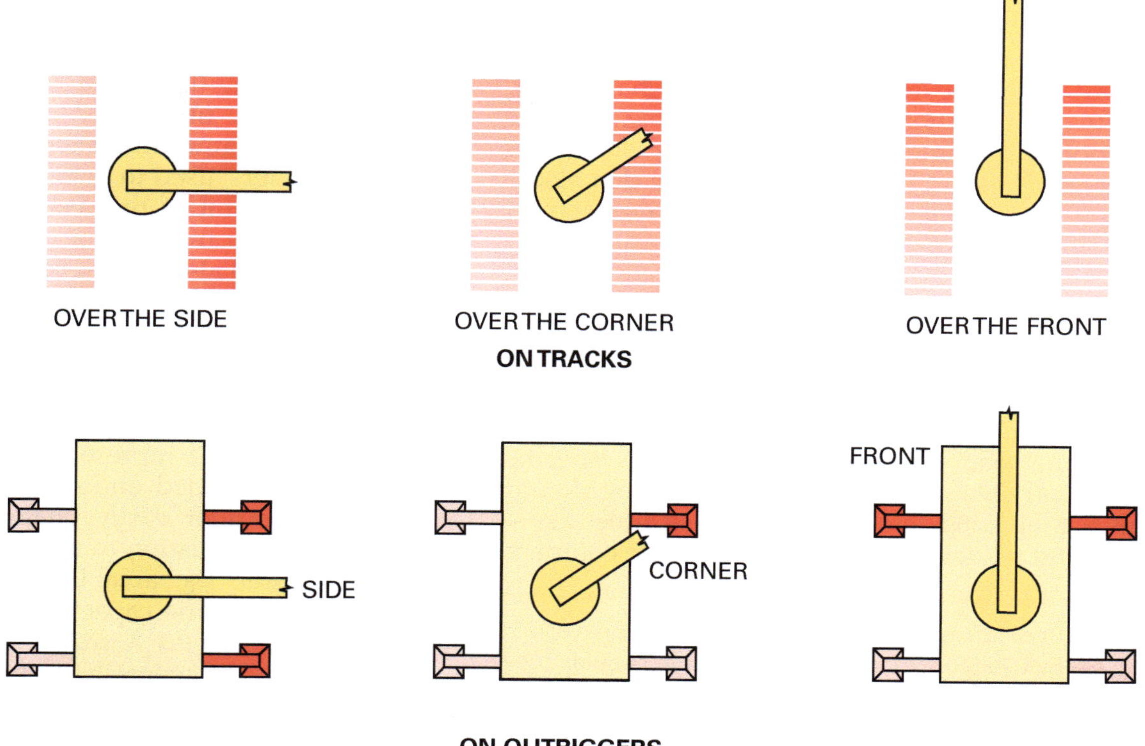

Figure 35 Relative ground pressures applied by track- and outrigger-supported cranes.

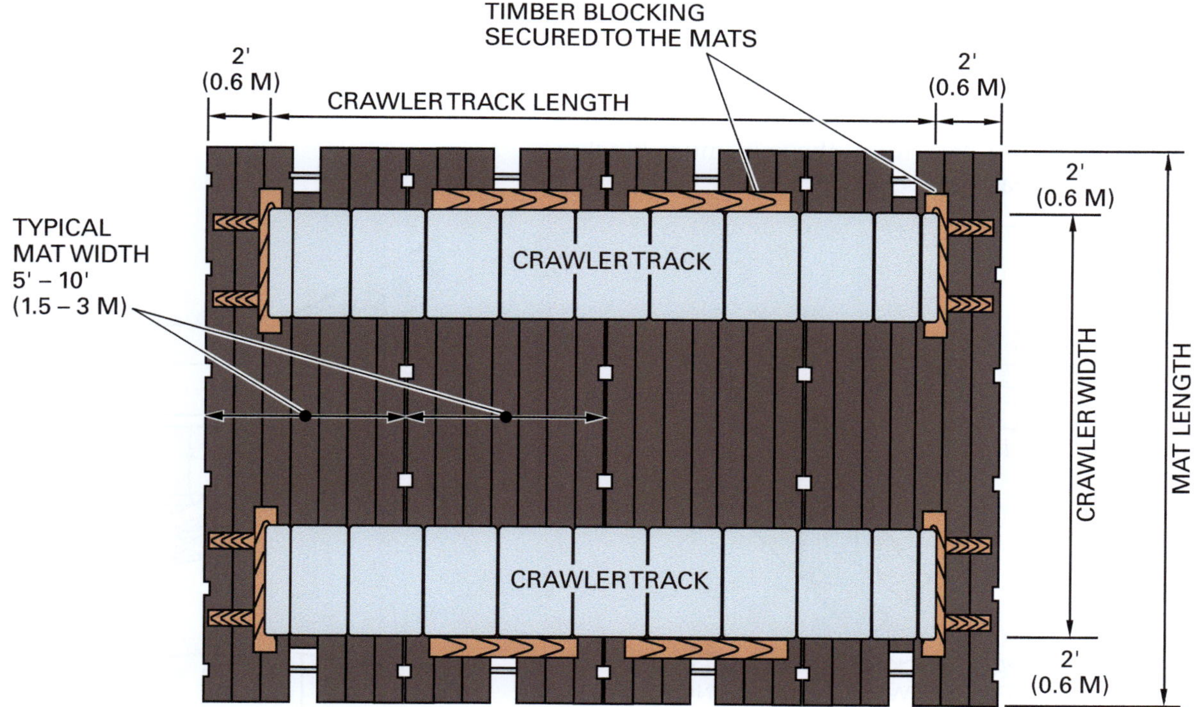

Figure 36 Typical crawler-crane mat arrangement.

they are often detached from the outrigger for travel. This limits the size of the float somewhat.

The blocking beneath the float must be made from a suitable material such as a hardwood. Softwoods are not generally acceptable. Heavier cranes often require blocking that is stacked to increase the total thickness. When wooden blocking is stacked, the lumber in each layer should be positioned at 90-degree angles to each other. See *Figure 37* for several examples of correct and incorrect blocking. Although blocking is often made of wood, it is not a requirement. Blocking products are also fabricated from man-made materials and marketed by various manufacturers.

Never assume the supporting surface will support the load without blocking unless specific technical data shows that it will. Blocking should always be considered a requirement unless specific and reliable data is presented to the contrary.

Spreading the load means the surface area of the blocking should cover as much ground surface as possible, within reason. The area of the blocking, to do this effectively, must be larger than the area of the float. Either of two methods can be used to determine the minimum blocking area for average soil:

- *Method 1* – Method 1 is sometimes referred to as the *Rule of Three*. Multiply the area of the crane float by 3 to determine the required area of the blocking (float area × 3 = blocking area). Note that the units used can be imperial or metric. The same unit, such as square feet or square meters, used to determine the float area will be the unit of the result. For example, if square meters are used for the area of the float, the result will also represent square meters.

- *Method 2* – Method 2 is sometimes referred to as the *Rule of Five*. Divide the capacity of the crane (in tons) by 5 to determine the required area of the blocking in square feet [Crane Capacity (tons) ÷ 5 = blocking area (square feet)]. The result represents the total amount of blocking needed under all outriggers. Note that you cannot easily change the imperial units in this equation to metric units. It is best to determine the area in square feet, using tons for the crane capacity, and then convert the result area to a metric unit if desired. This method is generally preferred for smaller cranes that have relatively small floats.

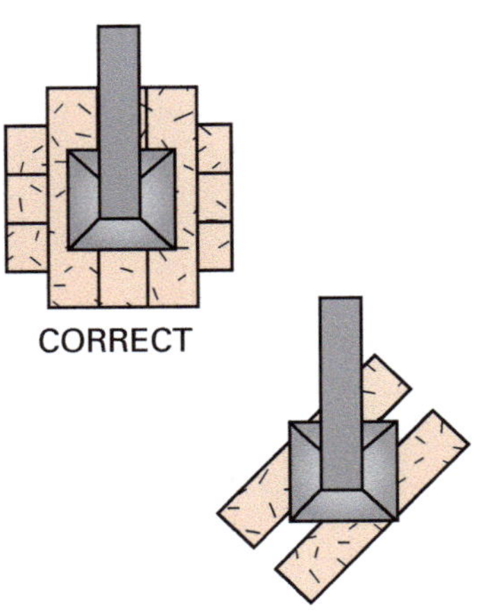

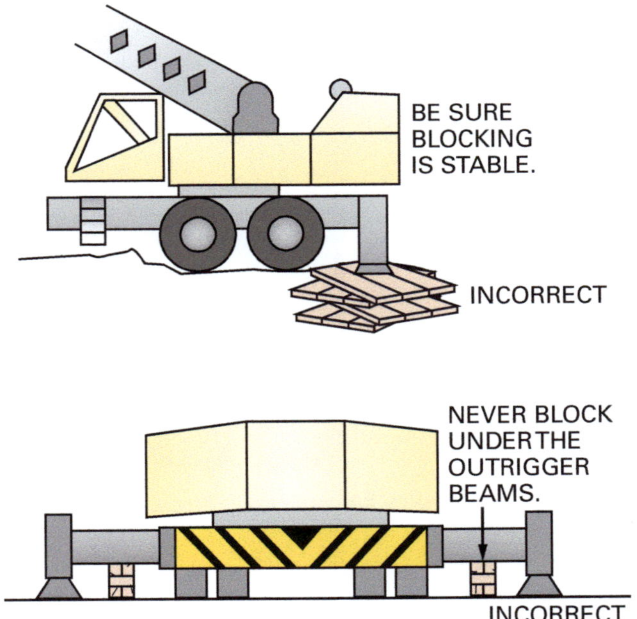

Figure 37 Correct and incorrect use of outrigger blocking.

NCCER – *Mobile Crane Operations*

Manufactured Blocking

Not all crane blocking and mats are made by the crane owner. Those that are made by the owner are usually made of wood. However, there are a number of manufacturers that specialize in the fabrication and sale of crane support materials. Although some manufacturers do make and sell wooden blocking and mats, others use materials that are more durable.

The crane blocking shown here can be made to order for a given crane to ensure it is properly sized. The fibrous material used is far more weather resistant than wood, which tends to crack and split over time, as well as absorb water. The material is said to have the strength of steel blocking, but weighs 70 percent less than steel. The manufacturer also makes smaller pads using a thermoplastic material that offers the same basic advantages. Crane blocking and mat manufacturers will also assist in the engineering and selection process.

Figure Credit: DICA Outrigger Pads

An example calculation for each method is provided below. Unless there are unique circumstances, the results from either method show the minimum size of the blocking for a single outrigger. There is no reason that the blocking can't be larger; indeed, larger is better. For this reason, it is best to always round the results of any calculation up when rounding is needed.

Example for Method 1 (Square Floats):

What is the minimum blocking area needed for a 27-inch square float?

The blocking area can be calculated as follows:

Step 1 Determine the area of the float:
$$27" \times 27" = 729 \text{ in}^2$$

Step 2 Multiply by 3 to determine the required blocking area:
$$729 \text{ in}^2 \times 3 = 2{,}187 \text{ in}^2$$

Step 3 If needed, the result can be converted to square feet or any other unit you find convenient to use. The result has been rounded up to the nearest tenth:
$$2{,}187 \text{ in}^2 \div 144 \text{ in}^2/\text{ft}^2 = 15.2 \text{ ft}^2$$

> **NOTE**
>
> When rounding blocking area calculations, always round up (not down) to ensure your result is not below the minimum blocking area.

There are several ways this information can be applied. For example, if square blocking is desired, finding the square root of 15.2 ft² will give you the dimensions of the square. Since the square root of 15.2 is approximately 3.9, then the blocking should be at least 3.9' × 3.9'. Note that this is very close to a 4-foot square.

Many outrigger floats are round rather than square. For round floats, you will need to determine the area of the circle. To find the area of a circle, the radius of the circle (which is half its diameter) is squared and then multiplied by the constant *pi*, represented by the symbol π. Rounded to the nearest hundredth, pi has a value of 3.14. The equation for determining the area of a circle appears as:
$$\text{Area} = \pi r^2$$

Example for Method 1 (Round Floats):

What is the minimum size of the blocking needed for a round float that is 50 cm in diameter?

The blocking area can be calculated as follows:

Step 1 Determine the radius of the round float:
$$\text{Radius} = \text{diameter} \div 2$$
$$\text{Radius} = 50 \text{ cm} \div 2$$
$$\text{Radius} = 25 \text{ cm}$$

Step 2 Determine the area of the float:
$$\text{Area} = \pi r^2$$
$$\text{Area} = 3.14 \times (25 \text{ cm})^2$$
$$\text{Area} = 3.14 \times 625 \text{ cm}^2$$
$$\text{Area} = 1{,}963 \text{ cm}^2$$

Step 3 Multiply by 3 to determine the required blocking area:

$$1,963 \text{ cm}^2 \times 3 =$$
$$5,889 \text{ cm}^2 \text{ of blocking surface area required}$$

Step 4 Determine the diameter of the round blocking needed, using the area. The same equation can be used to first find the radius of the round blocking:

$$\text{Area} = \pi r^2$$
$$5,889 \text{ cm}^2 = \pi r^2$$
$$(5,889 \text{ cm}^2 \div 3.14) = (3.14 \times r^2) \div 3.14$$
$$1,876 \text{ cm}^2 = r^2$$
$$\sqrt{1,876 \text{ cm}^2} = \sqrt{r^2}$$
$$43.3 \text{ cm} = r$$

Step 5 Now determine the diameter of the round blocking needed by simply multiplying the radius by 2, as follows:

$$43.3 \text{ cm} \times 2 = 86.6 \text{ cm}$$

Therefore, a section of round blocking that is 86.6 cm in diameter is the minimum needed for the 50-cm round float.

Note that the blocking can be square even if the float is round. Square blocking is easier to construct from lumber. Again, you can determine the size of the square by finding the square root of the required area. Since the area needed is 5,889 cm², use a calculator to find the square root. The square root of 5,889 is 76.7. Therefore, a square section of blocking 77 cm × 77 cm will also work for the 50-cm round float.

Example for Method 2:
What is the blocking area required for a crane with a capacity of 75 tons?

The blocking area can be calculated as follows:

Step 1 Divide the capacity of the crane (in tons) by five, which represents the number of square feet of blocking area per ton of weight. Remember that the Rule of Five only works with square feet:

$$75 \text{ tons} \div 5 \text{ tons/ft}^2 = 15 \text{ ft}^2$$

Step 2 If desired, the result can be converted to square meters or any other unit you find convenient to use. To convert square feet to square meters, for example, multiply square feet by the conversion factor of 0.093 m²/ft²:

$$15 \text{ ft}^2 \times 0.093 \text{ m}^2/\text{ft}^2 = 1.4 \text{ m}^2$$

Since 15 ft² represents the total amount of blocking needed for the crane, each of four outrigger pads would require 3.75 ft² of blocking beneath them. Therefore, a 2' × 2' square section of blocking under each pad would work nicely and provide a bit of extra area.

When tasked with constructing blocking that must cover a given area, lumber sizes are a factor. First, remember that lumber, such as a 4 × 4 timber, is not made to its nominal size. The actual dimensions of a common 4 × 4 are 3.5" × 3.5". This must be taken into consideration as the actual size of the blocking is planned. Square blocking does not need to be perfectly square. However, it is important to ensure that the final dimensions result in a section of blocking that does not allow any part of the float to hang over the edge, and that has an area no less than the calculated area.

> **WARNING!**
>
> These calculations are for average soil, which will not provide enough support for all soil conditions.

2.3.0 Center of Gravity and Operating Quadrants

Mobile crane operators must have a clear understanding of a crane's center of gravity and its **tipping fulcrum**. The location of and relationship between these two points determines if a crane will stay upright or tip over. These concepts are important to all members of a lift team, such as signal persons and riggers who work in the vicinity of the crane and directly support the effort.

2.3.1 Center of Gravity

The center of gravity can be defined as the point at which the entire weight of an object is considered to be concentrated. Supporting an entire object at its precise center of gravity results in the object remaining balanced in position. The location of a crane's center of gravity in relation to its tipping fulcrum directly affects its **leverage** and stability. During a lift, the crane's center of gravity must remain between the tipping fulcrum and the other point of crane support (tracks, tires, or outriggers opposite the load). If the center of gravity moves outside of this area toward the load, the crane will tip over.

To understand the relationship between a crane's center of gravity and stability, it is first necessary to understand the principle of leverage. Leverage can be illustrated with a simple teeter-totter (*Figure 38*). When the weight of the heavy load, multiplied by the distance (X) from its center of gravity to the tipping fulcrum, is equal to the weight of the lighter load multiplied by the distance (Y) from its center of gravity to the tipping fulcrum, a condition of balance has been reached. This relationship is further illustrated with a long lever bent upwards in *Figure 39*.

Another example shows the relationship with the lighter load being suspended below the long lever in *Figure 40*. This figure represents the profile of a crane. Note that in all three figures, the loads will remain balanced as long as the load weights remain the same and the horizontal distances X and Y remain the same.

2.3.2 Quadrant of Operation

The center of gravity usually changes depending on the crane's quadrant of operation. When the crane boom shifts from one quadrant of operation, such as over the side, to another quadrant, the center of gravity shifts closer to, or farther away from, the tipping fulcrum. This, in turn, may increase or decrease the crane's leverage. For a rough-terrain crane, maximum lifting capacity (with the greatest stability margin) is accessed when the crane boom is operated directly over the front. Maximum lifting capacity for other wheeled cranes is accessed when the boom is over the rear.

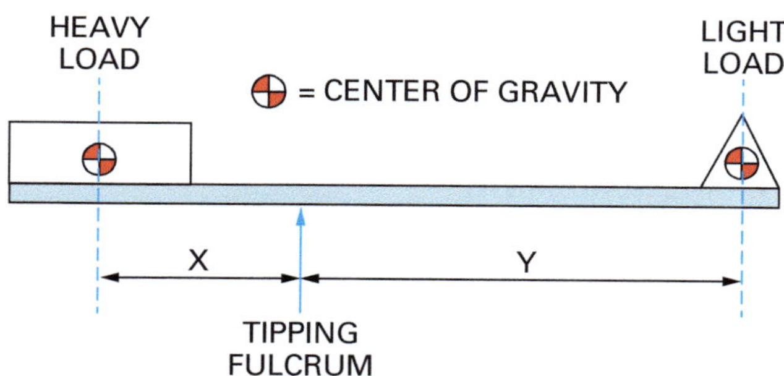

Figure 38 Teeter-totter demonstrating leverage.

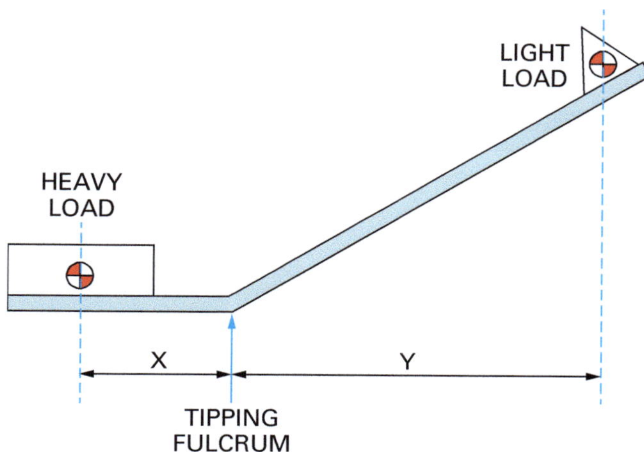

Figure 39 Leverage demonstrated on a bent lever.

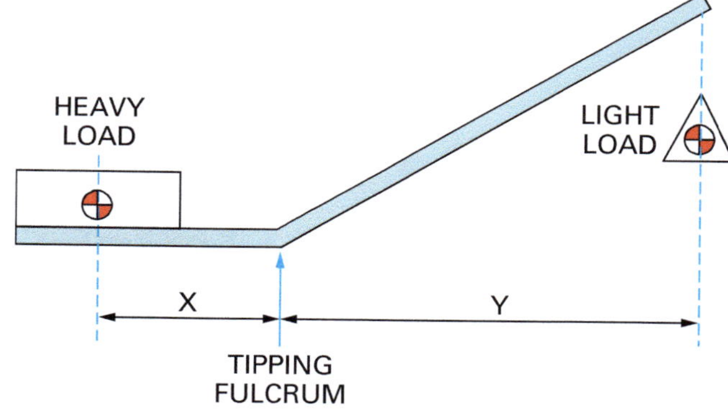

Figure 40 Crane-shaped lever.

The individual quadrants of operation are typically defined as follows:

- 360 degrees
- Over the front
- Defined arc over the front
- Over the rear
- Over the side

Figure 41 shows examples of defined quadrant boundaries for rough-terrain, wheeled truck, and crawler cranes. Note that the quadrant boundaries differ from crane to crane; always check the manufacturer's load charts and operating manual for the applicable quadrant boundaries.

Load charts are available for each quadrant of operation for which the crane is approved. The capacity for each quadrant can vary widely. Lifting from some quadrants may be prohibited entirely. This means that the lift and the position of the crane must be considered carefully beforehand. In addition, it means that operators must be cautious anytime the load must swing from one quadrant to another. In most situations, rotating the crane with a suspended load necessary in order to move the load.

If there is no load chart provided for a given quadrant, handling a load in that quadrant is not approved. This restriction means that no lifts, without exception, can be made from that quadrant, and loads cannot be swung into or through that quadrant. Most truck cranes (other than rough-terrain models) do not allow lifting over the front of the crane. Some are equipped with special bumper outriggers up front, allowing over-the-front operations. However, these are the exception rather than the rule. Check the charts available with each crane you operate to determine the approved operating quadrants and their boundaries. Load charts often provide pictures of the quadrant areas for clarity.

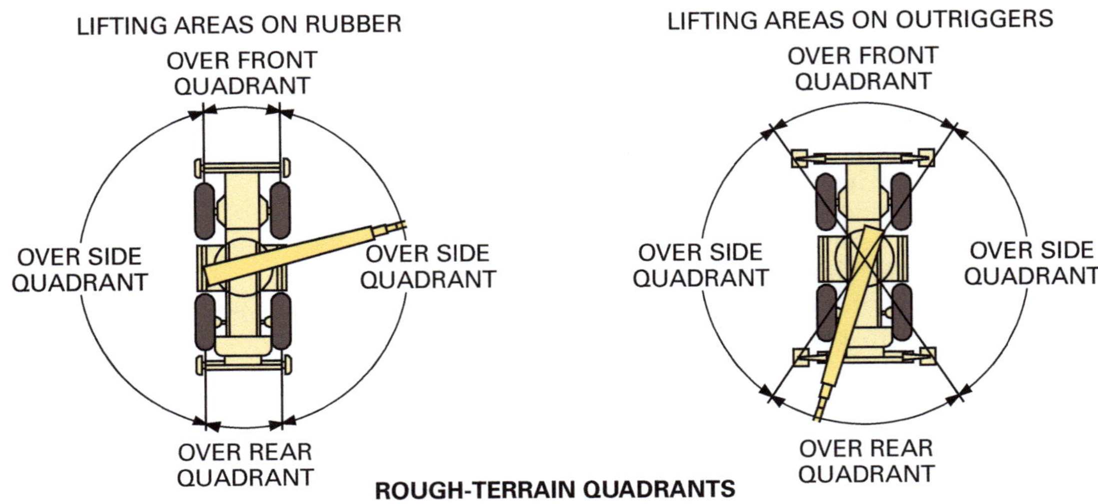

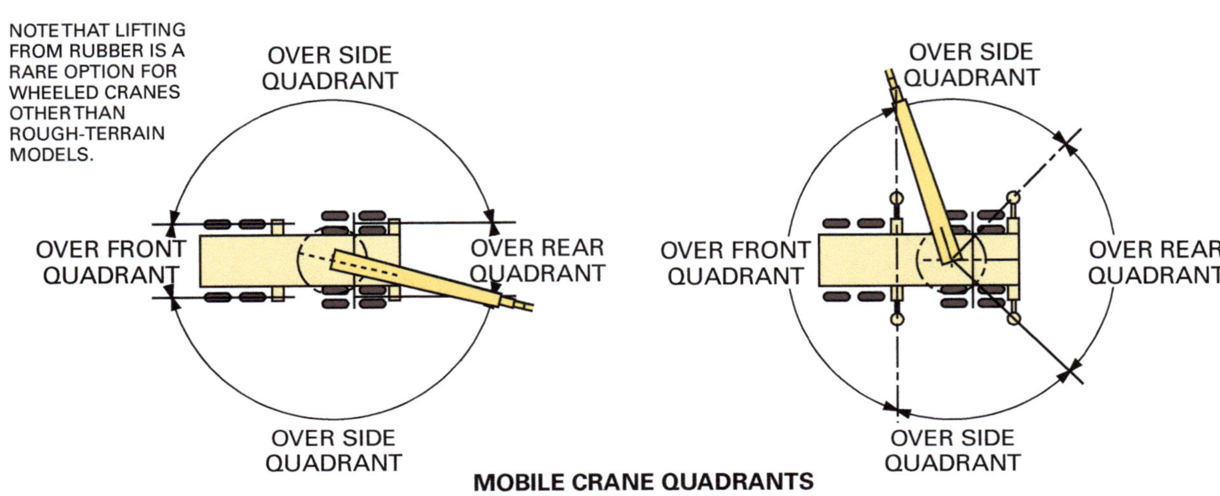

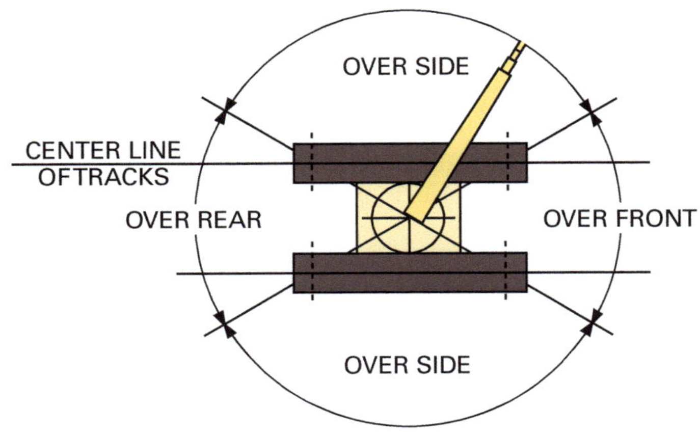

Figure 41 Quadrant definition drawings for rough-terrain, wheeled truck, and crawler cranes.

As explained earlier, the crane's stability depends on the relationship between the load leverage and the leverage of the crane. The crane leverage is determined by the distance between the crane's center of gravity and the tipping fulcrum. The load leverage is determined by the distance between the load's center of gravity and the same tipping fulcrum. Remember that the tipping fulcrum changes as a load swings from one quadrant to another, changing the relationship between it and the center of gravity.

In addition to changes in capacity and stability that occur as the boom moves from one quadrant to another, the leverage applied by the load changes as the angle of the boom changes. When the boom is near vertical, the leverage of the load is relatively small. As the boom is lowered towards a horizontal position, the load leverage increases due to the increased horizontal distance between it and the tipping fulcrum. Note that the safe lifting capacity of the crane is being reduced as the boom is lowered, as a direct result of this change in leverage. The crane remains stable as long as the crane's leverage is greater than the load's leverage.

The change in leverage that occurs as a boom is lowered is illustrated in *Figure 42*. Pay close attention to the changes in the length of the Y line, representing load leverage, as the boom drops toward horizontal. Once a crane starts to tip over, the crane leverage will rapidly decrease further due to the reduced distance to the tipping fulcrum (X_2), and tipping will accelerate.

The crane leverage for a wheeled crane (other than most rough-terrain models) in different quadrants varies more widely than a crawler crane. This is especially true for crawler cranes that can extend their crawler frames to increase stability. This is due to the large changes in the distance between the center of gravity of the crane and the tipping fulcrum of wheeled cranes.

The greatest leverage, and thus the greatest lifting capacity, is developed when the boom is over the rear of a wheeled crane. Even with the outriggers extended, the crane leverage with the boom over the side is less than over the rear, due to the shorter distance to the tipping fulcrum.

The least capacity is with the boom over the front, because the crane leverage is at its least and the weight of the cab and front drive train adds to the leverage of the load. Lifts over the front are not permitted with most wheeled truck cranes. *Figure 43* shows how the crane leverage changes for each position. Note that the crane leverage, represented by the line X, gets shorter from the top drawing to the bottom drawing.

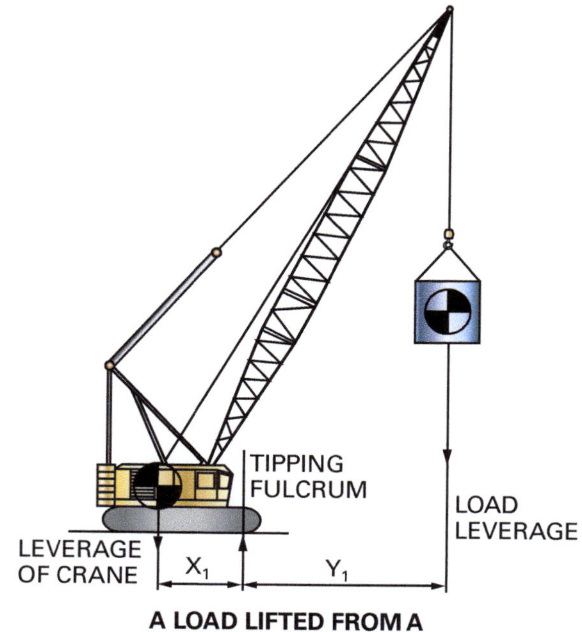

A LOAD LIFTED FROM A GIVEN BOOM ANGLE

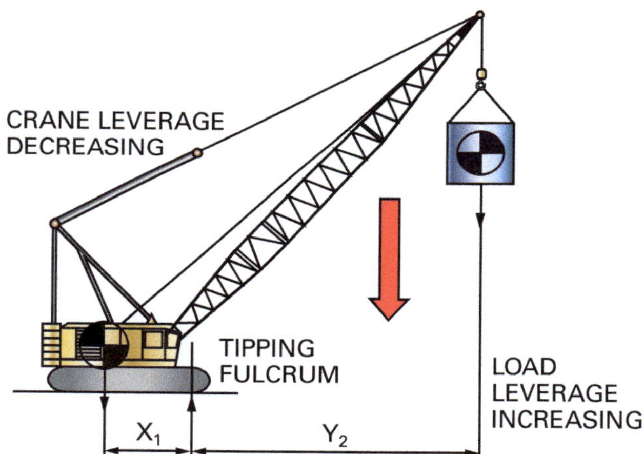

AS THE BOOM IS LOWERED, THE LENGTH OF Y AND LOAD LEVERAGE INCREASE, WHILE CRANE LEVERAGE DECREASES.

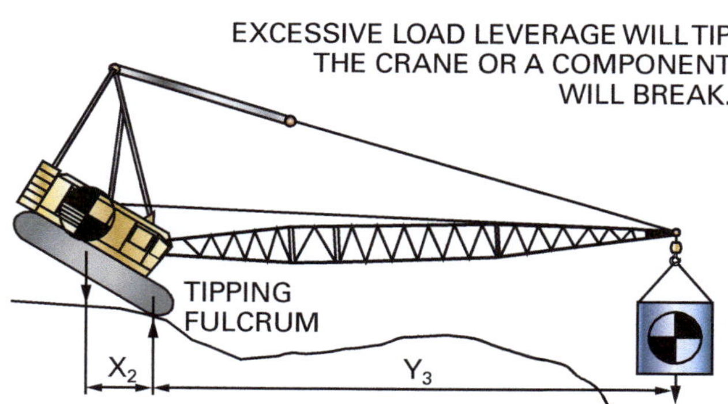

IF THE LOAD LEVERAGE INCREASES TO AN EXCESSIVE LEVEL, TIPPING BEGINS. ONCE TIPPING BEGINS, X IS SHORTENED AND TIPPING ACCELERATES.

Figure 42 Relative tipping forces as a boom is lowered.

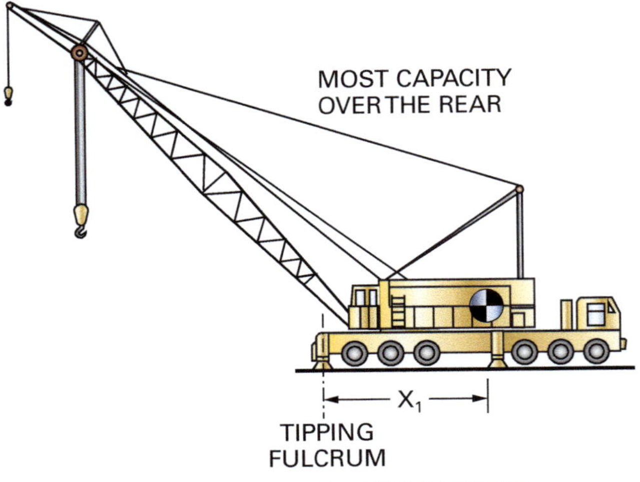

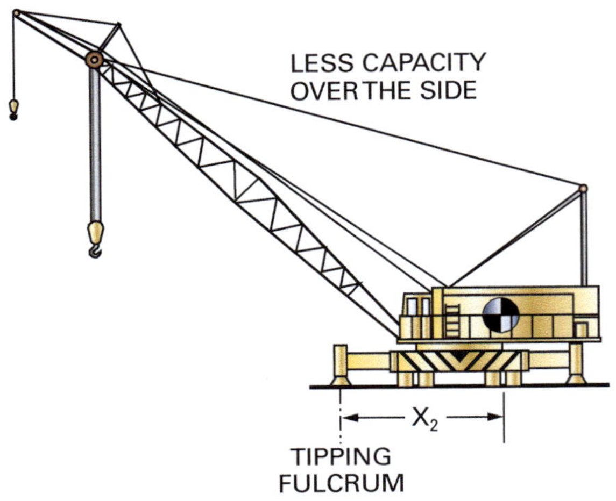

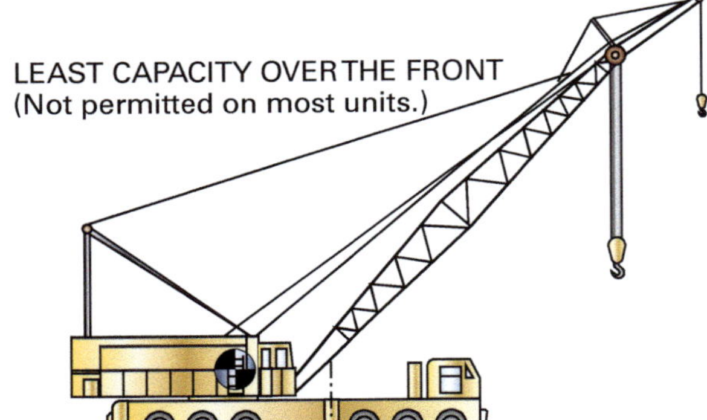

Figure 43 Tipping fulcrums for various quadrants.

To avoid tipping, cranes are rated with a built-in safety factor. The maximum rated capacity of any crane represents a percentage of the tipping load. *Table 4* shows the typical rated capacities of cranes as a percentage of their tipping load. Although these values are commonly used, they do not represent a standard shared by all manufacturers. It is best to know how the load capacity of the specific crane you are operating was determined.

2.4.0 Boom Position and Characteristics

Boom length, boom angle, operating radius, and boom-point elevation each have an effect on the capacity of a given crane. These characteristics are all within the control of the operator and are constantly changing (dynamic) as the boom position changes. *Figure 44* shows each of these measurements.

Boom length is the distance measured from the center of the boom hinge pin to the center of the sheaves at the boom tip. Boom length includes the jib and/or other boom extension when equipped. Boom length directly affects both crane capacity and reach, horizontally and vertically.

The boom angle is the angle existing between the horizontal plane and the center line of the boom in its present position. Generally, the smaller the angle (closer to the horizontal plane), the lower the capacity of the crane. As the boom angle increases, so does the crane's capacity. Smaller angles also place additional stress on the boom.

When a lattice boom is lowered to a smaller angle, it begins to sag in the middle from its own weight. This sag increases the pull on the pendants. These forces can exceed the strength

Table 4 Capacities of Cranes as a Percentage of Their Tipping Load

TYPE OF CRANE	CAPACITY
Locomotive	85%
Crawler	75%

Wheeled truck crane
    on outriggers
    on tires
	85%
	75%

Commercial boom truck on stabilizers	85%

* Check your crane. The capacities shown
  above are not used by all manufacturers.

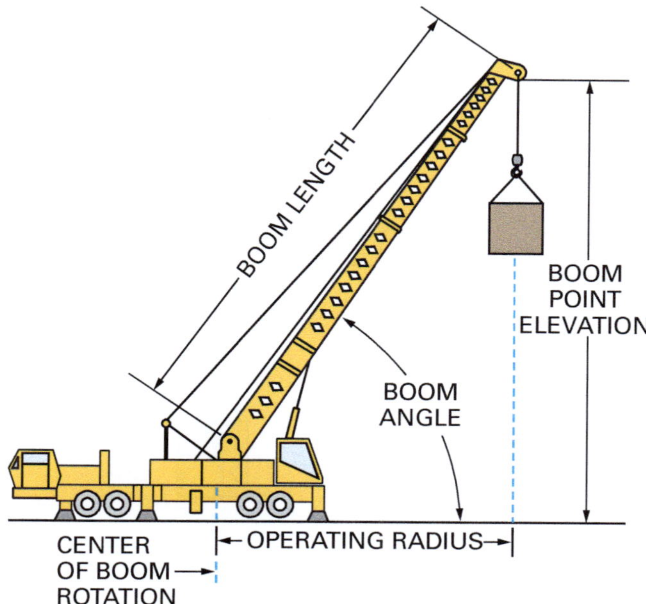

Figure 44 Boom length, boom angle, operating radius, and boom-point elevation.

compression of the earth directly beneath the tipping fulcrum (where most of the load is concentrated) will also cause the crane to lean toward the load. Although generally minor, any change in the operating radius as the lift progresses can become a major factor in a lift that is at or near the crane's capacity. In addition, the combined effect of all three of these factors in a single lift can interrupt what appears to be a safe and simple lift that was well within the crane's capacity during the planning stages.

Boom-point elevation is the vertical distance from the ground to the tip of the boom, boom extension, or jib point sheave. This dimension changes with the boom length (on telescopic booms) and the boom angle. It is used to determine the ability of the crane to reach a certain height and/or to determine clearances from power lines, buildings, or other hazards in the operating area.

2.4.1 Swing Out, Side Loading, and Dynamic Loads

The movement of the load in relation to the crane boom can cause additional stresses on the boom, resulting in failure. Swing out, side loading, and dynamic loads can cause potentially dangerous stress on the boom.

If the crane or boom is moved rapidly with a suspended load, the load will not initially follow the boom. When the swing begins, the load will lag behind the boom. As the load begins to swing to catch up, the centrifugal force of the swing will cause the load to swing out beyond the boom in the opposite direction. When the boom stops moving, the load will continue to swing. Each swing causes strain on the equipment and rigging that can result in failure. More importantly, swing out may also result in the load striking workers or equipment that were originally thought to be well outside the operating radius. For this reason, any movement of the crane or boom with a suspended load must be done gently and slowly. *Figure 46* demonstrates swing out.

Boom side-loading occurs when the load is positioned to one side of the boom due to the following factors:

- An off-level crane setup
- Rapid movement
- Sudden starting or stopping of boom rotation (swing out)
- Dragging of a load along the ground
- Lifting without first positioning the boom tip directly over the load
- Improper reeving

of the lattice boom structure. Helper cranes are often needed to assist in assembling long booms to prevent lattice failure. With a telescopic boom, the forces of the lower angle typically result in the bending of one or more boom sections. The degree of bending depends on how far the individual sections are extended. There is less stress when all of the sections are retracted, even when the boom is in the horizontal position (a 0-degree boom angle). When the sections are fully extended, a great deal of stress is placed on the end of each section.

The operating radius is defined as the horizontal distance from the axis of boom rotation to the center of the load block with the load suspended. As the operating radius increases, capacity decreases. Cranes have both minimum and maximum operating-radius specifications. These specifications are directly dependent on boom length. Unless a longer boom is installed, the only way to increase the operating radius of the crane is to lower the boom. It makes sense then, that the crane's capacity decreases as the operating radius increases.

It is not unusual for the operating radius to increase as the load is lifted and the weight transfers to the crane, due to the deflection of the boom and crane structure (*Figure 45*). If the crane is lifting from rubber, tire compression nearest the tipping fulcrum will also cause the operating radius to increase as the crane leans slightly toward the load.

A third factor is related to the compression of the supporting surface. Although the crane may be perfectly level before the load is lifted,

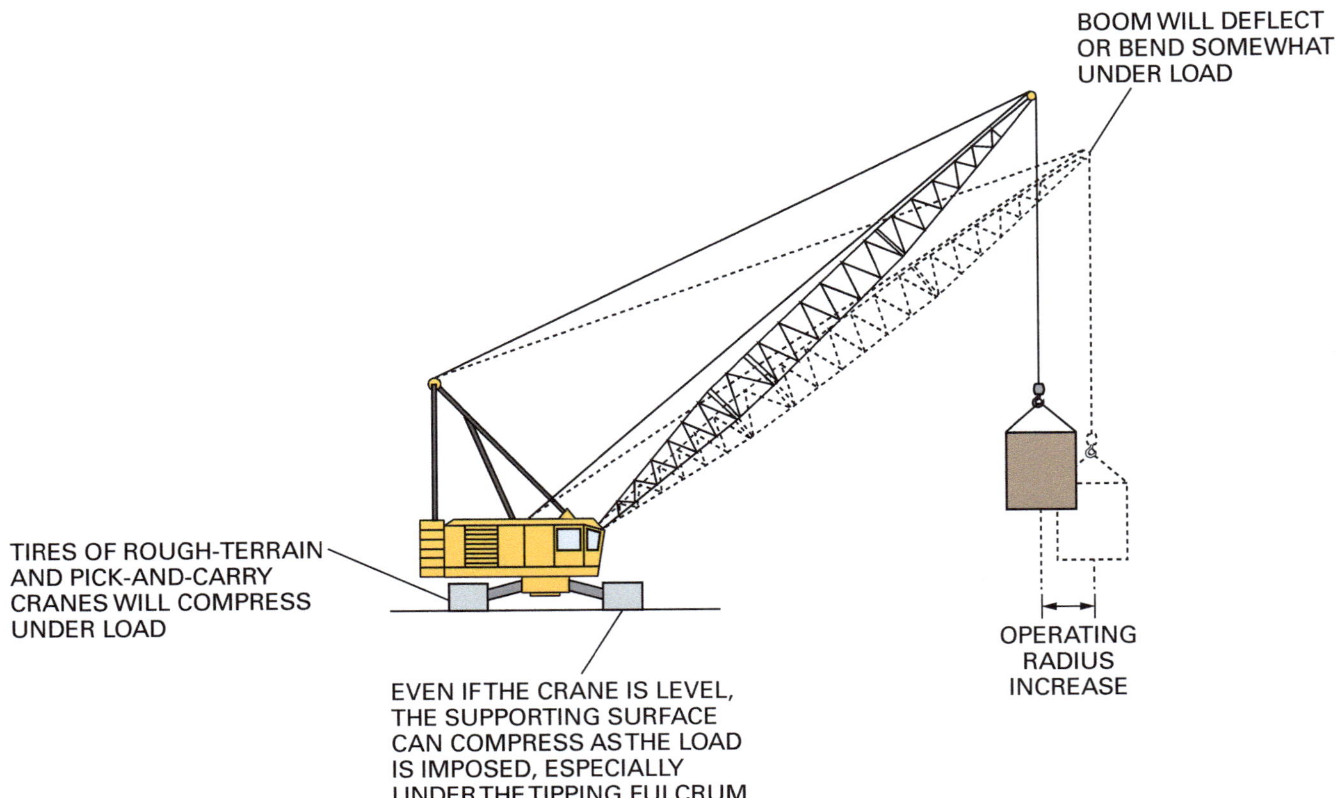

Figure 45 Increased operating radius due to various factors.

Dynamic loading, also called *shock loading*, refers to any changing stress placed on the crane by actions such as hoisting, lowering, stopping, or swinging the crane, particularly when these actions are sudden. When a crane is at rest with a suspended load on a firm, uncompromising surface and there is no wind, the load is said to be *static*. That is, a certain load is imposed on the crane structure and it remains relatively constant. However, the lightest breeze or any crane movement whatsoever can cause the load to fluctuate, making it *dynamic*. Sudden or rapid movements can easily increase the load by 30 to 100 percent; therefore, such movements must be avoided. Swing out and boom side-loading are two ways that dynamic loads are generated in a crane. It is important to note, however, that not all sources of dynamic loads are within the control of the operator. Wind, for example, is an uncontrollable factor.

2.5.0 Load Charts

A crane's load chart is as important to the mobile crane operator as the crane itself. It is a critical piece of the equipment. *ASME Standard B30.5, Section 5-1.1* says that "the crane manufacturer shall provide load rating charts and information for all crane configurations for which lifting is permit-

ted." Many manufacturers physically attach the load chart in a readily accessible location in the cab of the crane. Load charts are also built into the crane's software in models with a graphical display. The load chart indicates the maximum capacity of the crane under every permissible configuration. The understanding and use of load charts by crane operators and lift planners is critical to the safe operation of the crane.

The following are three basic load chart configurations. Note that there are many possible configurations for each crane, and each requires a different load chart. However, most charts fall into one of these three broad categories:

- Lifting from the boom with no extensions or jibs installed
- Lifting from the main load line with a boom extension and/or jib installed
- Lifting from the boom extension or jib

Remember that the advertised maximum capacity of a crane is calculated with the shortest boom option and lifting at the minimum operating radius. Do not think that just because a crane is rated at 50 tons (45.4 metric tons), it can lift that load in all situations. The advertised maximum capacity is likely more valuable in marketing efforts than it is to the crane operator. Experienced crane operators know that the load chart determines what

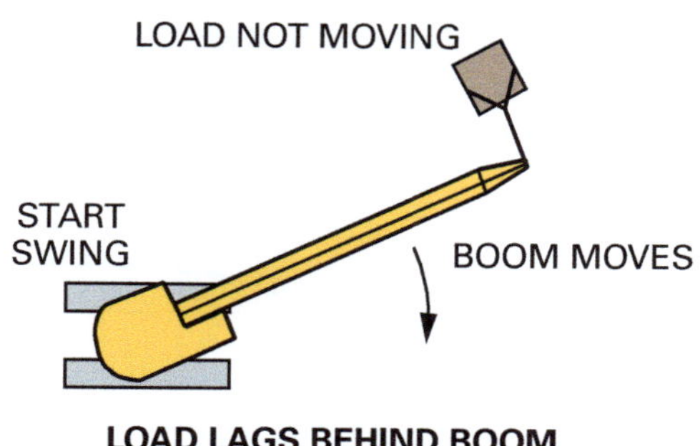

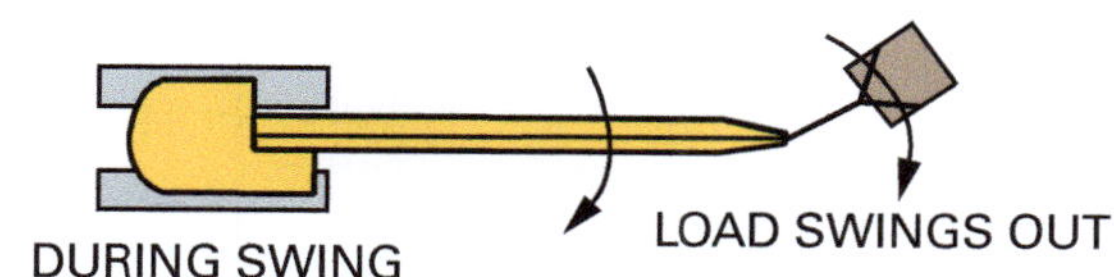

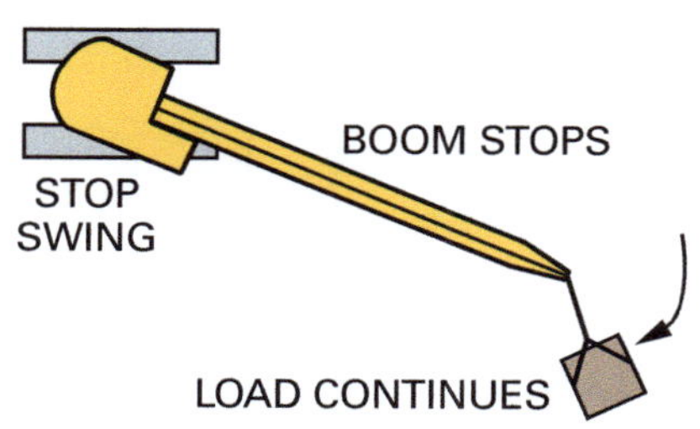

Figure 46 Operating sequence that leads to swing out.

the crane is capable of in a given set of conditions, and that the advertised maximum capacity is substantially higher than its actual capacity in most situations and configurations.

> **WARNING!**
>
> Load charts are based on a level crane that is properly supported on an appropriately prepared surface. Any deviation from the conditions that the load chart is based on, including an out-of-level position, can result in property damage and personal injury.

Each crane manufacturer has their own way of displaying a load chart, but all contain the same type of information, which includes the following:

- Type of crane base
- Type of crane configuration
- Quadrant of operation
- Length of boom
- Angle of boom
- Load radius

- Allowed deductions for additional equipment from the **gross capacity** to arrive at the **net capacity**

Figure 47 shows a sample working-range diagram for a National Crane Series NBT40 boom truck with a five-section, 142-foot (43.3 m) boom and a 26-foot (7.9 m) folding jib. *Figure 48* shows the matching load chart. Notice the heavy, dark line on the load chart, separating the top from the bottom. The capacities listed above this line are based on the structural strength of the boom extension as the weakest link, rather than on the tipping force. Weights below the line are based on the tipping force of the load at the designated angle and boom length. Other manufacturers may identify those values by shading the capacity numbers in the appropriate area, or placing an asterisk beside them.

Many cranes also have range diagrams and load charts in metric units, depending upon the distribution of that particular model. Metric data is not available for this particular crane model. Note that the load chart reports the gross capacity for a given set of conditions. The net capacity is calculated from this figure to determine the maximum weight of the load to be lifted.

Many cranes are designed to be used with different styles of boom tops, such as hammer-head, tapered top, boom extensions, or jibs. Each style of top has its own capacity chart or deductions for the extensions and jibs. This particular range diagram and load chart are based on the jib being in its stowed position. When it is stowed rather than removed, it adds to the weight managed by the boom.

Boom angle, operating radius, and boom length are critical factors in calculating the capacity of a crane. If, in determining loads from a chart, the chart increments do not precisely match the lift plan, use the next longer boom length, operating radius, or the next lower boom angle on the chart. For example, a lift requires a boom length of 105 feet (32 meters) and an operating radius of 60 feet (18.3 meters). Referring to the chart shown in *Figure 48*, note that there is no listing for a 105-foot boom across the top. The next longest boom length, 115 feet (35 meters), must be used. This shows that the maximum allowable load is 6,200 pounds when the outriggers are set and the jib is stowed. For this particular crane, this capacity is also valid around the entire 360-degree circle, rather than only in one or two quadrants. The point here however, is to avoid trying to calculate a value between the 101-foot boom and the 115-foot boom. This is known as **interpolation**, and it is not allowed. Trying to interpolate values can result in an unsafe lift.

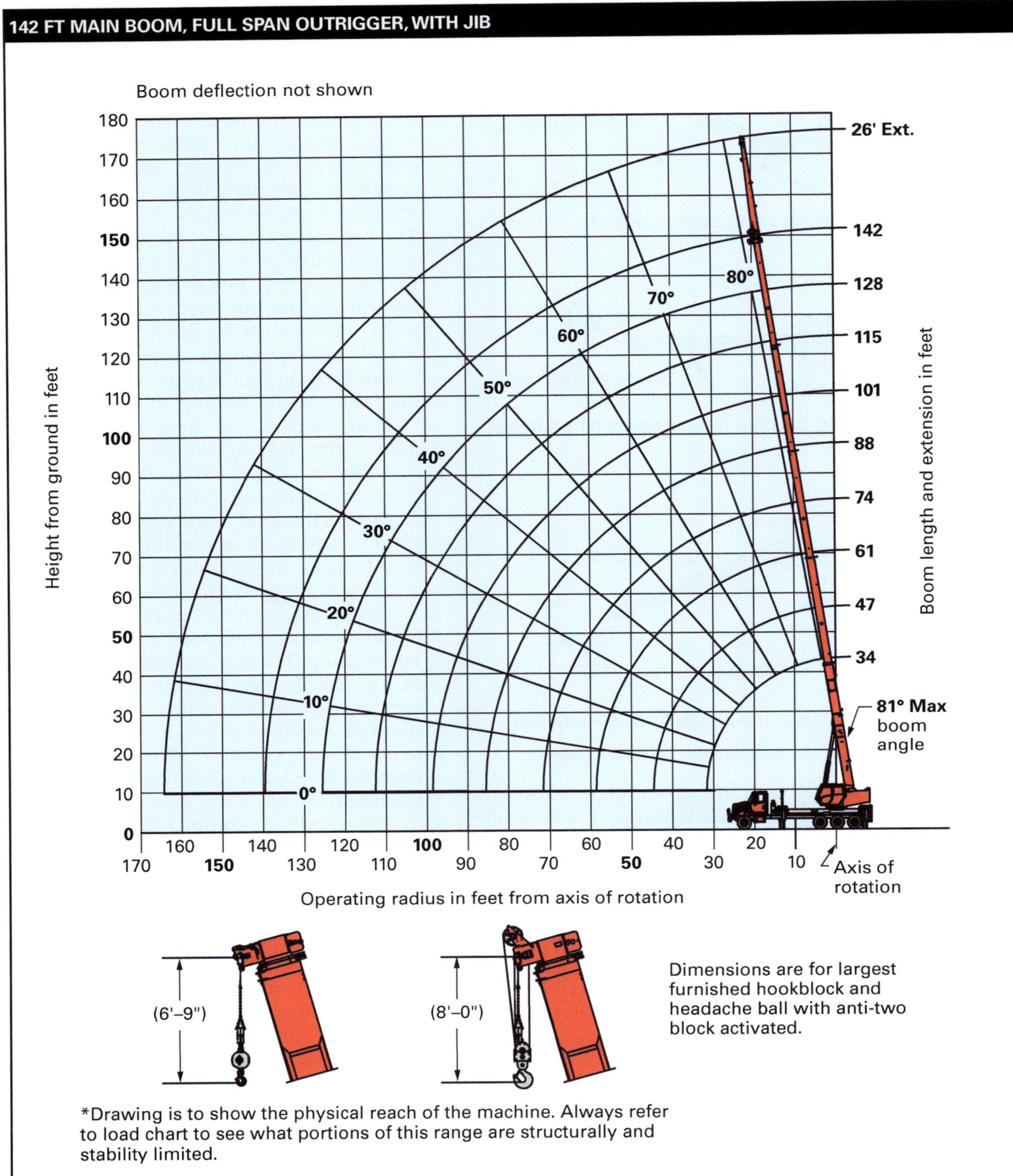

Figure 47 Sample range diagram.

Radius in feet	#02								
	Main boom length in feet								
	34	47-A	61-B	74-C	88-D	101-E	115-F	128-G	142
7	79,475 (74.9)								
8	74,475 (73.1)								
10	65,975 (69.4)	39,600 (75.6)							
12	54,475 (65.7)	39,600 (73.1)	39,700 (77.4)						
15	42,475 (59.7)	39,600 (69.2)	37,700 (74.5)	33,750 (77.7)					
20	30,225 (48.9)	31,000 (62.3)	31,500 (69.5)	29,750 (73.7)	22,850 (76.7)	17,200 (78.8)			
25	22,725 (35.7)	23,450 (55)	23,950 (64.2)	24,250 (69.5)	20,500 (73.4)	15,550 (75.9)	12,850 (78.3)		
30	17,475 (13.5)	18,400 (46.9)	18,900 (58.8)	19,200 (65.2)	18,550 (70)	14,100 (73.1)	12,000 (75.8)	9900 (78)	7875 (79.5)
35		14,750 (37.5)	15,250 (52.9)	15,550 (60.7)	15,800 (66.4)	13,000 (70.1)	11,000 (73.5)	9400 (75.8)	7475 (77.7)
40		11,800 (25.2)	12,350 (46.6)	12,650 (56)	12,900 (62.6)	12,000 (67.1)	10,250 (71)	8900 (73.7)	7325 (75.9)
45			10,000 (40.1)	10,300 (51.5)	10,550 (59.1)	10,700 (64.2)	9600 (68.4)	8400 (71.4)	7075 (74)
50			8100 (31.8)	8400 (46.2)	8600 (55)	8800 (60.8)	8950 (65.7)	7900 (69.1)	6675 (72)
55			6550 (20.6)	6900 (40.3)	7100 (50.8)	7250 (57.3)	7450 (62.8)	7450 (66.7)	6425 (70)
60				5650 (33.6)	5850 (46.3)	6000 (53.7)	6200 (59.7)	6350 (64.1)	6075 (67.9)
65				4600 (25.4)	4850 (41.4)	5000 (49.9)	5150 (56.6)	5300 (61.4)	5425 (65.6)
70				3750 (12.6)	4000 (35.9)	4100 (45.9)	4300 (53.3)	4400 (58.6)	4525 (63.1)
75					3250 (29.6)	3350 (41.6)	3550 (49.9)	3650 (55.7)	3775 (60.6)
80					2600 (21.6)	2700 (36.9)	2850 (46.4)	2950 (52.7)	3125 (58)
85					2100 (7.2)	2100 (31.5)	2250 (42.6)	2350 (49.6)	2475 (55.3)
90						1650 (25.1)	1350 (38.5)	1850 (46.3)	1925 (52.6)
95						1250 (16.5)	950 (34)	1350 (42.8)	1475 (49.7)
100							600 (28.8)	950 (39.2)	1075 (46.7)
105							500 (22.6)	650 (35.1)	675 (43.6)
110									525 (41.6)
Minimum boom angle (°) for indicated length (no load)						0	22.5	35	43.4
Maximum boom length (ft) at 0° boom angle (no load)					101				

NOTE: Loads displayed in pounds. ( ) Boom angles are in degrees.
#LMI operating code. Refer to LMI manual for operating instructions.

Lifting capacities at zero degree boom angle						
Boom angle	Main boom length in feet					
	34	47-A	61-B	74-C	88-D	101-E
0°	16,825 (31.5)	9550 (44.5)	5600 (58.5)	3550 (71.5)	2000 (85.5)	900 (98.5)

NOTE: ( ) Reference radii in feet.

80027135

Radius in feet	#03
33	4000 (80)
50	3800 (75)
65	3200 (70)
78	2450 (65)
90	1800 (60)
101	1250 (55)
112	650 (50)
Min. boom angle for indicated length (no load)	48°
Max. boom length at 0° boom angle (no load)	88 ft

80027138

NOTE: Loads displayed in pounds
( ) Boom angles are in degrees.
#LMI operating code. Refer to LMI manual for operating instructions.

Boom extension capacity notes:
1. All capacities above the bold line are based on structural strength of boom extension.
2. 26 ft extension length may be used for single line lifting service
3. Radii listed are for a fully extended boom with the boom extension erected. For main boom lengths less than fully extended, the rated loads are determined by boom angle. For boom angles not shown, use the rating of the next lower angle.
Warning: Operation of this machine with heavier loads than the capacities listed is strictly prohibited. Machine tipping with boom extension occurs rapidly and without advance warning.
4. Boom angle is the angle above or below horizontal of the longitudinal axis of the boom base section after lifting rated load.
5. Capacities listed are with outriggers properly extended and vertical jacks set.
6. When lifting over the main boom nose with 26 ft extension erected, the outriggers must be fully extended or 50% (17.5 ft) spread.

THIS CHART IS ONLY A GUIDE AND SHOULD NOT BE USED TO OPERATE THE CRANE.
The individual crane's load chart, operating instructions and other instructional plates must be read and understood prior to operating the crane.

Figure 48 Sample load chart.

Note that this section merely introduces load charts and doesn't represent a full presentation of the topic. Additional information on load charts is presented in future modules.

2.5.1 Capacity with No Attachments

All cranes have their greatest lifting capacity when they have no attachments. This is because attachments add to the lifted load. To calculate the capacity with no attachments, use the following steps:

Step 1   Use the load chart to find the gross capacity for the boom length, boom angle, load radius, and quadrant of operation.

Step 2   Use the load chart to determine the parts of line needed for the lift and to determine the traveling block needed.

Step 3   Use the load chart to determine any load deductions. This may include anything below the boom tip, such as the traveling block, headache ball, lifting rope, and all load rigging.

Step 4   Subtract all noted deductions from the gross capacity to determine the net capacity.

2.5.2 Capacity with Attachments

To calculate the capacity with attachments, use the following steps:

Step 1   Use the load chart to find the gross capacity for the boom length, boom angle, load radius, and quadrant of operation.

Step 2   Use the load chart to determine the parts of line needed for the lift and to determine the traveling block needed.

Step 3   Use the load chart to determine the **effective weight** of the extension or jib and attachments. This applies to any extensions or jibs in a stored position on the main boom. It also applies to installed extensions or jibs when the main boom (and not the extension or jib) is being used for the lift. When installed, the effective weight of the extension or jib may be as much as twice its actual weight. Deduct the effective weight from the boom capacity.

Step 4   Use the load chart to determine any load deductions. This may include anything below the boom tip, such as the traveling block, headache ball, lifting rope, and all load rigging.

Load Charts

Each crane can have many different load charts to consult, depending primarily on the crane's configuration. For example, the Link-Belt 218 HSL lattice-boom crawler crane (shown here) is a fairly common model rated at 110 tons (99.8 metric tons). The technical specifications for this crane include 18 different load charts.

Figure Credit: Link-Belt Construction Equipment Company

NCCER – *Mobile Crane Operations*

Step 5 Subtract all noted deductions from the gross capacity to determine the net capacity.

2.5.3 Capacity Using Extensions or Jibs

To calculate the capacity of a crane using extensions or jibs, perform the following:

Step 1 Use the jib chart, boom chart, and chart notes to find the jib offset, boom angle, or jib-to-ground angle.

Step 2 Use the jib chart and/or the boom chart to determine the gross capacity.

Step 3 Use the jib chart to determine which headache ball or block to use. Extension or jib lifts are usually made with a single part of line.

Step 4 Use the jib chart to determine the load deductions. This may include anything below the boom tip, such as the headache ball or block, lifting rope, and all load rigging.

Step 5 Subtract all noted deductions from the gross capacity to determine the net capacity.

2.5.4 Critical Lifts

There are a number of factors in a lift that can place it into the category of a critical lift. Lifting hazardous materials, working in close proximity to power lines, or swinging a load in a congested work area are all possible reasons to classify a lift as critical. In addition, any lift that exceeds 75 percent of a crane's capacity has been historically considered a critical lift by many organizations. A critical lift is defined the same way in 29 *CFR 1926.751, Steel Erection*. Per *ASME Standard B30.5*, "the criteria to categorize a lift as critical on this basis are established by site supervision, project management, a qualified person, or company policies.

Lift planning and oversight shall be tailored to each hoisting operation and shall be sufficient to manage varying conditions and their associated hazards." *ASME Standard B30.5* also has an appendix entitled *Nonmandatory Appendix A, Critical Lifts* that provides guidance, including a number of scenarios that typically indicate the need for a critical-lift classification.

Critical lifts require a more detailed lift plan than a standard lift plan, and the plan must be documented in writing. A critical lift plan should include but not be limited to the following:

- The total weight to be lifted
- Crane placement location
- Identification of the crane(s) to be used, including its required configuration, and the percentage of its capacity the load represents in that configuration
- Sling and rigging component selection
- A dimensioned diagram of the lifting area
- A diagram of the rigging configuration

Once a plan has been assembled, all responsible job-site personnel, including the crane operator, must review, approve, and sign the plan before it can be implemented. A pre-lift meeting is also called to ensure all parties are aware of the concerns before the lift begins.

Additional Resources

ASME Standard B30.5, Mobile and Locomotive Cranes. Current edition. New York, NY: American Society of Mechanical Engineers.

29 *CFR* 1926.1400, **www.ecfr.gov**

Mobile Crane Safety Manual. 2014. Milwaukee, WI: Association of Equipment Manufacturers.

Cranes: Design, Practice, and Maintenance, Ing J. Verschoof. Second Edition. 2002. Hoboken, NJ: John Wiley and Sons, Inc.

IPT's Crane and Rigging Handbook, Ronald G. Garby. Current Edition. Spruce Grove, Alberta, Canada: IPT Publishing and Training Ltd.

North American Crane Bureau, Inc. website offers resources for products and training, **www.cranesafe.com**

2.0.0 Section Review

1. When working near an excavation, any weight-bearing component, such as an outrigger, must be positioned at least ______.

    a. 1.5 times the depth of the excavation away
    b. 2 times the depth of the excavation away
    c. 2.5 times the depth of the excavation away
    d. 3 times the depth of the excavation away

2. When using crane mats to support a crawler crane, how far should the mat extend beyond the ends and edges of the tracks?

    a. 18 inches (46 centimeters)
    b. 2 feet (0.6 meters)
    c. 4 feet (1.2 meters)
    d. 5 feet (1.5 meters)

3. The rated capacity of a crawler crane is typically ______.

    a. 65 percent of its tipping load
    b. 75 percent of its tipping load
    c. 85 percent of its tipping load
    d. 95 percent of its tipping load

4. Dragging a load along the ground from one side of the crane to the other will cause ______.

    a. load lag
    b. swing out
    c. effective loading
    d. side loading

5. Manufacturers use bold lines, shading, or asterisks to identify load-chart capacity figures that are based on ______.

    a. the quadrant of operation
    b. lifts with and without a jib
    c. the structural strength of the boom extension
    d. the structural strength of the ring gear drive

SUMMARY

Mobile cranes are common on modern construction sites and many other environments. Some cranes are used to lift and carry materials and equipment around the site. Others are used to lift and place materials during construction. Mobile crane operators must be familiar with the components of the crane in use and the factors affecting its lifting capacity. Operators must also be aware of overhead, ground level, and below-ground conditions that may present a danger to workers in the area and nearby property.

Range diagrams and load charts provide essential information for each and every crane. No crane should be operated without the appropriate load chart on hand. These critical data resources will be explored further in future training modules, since reading and interpreting the information on load charts is a vital operator skill.

1. Rough-terrain cranes generally travel at or below a speed of roughly ______.
    a. 10 mph (16 kph)
    b. 20 mph (32 kph)
    c. 30 mph (48 kph)
    d. 45 mph (72 kph)

2. Wheeled truck cranes without a load can typically climb a job site ramp with a ______.
    a. 40-percent grade
    b. 50-percent grade
    c. 60-percent grade
    d. 70-percent grade

3. In addition to the jib itself, the jib forestay and jib backstay are also attached to the ______.
    a. hoist line
    b. load moment
    c. hoist drum
    d. jib mast

4. The lowest portion of a lattice boom that is connected directly to the upperworks is called the ______.
    a. lead
    b. boom base
    c. jib
    d. boom butt

5. To move loose material and unload it easily, without the need for ropes or cables to operate the device, a crane operator can use a(n) ______.
    a. clamshell bucket
    b. self-dumping bin
    c. grapple
    d. rehandling clamshell bucket

6. The two primary types of pile-driving equipment handled by a crane are ______.
    a. pile hammers and pile drills
    b. pile drills and vibratory pile drivers
    c. pile hammers and sheet hammers
    d. vibratory pile drivers and pile hammers

7. An example of a Category II operational aid is a(n) ______.
    a. boom hoist limiting device
    b. luffing jib limiting device
    c. anti-two-blocking device
    d. jib angle indicator (for luffing jibs)

8. If a crane hoist sheave has four-part reeving, the hoist line is terminated at the ______.
    a. jib
    b. actuator
    c. boom
    d. hoist sheave

9. Cranes are required to be no more than one percent out of level, which equals ______.
    a. 3.6 degrees
    b. 2.31 degrees
    c. 1.14 degrees
    d. 0.57 degrees

10. When a crane's level is missing or unreliable, the crane can be leveled using the headache ball as a ______.
    a. plumb bob
    b. level
    c. counterweight
    d. protractor

11. For a rough-terrain crane, the maximum lift capacity is accessed when the boom is in which quadrant?
    a. Over the front
    b. Over the rear
    c. Over the side
    d. 360 degrees

12. The rated capacity of a wheeled truck crane when on outriggers is typically ______.
    a. 65 percent of its tipping load
    b. 75 percent of its tipping load
    c. 85 percent of its tipping load
    d. 95 percent of its tipping load

13. When a lattice boom is lowered to a smaller angle (closer to the horizontal plane), ______.

    a. it begins to sag in the middle
    b. the leverage applied by a load is reduced
    c. its capacity becomes nearly infinite
    d. a whip line cannot be used at all

14. When using a specific load chart for a crane, ______.

    a. you can assume the quadrant of operation is 360 degrees
    b. the effective weight of a jib will be the same as its actual weight
    c. interpolation is not allowed
    d. the net capacity is read directly from the chart

15. A critical lift plan must be reviewed and signed by ______.

    a. the site supervisor
    b. the crane manufacturer
    c. the lift director
    d. all responsible job site personnel

Trade Terms Quiz

Fill in the blank with the correct term that you learned from your study of this module.

1. A(n) __________ is a portable platform, typically made of large wooden timbers bolted together to support and spread the weight of a crane over a larger ground area.

2. A(n) __________ is a piece of standing rigging that is routed from the jib mast back to the main boom to help support the jib.

3. To pick up items such as a downed tree, a stack of tree limbs, or a large rock, the best attachments for the job would be __________.

4. The crawler crane assembly that consists of the carbody, ring gear drive, crawler frames, and tracks is called the __________.

5. Steel structures called __________ provide support for a pile hammer and help to align the hammer with the pile to be driven.

6. To prevent the load block or hook assembly from coming into contact with the boom tip, a(n) __________ is used.

7. A system of two or more pulleys with a rope or cable threaded between them to use as a lifting aid is called a(n) __________.

8. Loads that are not constant, but instead are consistently changing due to various factors in a lift are called __________.

9. The point at which the entire weight of an object is considered to be concentrated is called its __________.

10. The __________ is the lowest portion of a telescopic boom that houses the other telescopic sections, but does not itself extend.

11. Component used to guide or place tension on a belt or crawler-crane track are called __________.

12. To help oppose the weight of the load and improve stability, __________ are added to a crane.

13. Improper reeving of the boom tip sheave, where the descending rope is placed to one side of the sheave rather than in the center, leads to __________.

14. If crane documentation refers to a percentage of time that the crane is designed to operate over a 24-hour period, it is referring to the crane's __________.

15. Crane assemblies that include crawler tracks, track idlers, and track power sources of a crawler crane are called __________.

16. Soil and rock is used as __________ to level an area or fill voids, such as the perimeter of building foundations or trenches.

17. To determine an unknown value that lies between two known values, __________ is needed.

18. A(n) __________ is used to wind or unwind the rope for hoisting or lowering the load.

19. A device designed to allow flow in one direction but prevents fluid flow in the opposite direction is a(n) __________.

20. The __________ of an accessory such as a boom extension or jib reflects the effect of its weight on the lift, usually based on its position, rather than its actual weight.

21. __________ refers to a hard, compacted layer of subsoil, usually with a major clay component.

22. If a lifting operation will present a significantly increased level of risk beyond normal lifting activities, it will likely be identified as a(n) __________.

23. A(n) __________ extends or retracts using pressure from the hydraulic system.

24. A(n) __________ refers to the base of a wheeled crane that provides crane movement.

25. The feet of outriggers on wheeled truck cranes are called __________.

26. Motors powered by fluid pressure, called __________, are often used to power crawler-crane tracks.

27. Extendable or fixed members known as __________ are attached to a crane base to stabilize and support a crane.

28. The dynamic effects on a stationary or mobile body as imparted by the forcible contact of another moving body or the sudden stop of a fall are called __________.

29. When you apply a lever, you are using the mechanical advantage in power known as __________.

30. Extension attached to booms point to provide added length for reaching and lifting loads are called __________.

31. A heavy, round weight called a(n) __________ is often attached to a load line to provide sufficient weight to allow the load line to unspool from the drum when there is no live load.

32. When reeving, the space between the sheave and the frame of a block through which the rope is passed is called the __________.

33. The __________ is a piece of standing rigging that is routed from the far tip of the jib back to the jib mast, holding the tip of the jib up.

34. According to 29 *CFR* 1926.1401, the __________ of a crane can also be referred to as the superstructure.

35. A(n) __________ is a secondary hoisting rope usually of lower capacity than that provided by the main hoisting system.

36. A structure mounted on the main boom that provides a point of connection for the jib forestay and jib backstay on lattice-boom cranes is called the __________.

37. A boom constructed of steel angles or tubing to create a relatively lightweight but strong, rigid structure is referred to as a(n) __________.

38. Ropes or strands of a specified length with fixed end connections, used to support a lattice boom or boom components, are called __________.

39. The __________ is usually reported to the operator as a percentage of the crane's capacity at the present set of conditions.

40. The part of a crawler-crane base mounting that carries the rotating upperworks is called the __________.

41. The total amount a crane can safely lift under a given set of conditions defines its __________.

42. When you adjust the boom angle by varying the length of the suspension ropes, you are __________ the boom.

43. A lattice boom with an opening in the boom structure near the far end, allowing the hoist lines to drop through the boom rather than over the end of the boom, is called a(n) __________.

44. The __________ of a crane is its gross capacity minus all noted capacity deductions.

45. A process known as __________ is used to multiply pulling or lifting capability by using wire rope routed through multiple pulleys or sheaves a number of times.

46. A metal that contains no iron, such as aluminum, is considered __________.

47. The distance from the center of the boom's mounting point (usually the ring gear drive) to the center of gravity of the load defines a crane's __________.

48. A trailer with a low frame for transporting very tall or heavy loads, such as large cranes, is called a(n) __________.

49. Single- or multiple-line __________ is used for whip, boom, and jib lines.

50. If the capacity of a crane is exceeded, the crane will turn over at a point called the __________.

51. When a line is reeved on the hoist sheaves, the resulting number of lines that are supporting the load block are referred to as __________.

52. The combined operating height and operating radius of a boom determine a crane's __________.

53. The direction of the boom relative to the base mounting or carrier body determines its __________.

54. A jib mounted on the end of a boom that can be positioned at different angles relative to the main boom is called a(n) __________.

55. The upperworks of a crane are mounted to the __________, which provides the pivot point for the entire assembly. It is sometimes referred to as the swing circle.

56. Wheels that have a groove for a belt, rope, or cable to run in re called __________.

57. The distance between the front and rear axles of a vehicle is called the __________.

58. Cribbing is another word for __________.

Trade terms

Anti-two-blocking device	Duty cycle	Jib backstay	Parts of line
Backfill	Dynamic loads	Jib forestay	Pendants
Base mounting	Effective weight	Jib mast	Quadrant of operation
Base section	Floats	Lattice boom	Reach
Block and tackle	Grapples	Leads	Reeving
Blocking	Gross capacity	Leverage	Ring gear drive
Boom torque	Hardpan	Load moment	Sheaves
Carbody	Headache ball	Lowboy	Swallow
Carrier	Hoist drum	Luffing	Telescopic boom
Center of gravity	Hoist reeving	Luffing jib	Tipping fulcrum
Check valve	Hydraulic motors	Net capacity	Upperworks
Counterweights	Idlers	Non-ferrous	Wheelbase
Crane mat	Impact loads	Open-throat boom	Whip line
Crawler frames	Interpolation	Operating radius	
Critical lift	Jibs	Outriggers	

Richard Laird
Mobile Crane and Heavy Equipment Lead Instructor

Associated Builders and Contractors Pelican Chapter

Building upon a strong work ethic and an expansive base of experience in the petrochemical industry, Richard maintains his excellence by continually improving his own skills as well as training and certifying the next generation of Mobile Crane and Rigging/Signal persons, with a vigilant focus on worker safety.

Please give a brief synopsis of your construction career and your current position.

After working 34 years in the petroleum chemical industry, I retired from ExxonMobil as the refinery lift specialist in the rigging and mobile crane operation. My work also included procuring cranes and negotiating work orders with the major crane rental companies in the area. I now provide training and certification to companies, and work with Associated Builders and Contractors (ABC) in Baton Rouge as the Lead Instructor for mobile crane and heavy equipment operators. I am also involved with NCCER as a Mobile Crane Practical Examiner (CPE), and a Rigging/Signal Person Practical Examiner.

How did you get started in the construction industry?

Like every young man wanting to have a steady job and money in the bank, I successfully launched my career by answering an ad in the paper, and began working with Humble Oil and Refining Company in Baton Rouge, now known as ExxonMobil Refinery.

Who or what inspired you to enter the construction industry?

I enjoyed watching cranes and riggers work together to move equipment. The more I saw this, the more I became inspired and knew this was what I wanted to do every day.

How has training in construction impacted your life and career? What types of training have you completed?

As I progressed through my own training and mastered skills, I was offered the opportunity to do tutoring work for ABC. As a result, I became an instructor for mobile crane operators and, eventually, a Practical Examiner.

Why do you think credentials are important in the construction industry?

There is a great need in the industry today for construction people to be credentialed. Not only does this bring a more well-rounded and verifiably skilled person to the craft, but it also encourages a safety-focused culture and promotes worker awareness of the need to protect themselves and others on the job.

What do you enjoy most about your career?

The most rewarding part of my career is being able to encourage and watch both young men and women making career changes that generate job satisfaction and improved financial positions in their lives.

Would you recommend construction as a career to others? Why?

The construction environment presents an opportunity for people to find well-paying jobs that bring great personal satisfaction, and being able to provide a service that is of value and a benefit to others.

What advice would you give to someone who is new to the construction industry?

I would advise anyone beginning to work in the construction trades to be aware that there are many opportunities available for continued training and competency that can fast-track their career goals.

How do you define craftsmanship?

Craftsmanship is what happens when a tradesman advances the quality of his or her work to the next level and then beyond, achieving the goals of being well-trained and proficient in a given craft. A craft professional consistently produces excellence in their work by not short-cutting tasks and always maintaining a safe work environment for themselves and others.

Trade Terms Introduced in This Module

Anti-two-blocking device: Two-blocking refers to a condition in which the lower load block or hook assembly comes in contact with the boom tip, boom tip sheave assembly or any other component above it as it is being raised. If this occurs, continuing to apply lifting power to the cable can result is serious equipment damage and/or failure of the hoist line. An anti-two-blocking device, therefore, prevents this condition from occurring.

Backfill: Soil and rock used to level an area or fill voids, such as the perimeter of building foundations or trenches. Areas with fresh backfill may not be stable enough to support a crane.

Base mounting: A crawler crane assembly consisting primarily of the carbody, ring gear drive, crawler frames, and tracks.

Base section: The lowest portion of a telescopic boom that houses the other telescopic sections but does not extend.

Block and tackle: A system of two or more pulleys, which form a block, with a rope or cable threaded between them, reducing the force needed to lift or pull heavy loads.

Blocking: Wood or a similar material used under outrigger floats to support and distribute loads to the ground. Also referred to as *cribbing*.

Boom torque: A twisting force applied to the crane boom, typically resulting from imbalanced reeving of the boom tip sheave assembly ropes.

Carbody: The part of a crawler-crane base mounting that carries the rotating upperworks.

Carrier: The base of a wheeled crane that provides crane movement and supports the upperworks.

Center of gravity (CG): The point at which the entire weight of an object is considered to be concentrated, such that supporting the object at this specific point would result in its remaining balanced in position.

Check valve: A valve designed to allow flow in one direction but closes as necessary to prevent flow reversal.

Counterweights: Weights added to the crane, usually on the end opposite the boom, to help counter the weight of the load and improve stability.

Crane mat: A portable platform, typically made of large wooden timbers bolted together, used to support and spread the weight of a crane over a larger ground area.

Crawler frames: Crane assemblies comprised of the crawler tracks, track idlers, and track power sources of a crawler crane. Also called *tread members* or *track assemblies*.

Critical lift: Defined in *ASME Standard B30.5* as a hoisting or lifting operation that has been determined to present an increased level of risk beyond normal lifting activities. For example, increased risk may relate to personnel injury, damage to property, interruption of plant production, delays in schedule, release of hazards to the environment, or other significant factors.

Duty cycle: An expression of equipment use over time. In the case of mobile cranes, an 8-, 16-, or 24-hour rating expressed as a percentage.

Dynamic loads: A load on a structure (in this case, a crane) that is not constant, but consistently changing as the result of one or more changes in various factors. Also referred to as shock loading, significant dynamic loads can be applied to a crane through abrupt motions and lifting a load from its support too quickly.

Effective weight: The weight of an accessory such as a boom extension or jib that reflects the effect of its weight on the lift, usually based on its position, rather than its actual weight. For example, a jib folded and stored on the main boom will have different effective weight than when it is installed on the main boom tip.

Floats: The portion of outriggers that touches the ground; the feet of the outriggers.

Grapples: Devices used to pick up bulk items, containers, rocks, trees and tree limbs, etc. Grapples typically have several jaws that operate like fingers to pick up material, using mechanical or hydraulic power.

Gross capacity: The total amount a crane can safely lift under a given set of conditions. The gross capacity includes but is not limited to the load block, ropes, and rigging as well the primary load.

Hardpan: A hard, compacted layer of subsoil, usually with a major clay component.

Headache ball: A heavy round weight often attached to a load line to provide sufficient weight to allow the load line to unspool from the drum when there is no live load. Larger versions of headache balls are used to swing into structures to demolish them.

Hoist drum: A drum is a cylindrical component around which a rope is wound. The hoist drum is used to wind or unwind the rope for hoisting or lowering the load; the part of a crane that spools and unspools the lifting line.

Hoist reeving: The reeving pattern applied to the hoist sheaves. Single- or multiple-line hoist reeving is used for whip, boom, and jib lines.

Hydraulic motors: Motors powered by hydraulic pressure provided by an external pump. Hydraulic motors are often used to power the tracks of crawler cranes, instead of complex drive systems connected directly to the diesel engine.

Idlers: Pulleys, wheels, or rollers that do not transmit power, but guide or place tension on a belt or crawler-crane track.

Impact loads: The dynamic effects on a stationary or mobile body as imparted by the forcible contact of another moving body or the sudden stop of a fall.

Interpolation: The process of estimating or calculating unknown values between two known values.

Jibs: Extensions attached to the boom point to provide added boom length for reaching and lifting loads. Jibs may be in line with the boom, offset to another angle, or adjustable to a variety of angles. A jib is sometimes referred to as a *fly*.

Jib backstay: A piece of standing rigging that is routed from the jib mast back to the main boom to help support the jib.

Jib forestay: A piece of standing rigging that is routed from the far tip of the jib back to the jib mast, holding the tip of the jib up.

Jib mast: A structure mounted on the main boom that provides a fixed distance for the point of connection of the jib forestay and jib backstay. Also referred to as a *jib strut*.

Lattice boom: A boom constructed of steel angles or tubing to create a relatively lightweight but strong, rigid structure.

Leads: Steel structures that provide support for a pile hammer and help to align and position the hammer with the pile to be driven. The hammer can travel up or down in the leads as necessary.

Leverage: The mechanical advantage in power gained by using a lever.

Load moment: The force applied to the crane by the load; the leverage of the load, opposing the leverage of the crane. The load moment is calculated by multiplying the gross load weight by the horizontal distance from the tipping fulcrum to the center of gravity of the suspended load. The load moment is usually reported to the operator as a percentage of the crane's capacity at the present set of conditions. As those conditions change, such as the boom angle, the load moment changes as well.

Lowboy: A trailer with a low frame for transporting very tall or heavy loads. A typical lowboy has two drops in deck height: one right after the gooseneck connecting it to the tractor, and one right before the wheels. This allows the trailer deck to be extremely low compared with common trailers.

Luffing: Changing a boom angle by varying the length of the suspension ropes.

Luffing jib: A jib mounted on the end of a boom that can be positioned at different angles relative to the main boom.

Net capacity: The weight of the item(s) that can be lifted by the crane; the gross capacity of a crane minus all noted capacity deductions.

Non-ferrous: Having no iron. Ferrous metals, such as steel, contain iron and are magnetic as a result.

Open-throat boom: A lattice boom with an opening in the boom structure near the far end, allowing the hoist lines to drop through the boom rather than over the end of the boom.

Operating radius: The distance from the center of the boom's mounting point (usually the ring gear drive) to the center of gravity of the load.

Outriggers: Extendable or fixed members attached to a crane base that rest on ground supports at the outer end to stabilize and support the crane.

Parts of line: When a line is reeved more than once, the resulting number of lines that are supporting the load block.

Pendants: Ropes or strands of a specified length with fixed end connections, used to support a lattice boom or boom components. According to 29 *CFR* 1926.1401, a pendant may also consist of a solid bar.

Quadrant of operation: The direction of the boom relative to the base mounting or carrier body.

Reach: The combined operating height and radius of a boom, or the combination of boom and jib.

Reeving: A method often used to multiply the pulling or lifting capability by using wire rope routed through multiple pulleys or sheaves a number of times.

Ring gear drive: Sometimes referred to as the swing circle. An assembly that provides the point of attachment and pivot point for the upperworks of a crane. The ring gear is typically driven by hydraulic pressure, allowing the upperworks to rotate on a set of bearings that reduce friction and transfer the weight of the upperworks (and any load) to the carbody.

Sheaves: Wheels that have a groove for a belt, rope, or cable to run in. The terms *sheave* and *pulley* are often used interchangeably.

Swallow: The space between the sheave and the frame of a block, through which the rope is passed.

Telescopic boom: A crane boom that extends and retracts in sections that slide in and out, powered by hydraulic pressure.

Tipping fulcrum: The point of crane contact with the ground where it would pivot if it were to tip over; the fulcrum of the leverage applied by the load. Depending on the attitude and type of crane, the tipping fulcrum may be the edge of one crawler assembly, one or more outriggers, or similar locations.

Upperworks: A term that refers to the assembly of components above the ring gear drive; the rotating collection of components on top of the base mounting or carrier; may also be referred to as the *house,* or as the *superstructure* as defined in 29 *CFR* 1926.1401.

Wheelbase: The distance between the front and rear axles of a vehicle.

Whip line: A secondary hoisting rope usually of lower capacity than that provided by the main hoisting system. When a whip line exists, it is typically out at the tip of a jib, while the main hoist line is closer to the crane and operated from the tip of the main boom.

NCCER – *Mobile Crane Operations*

Additional Resources

This module presents thorough resources for task training. The following reference material is recommended for further study.

ASME Standard B30.5, Mobile and Locomotive Cranes. Current edition. New York, NY: American Society of Mechanical Engineers.

29 CFR 1926.1400, **www.ecfr.gov**

Mobile Crane Safety Manual (AEM MC-1407). 2014. Milwaukee, WI: Association of Equipment Manufacturers.

Cranes: Design, Practice, and Maintenance, Ing J. Verschoof. Second Edition. 2002. Hoboken, NJ: John Wiley and Sons, Inc.

IPT's Crane and Rigging Handbook, Ronald G. Garby. Current Edition. Spruce Grove, Alberta, Canada: IPT Publishing and Training Ltd.

North American Crane Bureau, Inc. website offers resources for products and training, **www.cranesafe.com**

Figure Credits

Link-Belt Construction Equipment Company, Module opener, Figures 1, 5, 7–9, 10A, 14–16, 28, SA04

Topaz Publications, Inc., Figures 2, 4

The Manitowoc Company, Inc., Figures 6, 12, 13, 47, 48

Manitex, Inc., Figure 10B

© Hellen Sergeyeva/Shutterstock.com, Figure 11

Liebherr USA, Co., SA01

Smiley Lifting Solutions - www.spydercrane.com, SA02

© iStockphoto.com/dane-mo, Figure 19

© iStockphoto.com/Paul Vasarhelyi, Figure 20

Bigfoot Crane Company Inc., Figure 21

© iStockphoto.com/Bradford Martin, Figure 22

© AngelPet/Shutterstock.com, Figure 23

Kenco Construction Products, Inc., Figure 24

Lifting Technologies, Figure 25

Carolina Bridge Co., Figures 26, 27

SANY America, Inc., Figure 29

DICA Outrigger Pads, SA03

Section Review Answer Key

Answer	Section Reference	Objective
Section One		
1. a	1.1.2	1a
2. c	1.2.2	1b
3. d	1.3.1	1c
4. a	1.4.0	1d
5. c	1.5.0	1e
Section Two		
1. a	2.1.1	2a
2. b	2.2.1	2b
3. b	2.3.2; Table 4	2c
4. d	2.4.1	2d
5. c	2.5.0	2e

NCCER CURRICULA — USER UPDATE

NCCER makes every effort to keep its textbooks up-to-date and free of technical errors. We appreciate your help in this process. If you find an error, a typographical mistake, or an inaccuracy in NCCER's curricula, please fill out this form (or a photocopy), or complete the online form at **www.nccer.org/olf**. Be sure to include the exact module ID number, page number, a detailed description, and your recommended correction. Your input will be brought to the attention of the Authoring Team. Thank you for your assistance.

Instructors – If you have an idea for improving this textbook, or have found that additional materials were necessary to teach this module effectively, please let us know so that we may present your suggestions to the Authoring Team.

NCCER Product Development and Revision
13614 Progress Blvd., Alachua, FL 32615

Email: curriculum@nccer.org
Online: www.nccer.org/olf

❏ Trainee Guide ❏ Lesson Plans ❏ Exam ❏ PowerPoints Other ________________

Craft / Level: Copyright Date:

Module ID Number / Title:

Section Number(s):

Description:

____

____

____

____

Recommended Correction:

____

____

____

____

Your Name:

Address:

Email: Phone:

Rigging Practices

OVERVIEW

Rigging is the preparation of a load for movement, as well as preparation of the hardware and other components used to connect the load to a crane. Rigging is associated with all types of cranes, and rigging skills are also required to move and position equipment inside buildings and other areas where cranes are not involved. This module will provide insight into rigging hardware, lifting slings and their proper use, and various types of rigging equipment.

Module 38102

Rigging Practices

Objectives

When you have completed this module, you will be able to do the following:

1. Identify and describe various types of rigging hardware.
   a. Identify and describe various hooks, shackles, eyebolts, and clamps.
   b. Identify and describe various lugs, turnbuckles, plates, and spreader beams.
2. Identify and describe various types of slings and sling hitches.
   a. Identify and describe wire-rope slings and their proper care.
   b. Identify and describe synthetic slings and their proper care.
   c. Identify and describe chain slings and their proper care.
   d. Explain the significance of sling angles and describe common hitches.
   e. Describe how to properly rig and handle piping materials and rebar.
   f. Identify and describe how to use taglines and knots for load control.
   g. Identify common rigging-related safety precautions.
3. Identify and describe how to use various types of hoisting and jacking equipment.
   a. Identify and describe how to use manual and powered hoisting equipment.
   b. Identify and describe how to use jacks.

Performance Tasks

Under the supervision of your instructor, you should be able to do the following:

1. Inspect various types of rigging components and report on the condition and suitability for a task.
2. Configure a sling to produce a single-wrap basket hitch.
3. Configure a sling to produce a double-wrap basket hitch.
4. Configure a sling to produce a single-wrap choker hitch.
5. Configure a sling to produce a double-wrap choker hitch.
6. Select the correct tagline for a specified application.
7. Tie specific instructor-selected knots.
8. Select, inspect, and demonstrate the safe use of the following rigging equipment:
   - Block and tackle
   - Chain hoist
   - Ratchet-lever hoist
   - One or more types of jack

Industry Recognized Credentials

If you are training through an NCCER-accredited sponsor, you may be eligible for credentials from NCCER's Registry. The ID number for this module is 38102. Note that this module may have been used in other NCCER curricula and may apply to other level completions. Contact NCCER's Registry at 888.622.3720 or go to **www.nccer.org** for more information.

Note

This module provides instruction and information about common rigging equipment and hitch configurations, but it does not provide any level of rigging certification. Any questions about rigging procedures and/or certification should be directed to an instructor or supervisor.

Contents

1.0.0 RIGGING HARDWARE

Objective

Identify and describe various types of rigging hardware.

a. Identify and describe various hooks, shackles, eyebolts, and clamps.
b. Identify and describe various lugs, turnbuckles, plates, and spreader beams.

Performance Task

1. Inspect various types of rigging components and report on the condition and suitability for a task.

Trade Terms

Blind hole: A hole that does not penetrate the material completely, leaving a hole with a bottom.

Bridle hitch: A type of hitch that consists of two or more slings that support the load attached to a common lifting point.

Center of gravity (CG): The point at which the entire weight of an object is considered to be concentrated, such that supporting the object at this specific point would result in its remaining balanced in position.

Equalizer beams: Beams used to distribute the load weight on multi-crane lifts. The beam attaches to the load below, with two or more cranes attached to lifting eyes on the top.

Equalizer plates: A type of rigging plate that has three or more holes, used to level loads when sling lengths are unequal.

Minimum breaking strength (MBS): The amount of stress required to bring a rigging component to its breaking point. The MBS is a factor in determining a component's rated load capacity.

Rated load: The maximum working load permitted by a component manufacturer under a specific set of conditions. Alternate names for rated load include *working load limit (WLL), rated capacity,* and *safe working load (SWL).*

Rigging links: Links or plates with two holes used as termination hardware to appropriate lifting points.

Saddle: The portion of a hook directly below the center of the lifting eye.

Sling angle: The angle formed by the legs of a sling with respect to the horizontal plane when tension is placed on the rigging.

Spreader beams: Beams or bars used to distribute the load of a lift across more than one point to increase stability. Spreader beams are often used when the object being lifted is too long or large to be lifted from a single point, or when the use of slings around the load may crush the sides.

Hardware used in rigging includes items such as slings, hooks, shackles, eyebolts, spreader beams, equalizer beams, and blocks. There are also many unique pieces of hardware designed for specific applications. These hardware items must be carefully matched to the load to be lifted to ensure the safety of the load and all workers in the area. Careful inspection and maintenance of all lifting hardware is essential to safe and effective material movement. Rigging hardware should always be inspected before each use.

To understand rigging hardware and applications, it is important to be familiar with terms related to the capacity of an object to withstand weight or force. Many of these terms are used interchangeably. The term rated load for rigging components can be defined simply as the weight that a component can safely lift without fear of breaking, based on manufacturer testing. Other terms for rated load that are used by manufacturers in their product specifications include *rated capacity* and *working load limit (WLL).*

> **NOTE**
>
> *ASME Standard B30.10,* which is devoted exclusively to hooks, uses the term *rated load* to describe the maximum amount of weight or force that can be safely applied to a component. For this reason, rated load will be used primarily throughout this module.

To determine the weight a component can lift without fear of breaking and ensure that rigging hardware provides safe service, testing is done to determine the breaking point. This weight is referred to as the minimum breaking strength (MBS). The rated load for a component is determined by applying a factor to the MBS. It is not uncommon for the rated load to be only 20 percent of the MBS, resulting in a safety factor of 5.

For example, if the MBS of a hook is 5,400 lbs (2,449 kg) and the factor applied is 20 percent, the rated load would be 1,080 lbs (490 kg). As a result, the proper application of rigging hardware means that the load imposed should always be well below the breaking point. Safety factors may vary, so it is important not to assume that the rated load should be $\frac{1}{5}$ of the MBS. Always work within the specified rated load of the component.

CAUTION

Always refer to the manufacturer's instructions for all types of rigging equipment and its proper application.

1.1.0 Hooks, Shackles, Eyebolts, and Clamps

Hooks and shackles are essential items for rigging almost every load. Eyebolts and clamps are also commonly used components for lifts. These devices come in different types, and there are a variety of guidelines that must be observed for their safe operation.

1.1.1 Hooks

A rigging hook is typically used to attach a sling to a load. The eye hook (*Figure 1*) is the most common type. Hooks used for rigging must be equipped with safety latches to prevent a connection from slipping off of the hook if any slack develops in the sling. The capacity of a rigging hook is determined by its material of construction, size, and physical dimensions. Information about a specific hook's capacity is always available from the manufacturer or authorized distributor.

Always inspect hooks before each use. Look for wear in the saddle of the hook. Wear in a hook should not exceed 10 percent of the original hook

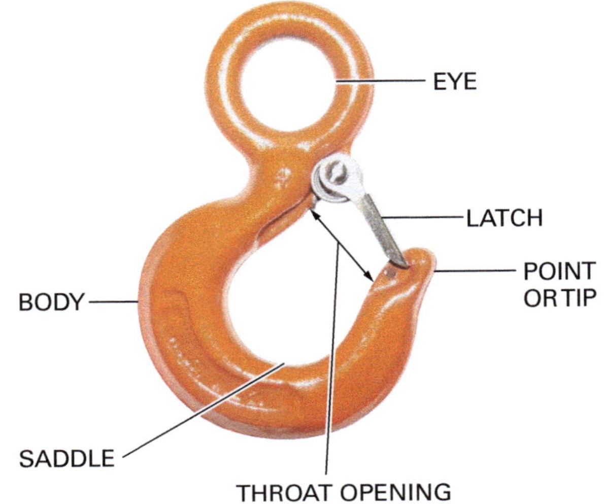

Figure 1 Typical eye hook.

dimensions. *ASME Standard B30.10* also requires that the hook be removed from service if the throat opening has enlarged 5 percent from its original size, to a maximum of $\frac{1}{4}''$ (6.4 mm). *Figure 2* shows some of the inspection points for a hook.

Some other problems that result in the hook being removed from service include the following:

- Missing or illegible manufacturer's identification or rated load information
- Any visually apparent amount of bending or twisting of the hook
- Cracks, nicks, or gouges that could compromise its strength
- Damaged or inoperative latch that does not properly close the throat of the hook
- Any sign of modifications such as grinding, drilling, or machining

Never use a sling on a hook if the sling eye is marginal in size and must be forced over the hook. The body diameter of the hook should fit easily into the sling eye. (Note that information regarding the proper fit of the sling eye to other

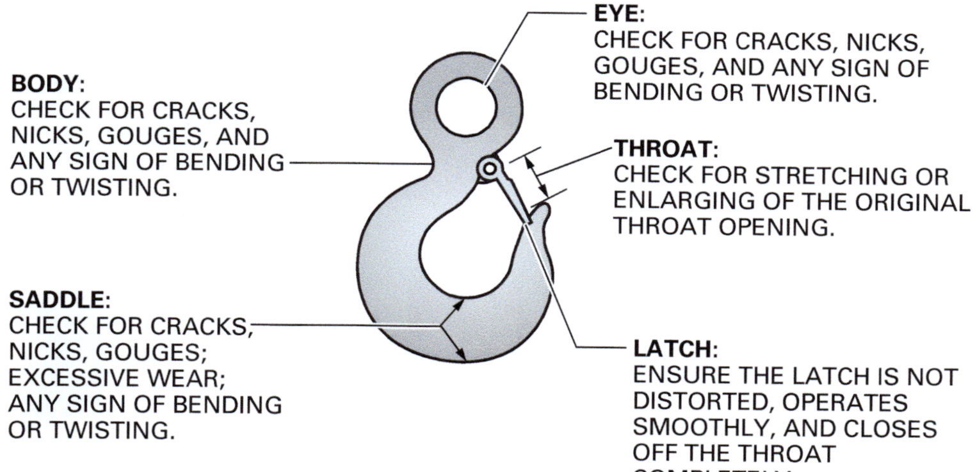

Figure 2 Rigging hook inspection points.

lifting hardware varies depending on the type of sling and hitch used. Slings and hitches will be discussed in more detail in later sections.)

The rated load of the hook is accurate only when the load is suspended from the saddle of the hook. The saddle is the portion of the hook that is directly beneath the center of the lifting eye. If the load is applied anywhere between the saddle and the hook tip, the rated load is reduced considerably, as shown in *Figure 3*. Point loading, also shown in the figure, is not acceptable.

1.1.2 Shackles

A shackle is used to attach an item to a load or to attach slings together. It can be used to attach the end of a wire rope to an eye fitting, hook, or other type of connector. Shackles used for lifting are made of forged steel and are sized by the diameter of the steel in the body, or bow section, of the shackle. However, the rated load capacity, pin size, and distance between the two lugs are often provided in the catalog data. Shackles are made with either screw pins or round pins as a means of safe closure, as shown in *Figure 4*. Screw-pin shackles, the most popular type, are threaded and have threaded pins that screw directly into the body of the shackle. No nut or cotter pin for security is required. A screw-pin shackle designed for synthetic web slings that allows the material to lie flat in the shackle throat is also shown in *Figure 4*.

A round-pin shackle body is not threaded. The round pin itself is threaded, but it is designed to pass completely through the shackle body. The threaded pin then receives a nut and a cotter pin to keep the nut from loosening. This type may also be referred to as a *safety shackle*, since the pin is more secure.

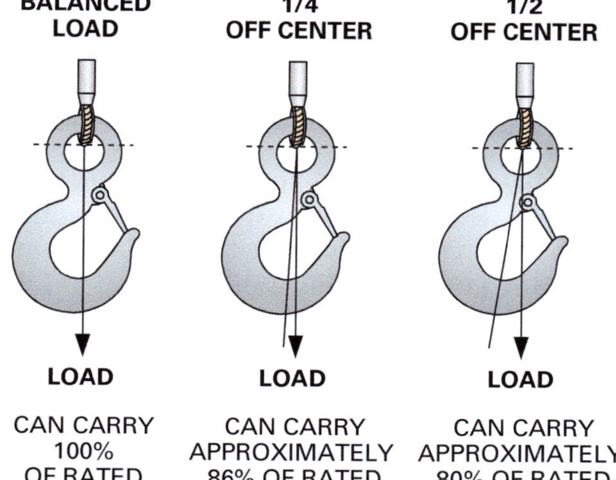

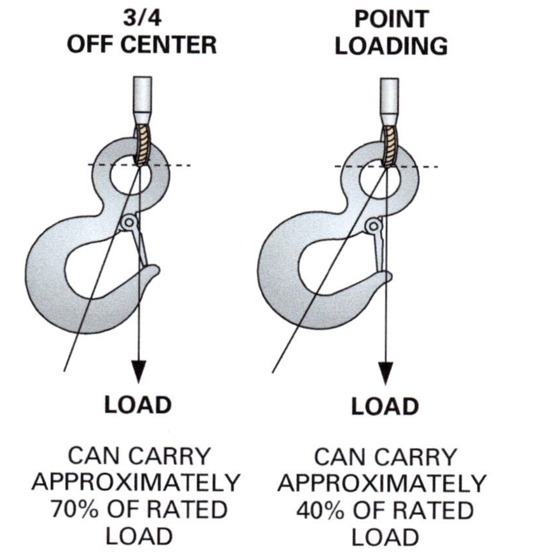

Figure 3 Balanced and off-center loads.

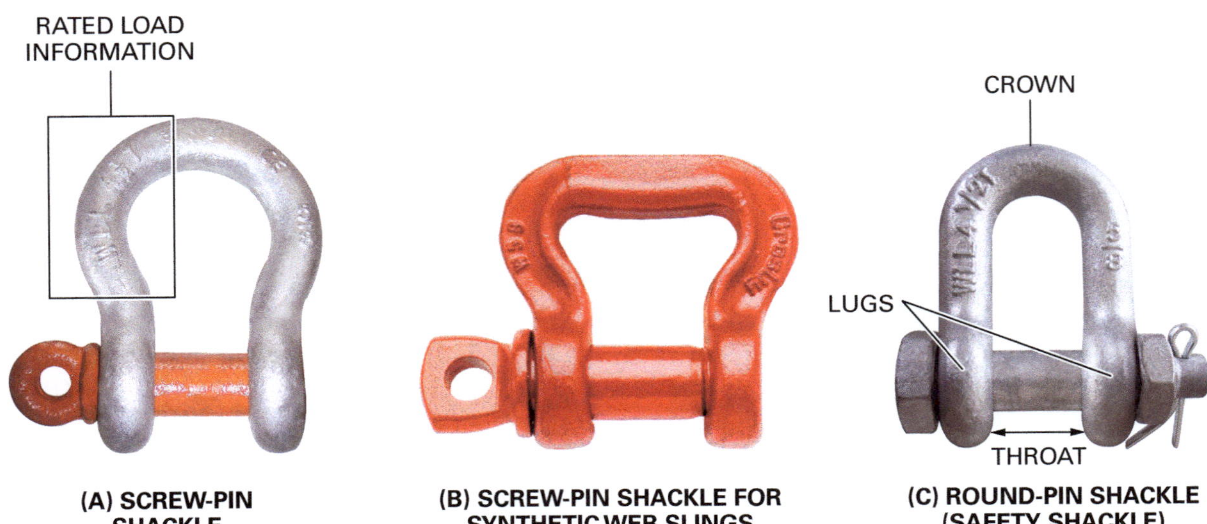

Figure 4 Screw-pin and round-pin shackles.

The threaded pin of a screw-pin shackle must be fully seated in order for the shackle to function at its full rated capacity. In the field, the threaded pin is often left loose to make it easier to remove later, since lifting the load tends to tighten the pin. When it is tight, a tool is needed to loosen it. However, capacity can be significantly affected by leaving the pin loose, and vibration may cause the pin to back out and fall away. The pin must be fully seated in all cases to be safe.

When using shackles, be certain that all pins are straight, all screw pins are completely seated, and cotter pins are used with all round-pin shackles. It is a good practice to replace cotter pins used with shackles after each use to ensure their integrity.

Shackle pins should never be replaced with a common bolt. Common bolts are not hard enough and cannot take the stress normally applied to a shackle pin. Shackles that are stretched, or that have crowns or pins worn more than 10 percent of their original size should be removed from service.

Only shackles with suitable load ratings can be used for lifting. Like hooks, shackles must have the rated load information on the body. When using a shackle on a hook, the pin of the shackle should be hung on the hook, while the load is placed on the bow of the shackle (*Figure 5*). Spacers, such as large washers, can be used on the pin, on each side of the hook, to keep the shackle centered on it. Never use a screw-pin shackle in a situation where the pin can roll as the load sways, as shown in *Figure 6*.

Hook Integrity

The crane hook is a crucial part of many lifts that is used over and over again. Hook failure can happen at any time and may be caused by a number of factors, including cumulative fatigue, overloading, and mechanical abuse such as a free fall to a hard surface.

Hooks can be visually inspected in the field, but they should also be periodically removed, disassembled, and tested to ensure their integrity. In addition to the thorough examination of areas that are not always visible to the operator, various nondestructive testing techniques such as dye-penetrant, magnetic-particle and magnetic-rubber tests can be conducted. These techniques can help detect fatigue and damage that lead to failure. This photo

Figure Credit Konecranes Americas, Inc.

shows a magnetic-particle test being conducted on a hook. Magnetic-particle, or magnaflux, testing can reveal surface flaws, as well as flaws slightly below the surface of ferrous metals.

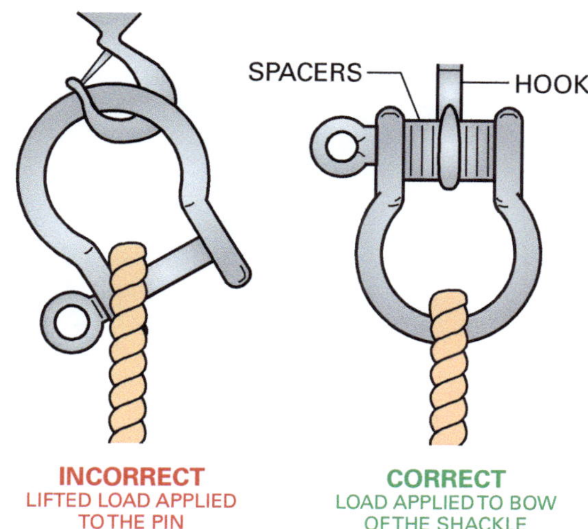

Figure 5 Shackle positioning.

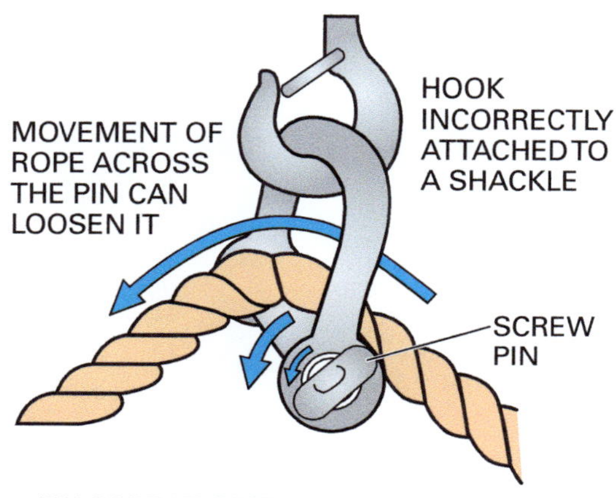

Figure 6 Incorrect attachment of shackle causing pin to roll.

1.1.3 Eyebolts

Eyebolts (*Figure 7*) are often attached to heavy loads to aid in their handling and hoisting. Eyebolts can either have shoulders or be shoulderless. The shouldered type is recommended for use in hoisting applications because it is stronger and resists bending when being pulled from an angle. The shoulder provides a broad surface area of contact between the bolt eye and the load, stabilizing the bolt. The shoulderless type is designed only for lifting a vertical load.

Regardless of the type used, the rated load of all eyebolts however, is reduced during angular loading. Loads should always be applied in the same plane as the eye to reduce the chance of bending. Aligning the eye of the bolt with the shackle or other connector is particularly important when a bridle hitch is used. Proper and improper orientation of the eyebolt is shown in the bottom left corner of *Figure 7*.

When installed, the shoulder surface of the eyebolt must make full contact with the load surface. Washers or other suitable spacers may be used to ensure that the shoulders are in firm contact with the working surface. A threaded blind hole used for an eyebolt should have a minimum depth of $1\frac{1}{2}$ times the bolt diameter to ensure sufficient thread engagement. However, this does not ensure that the threaded portion of the hole is deep enough to allow the shoulder to make firm contact with the surface. For shouldered eyebolts, a blind hole should be deep enough so that the threaded portion is deeper than the length of the threaded eyebolt shank. This prevents the eyebolt from bottoming out or reaching the end of the threads before it is properly seated.

Swivel hoist rings, also shown in *Figure 7*, may be used instead of eyebolts. These devices swivel to the desired lift position and therefore do not require any load-rating reduction due to an angular pull.

1.1.4 Beam and Plate Clamps

Beam clamps (*Figure 8*) are used to connect hoisting devices to beams so the beams can be lifted and positioned. Observe the following guidelines when using beam clamps:

- Do not use clamps unless they are designed, load tested, and stamped by an engineer. Homemade clamps should never be used.
- Ensure that the clamp fits the beam and has the necessary rated load capacity.
- Ensure that the clamp is securely fastened to the beam.
- Be careful when using beam clamps where angular lifts are required. Most are designed for straight vertical lifts only.
- Be certain the rated load appears on the beam clamp and is legible. Never load a beam clamp beyond its rated load capacity.
- Attach rigging to the beam clamp using a shackle; do not place a hoist hook directly in the beam clamp lifting eye.
- Examine the clamp for the following defects, and remove it from service if any are present:
  - Jaws of the beam clamp have been opened more than 15 percent of their normal opening
  - Lifting eye worn, bent, or elongated
  - Excessive rust or corrosion
  - Capacity and beam size information unreadable

Plate clamps (*Figure 9*) attach quickly to structural steel plates to allow for easier rigging attachment and handling of the plate. There are two basic types of plate clamps: serrated-jaw clamps and screw clamps.

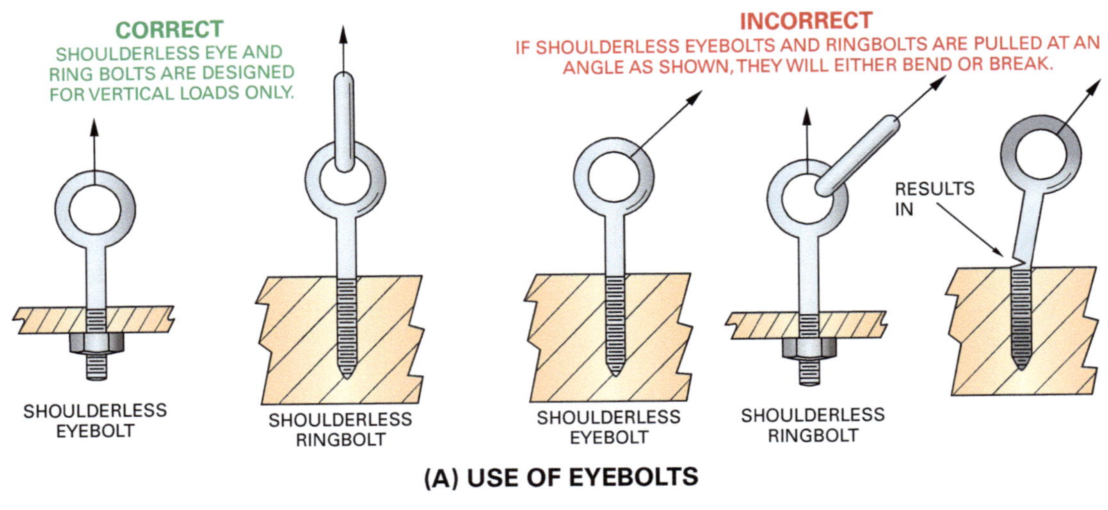

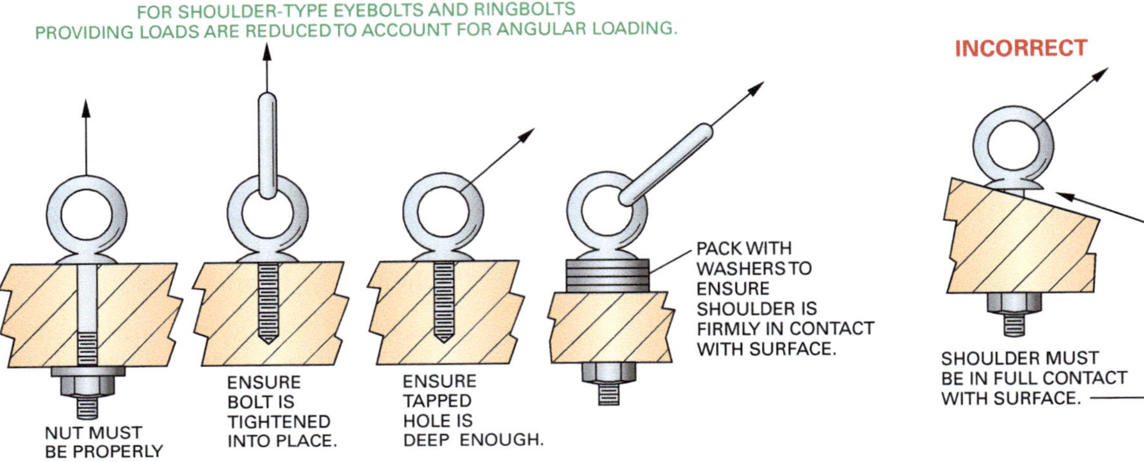

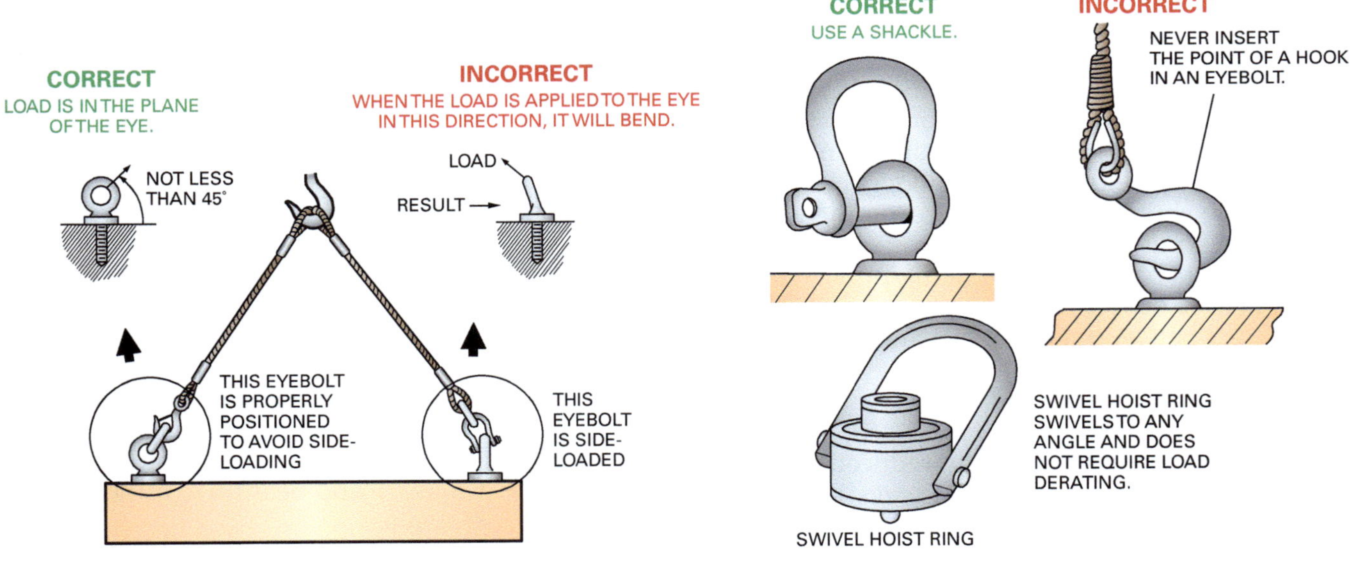

Figure 7 Eyebolt installation and lifting criteria.

Figure 8 Typical beam clamp application.

Serrated-jaw clamps are designed to grip a single plate for hoisting and are available with a locking device. The jaws of a serrated-jaw clamp are cam-operated; as the plate tries to escape the jaws, they rotate down slightly and press more tightly against the plate. Note that all plate clamps are designed to lift only one plate at a time. Always follow the manufacturer's recommendations for their use and remain within the rated load of the clamp.

Plate clamps must be examined before use and removed from service if any of the following defects are present:

- Identifying information and/or the rated load is absent or unreadable
- Distortion of the opening or wear of the jaw teeth
- Cracks in body
- Loose or damaged rivets

- Lifting eye worn, bent, or elongated
- Excessive rust or corrosion

1.2.0 Lugs, Turnbuckles, Plates, and Beams

There are many different types of hardware used for both common and unique rigging applications. Some of these include lifting lugs, turnbuckles, plates, and spreader beams. Riggers and crane operators must be familiar with different varieties of these components and know how to use them safely and effectively.

1.2.1 Lifting Lugs

Lifting lugs are often welded or bolted to an object by the manufacturer so that it can be lifted or moved more easily. A welded lifting lug is shown in *Figure 10*. They are typically designed and located to suspend a load near its **center of gravity (CG)** and support it safely. Lifting lugs should be used for straight, vertical lifting only. They are more likely to bend or fail if they are side-loaded. When lifting lugs are field-installed, consideration must be given to how they will be used in lifting and from what direction the stress will be applied.

1.2.2 Turnbuckles

Turnbuckles are available in a variety of sizes. They are used to adjust the length of rigging connections without twisting the cables or ropes. Two common types of turnbuckles use eyes and jaws as terminations (*Figure 11*). They can be used in any combination. Turnbuckles with hooks are also available but should not be used for rigging purposes as they can easily be disconnected and

(A) HORIZONTAL SERRATED-JAW CLAMP

(B) VERTICAL SERRATED-JAW CLAMP

(C) SCREW CLAMP

Figure 9 Plate clamps.

Figure 10 Welded lifting lug on a spreader beam.

have lower rated load capacities than the other types. The rated load for turnbuckles is based on the diameter of the threaded rods.

Observe the following guidelines and considerations when selecting turnbuckles:

- Turnbuckles should be made of forged steel and should not be welded.
- Do not use turnbuckles with open hooks for rigging loads.
- When using turnbuckles with multi-leg slings, do not use more than one turnbuckle per leg.

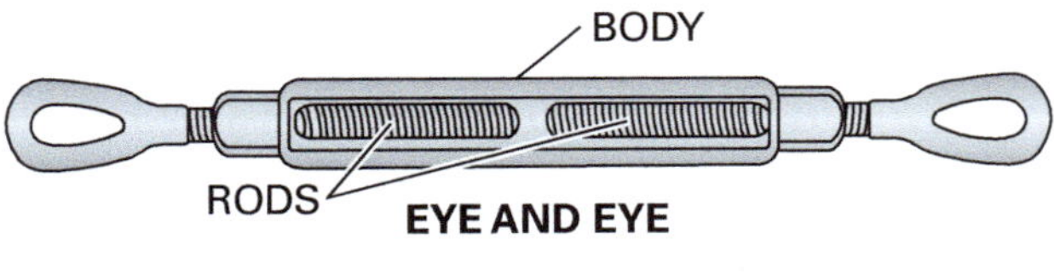

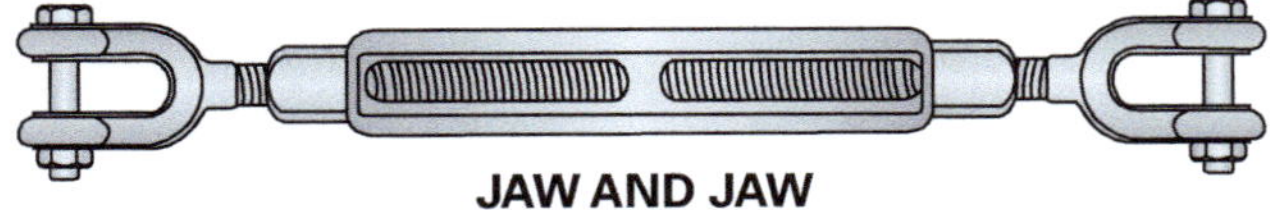

Figure 11 Turnbuckles.

- Do not use jamb nuts on turnbuckles that do not come equipped with them.
- Turnbuckles should not be overtightened. Perform tightening with a wrench of the proper size, using only as much force as a person can achieve by hand, without applying added leverage.

When inspecting turnbuckles, check for bent threaded rods and thread damage, and look for cracks in the body, threaded rods, and terminations.

1.2.3 Rigging Plates and Links

Rigging plates and links (*Figure 12*) are made for specific uses. The holes in the plates or links may be different sizes and may be placed in different locations in the plates. This creates a piece of connecting hardware that can be used for rigging the same type of objects repetitively. Plates with two holes are called **rigging links**. Plates with three or more holes are called **equalizer plates**. Equalizer plates can be used to level loads when the legs of a sling are unequal. Plates are attached to the rigging with high-strength pins or bolts.

Inspect rigging plates and links before using them, and remove them from service if any of the following are present:

- Cracks in body
- Worn or elongated lifting eye
- Excessive rust or corrosion

1.2.4 Spreader and Equalizer Beams

Spreader beams (*Figure 13*) are used to support and protect long loads. These devices prevent the load from tipping, sliding, or bending. They decrease the possibility of a low **sling angle** and help prevent the sling from crushing the load. Equalizer beams are used to balance the load on sling

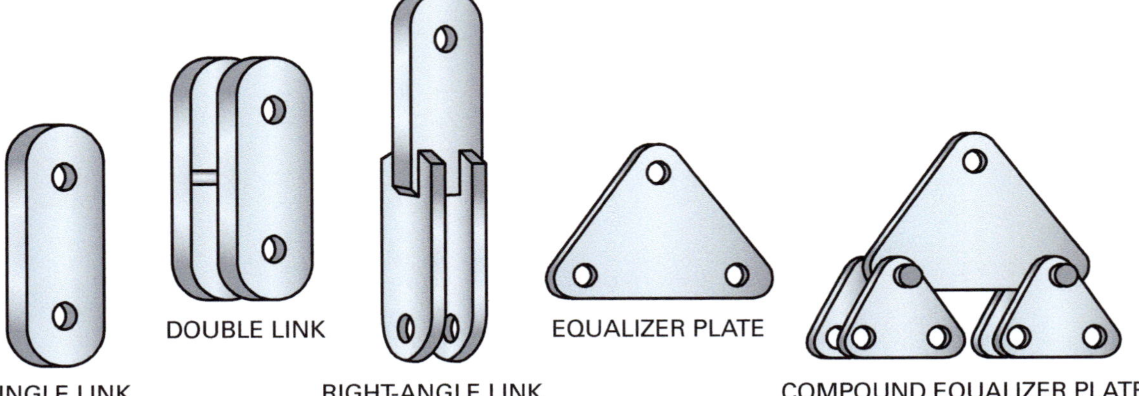

Figure 12 Rigging plates and links.

NCCER – *Mobile Crane Operations*

legs and to maintain equal loads on dual hoist lines when making multi-crane, or tandem, lifts.

> Spreader and equalizer beams are discussed in more detail in NCCER's *Mobile Crane Operations Level Two* and *Mobile Crane Operations Level Three*.

Both types of beams are often fabricated to suit a specific application. They are commonly made of heavy pipe, I-beams, or other suitable material. Custom-fabricated spreader or equalizer beams must be designed by a qualified person (often an engineer) and have their capacity clearly stamped on the side. All such beams should be tested at 125 percent of their rated load. Information on any beams in use should be kept on file.

The rated load capacity of beams designed for use with multiple attachment points depends upon the distance between attachment points. For example, if the distance between the attachment points is doubled, the beam capacity is typically cut in half.

Before use, a spreader or equalizer beam should be inspected for the following:

Figure 13 Typical use of a spreader beam.

- Beam is clearly marked with the following information:
    - Manufacturer's name and address
    - Serial number
    - Device weight, if over 100 pounds (45 kg)
    - Rated load
    - *ASME Standard BTH-1* Design Category
    - *ASME Standard BTH-1* Service Class

Moving Materials Using Vacuum Principles

There are many devices used by cranes and other heavy equipment to grab, lift, and position construction materials. However, one method is a little more unusual than the others.

Vacuum lifters are equipped with vacuum pumps, usually powered by a small diesel engine, that draw out the air between the lift mechanism and the load. Irregular surfaces that cannot provide a tight seal are not candidates for vacuum lifting because integrity of the seal is essential for them to function. Vacuum lifters have been specifically designed for pipe, slabs, beams, and similar items. An advantage of this method is the elimination of a great deal of rigging hardware and the labor required to assemble it prior to a lift, saving both time and money. These devices and their design criteria are covered by *ASME Standard B30.20, Below-The-Hook Lifting Devices*, and *ASME Standard BTH-1, Design of Below-The-Hook Lifting Devices*. The power of a vacuum is surprising— rated load capacities for the strongest vacuum lifters exceed 20 tons (18 metric tons).

(A) PLATE LIFTER

(B) SLAB LIFTER

Figure Credit: Vaculift ™ , Inc. d.b.a. Vacuworx®

- Welds are free of cracks or other significant flaws.
- No cracks, nicks, gouges, or corrosion are present.
- Attachment points are not damaged or distorted.

Additional Resources

ASME Standard B30.5, Mobile and Locomotive Cranes. Current edition. New York, NY: American Society of Mechanical Engineers.

ASME Standard B30.10, Hooks. Current edition. New York, NY: American Society of Mechanical Engineers.

ASME Standard B30.20, Below-The-Hook Lifting Devices. Current edition. New York, NY: American Society of Mechanical Engineers.

ASME Standard BTH-1, Design of Below-The-Hook Lifting Devices. Current edition, New York, NY: American Society of Mechanical Engineers.

29 *CFR* 1926, Subpart CC, **www.ecfr.gov**

29 *CFR* 1926.251, **www.ecfr.gov**

29 *CFR* 1926.753, **www.ecfr.gov**

Mobile Crane Safety Manual (AEM MC-1407). 2014. Milwaukee, WI: Association of Equipment Manufacturers.

Willy's Signal Person and Master Rigger Handbook, Ted L. Blanton, Sr. Current edition. Altamonte Springs, FL: NorAm Productions, Inc.

North American Crane Bureau, Inc. website offers resources for products and training, **www.cranesafe.com**

1.0.0   Section Review

1. Which of the following statements about shackles is true?

    a. Shackles with pins worn more than 10 percent of their original size should be removed from service.
    b. The size of a shackle is based on its pin size.
    c. When connecting a shackle to a hook, the body of the shackle is hung on the hook.
    d. If a shackle is pulled at an angle rather than a straight pull, its rated load increases.

2. Which rigging devices require the *ASME Standard BTH-1* Design Category and Service Class on the labeling?

    a. Spreader beams
    b. Shackles
    c. Hooks
    d. Eyebolts

2.0.0 SLINGS AND HITCHES

Objective

Identify and describe various types of slings and sling hitches.

a. Identify and describe wire-rope slings and their proper care.
b. Identify and describe synthetic slings and their proper care.
c. Identify and describe chain slings and their proper care.
d. Explain the significance of sling angles and describe common hitches.
e. Describe how to properly rig and handle piping materials and rebar.
f. Identify and describe how to use taglines and knots for load control.
g. Identify common rigging-related safety precautions.

Performance Tasks

1. Inspect various types of rigging components and report on the condition and suitability for a task.
2. Configure a sling to produce a single-wrap basket hitch.
3. Configure a sling to produce a double-wrap basket hitch.
4. Configure a sling to produce a single-wrap choker hitch.
5. Configure a sling to produce a double-wrap choker hitch.
6. Select the correct tagline for a specified application.
7. Tie specific instructor-selected knots.

Trade Terms

Basket hitch: A common hitch made by passing a sling around a load or through a connection and attaching both sling eyes to the hoist line.

Bird caging: A deformation of wire rope that causes the strands or lays to separate and balloon outward like the vertical bars of a bird cage.

Choker hitch: A hitch made by passing a sling around the load, and then passing one eye of the sling through the other. The one eye is then connected to the hoist line, creating a choke-hold on the load.

Independent wire rope core (IWRC): Wire rope with a core consisting of wire rope, as opposed to a fiber or single-stranded core; considered to be the most durable for rigging applications.

Spur track: A relatively short branch leading from a primary railroad track to a destination for loading or unloading. A spur is typically connected to the main at its origin only (a dead end).

Tagline: A rope attached to a lifted load for the purpose of controlling load spinning and swinging, or used to stabilize and control suspended attachments.

Vertical hitch: A simple hitch that uses one end of a sling to connect to a point on the load and the opposite end to connect to the hoist line. Also known as a *straight-line hitch*.

Slings are available in a variety of configurations. Some of these include wire rope, synthetic web, and chain slings, which are all presented in this section. Some slings are made from natural fibers or synthetic rope, but these are not typically used for rigging applications involving mobile cranes. All types of slings are likely to need protection to ensure they are not damaged by sharp edges, protrusions, and other potential sources of damage. In many cases, the load itself also benefits from such protection. Like hooks used in rigging, an ASME standard is devoted exclusively to slings. *ASME Standard B30.9, Slings,* provides a wealth of information about slings and their proper application. In addition, slings are also the topic of 29 *CFR* 1910.184.

2.1.0 Wire-Rope Slings

Wire-rope slings (*Figure 14*) are made of high-strength steel wires formed into strands and wrapped around a supporting core. They are lighter and easier to handle than chain slings, and can withstand substantial abuse and relatively high temperatures. However, because wire-rope slings can slip, the use of synthetic slings is often preferred. Wire-rope slings are still being used, so it is still important to learn the design, characteristics, applications, inspection, and maintenance of wire-rope slings.

2.1.1 Characteristics and Applications

Wire-rope slings usually consist of six strands, with each strand containing an average of 19 wires (written as 6 x 19), laid in a specific pattern

Figure 14 Typical wire-rope sling.

around a wire rope core. Cores other than wire are available, such as fiber cores, but independent wire rope core (IWRC) is the most common and considered the most durable. Wire rope will suffer the loss of only about 1 percent of its strength if a wire breaks. *Figure 15* shows examples of the most common wire-rope sling applications used in construction rigging.

Wire-rope slings should be protected where the slings are wrapped around the sharp edges of the object to be lifted (*Figure 16*). Of course, the load can often be damaged by the sling as well. Even if the edge of the load is a soft material that would not cut the sling, one or more individual wires may break if they are sharply bent. If wire-rope slings are kinked, the severe bending stress and displaced strands allow for unequal distribution of the live load, with some strands taking more than their normal load. Damage done to the rope by kinking is usually permanent, resulting in the disposal of many slings as well as some failures.

Protective material can include simple pieces of wood to separate the sling from the load. However, protective materials are also available in a variety of manufactured forms. *Figure 17* shows examples of some materials designed for the task. The plastic protectors shown here have magnets embedded in them, allowing them to remain in place on ferrous materials before the sling is applied. Others are made of metal alloys and are designed to slip around a wire-rope sling at any point along its length. Some type of protective material is often needed for any type of sling, depending upon the hitch used and the nature of the load being lifted. These protective materials are not used with wire rope only.

A special type of wire-rope sling, called a *braided-belt sling*, is made by braiding six or more small-diameter wire ropes together. This provides a sling with a wide, flat bearing surface of great strength and flexibility in all directions. They are especially well suited for a basket hitch or a choker hitch where sharp bends are encountered.

Rigging hardware placed into the eye of a sling must be appropriate to prevent failures and/or sling damage. *ASME Standard 30.9*, Section 2.10.4(p) states that "an object in the eye of a (wire-rope) sling should not be wider than one half the length of the eye nor less than the nominal sling diameter." For example, if the eye of wire-rope sling is 6" (15 cm) long, then nothing wider than 3" (7.5 cm) should be placed through the eye. (Half of 6" is 3", or half of 15 cm is 7.5 cm.) The second portion refers to the minimal width of hardware placed in the sling eye; it should never be smaller in diameter than the wire-rope sling itself. Sling manufacturers provide information about the application of their products that should be followed as well.

2.1.2 Storage and Inspection

Store slings in a rack to keep them off the ground. The rack should be in an area free of moisture and away from acid, acid fumes, or extreme heat. Both moisture and fumes can lead to corrosion. Never let slings lie on the ground in areas where heavy machinery may run over them, or where they can become filled with sand and other abrasives internally.

> **WARNING!**
>
> Broken wires in a wire-rope sling are extremely sharp and can easily cut or puncture the skin. Always wear gloves when handling wire-rope slings and when inspecting them by running your hand along their length.

Slings should be regularly inspected for broken wires, kinks, rust, or damaged fittings. A visual inspection should be made before each use. Inspections at the beginning of each shift or work day are required by *ASME Standard B30.9*. Any slings found to be defective should be removed from service for repair or disposal. Repairs should not be attempted by anyone other than the manufacturer or other qualified party.

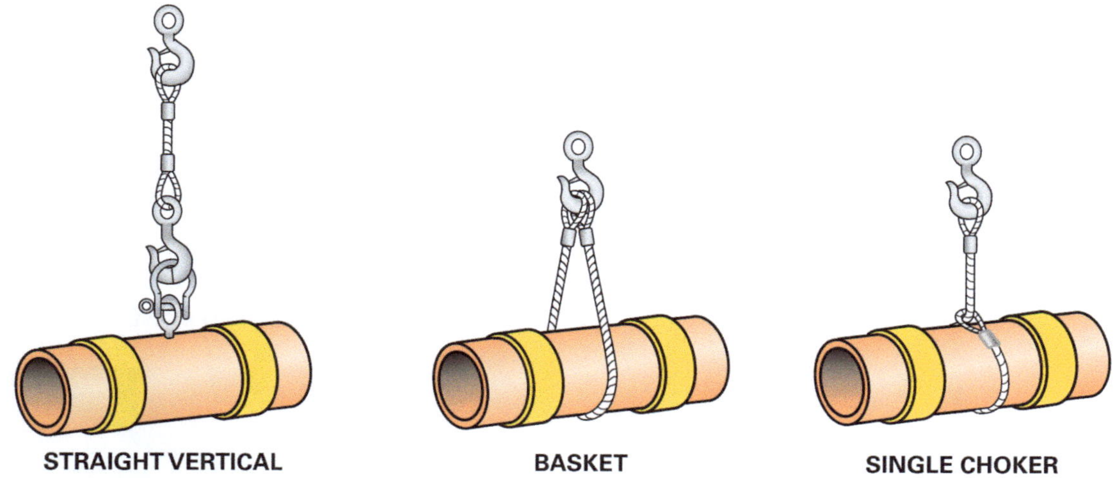

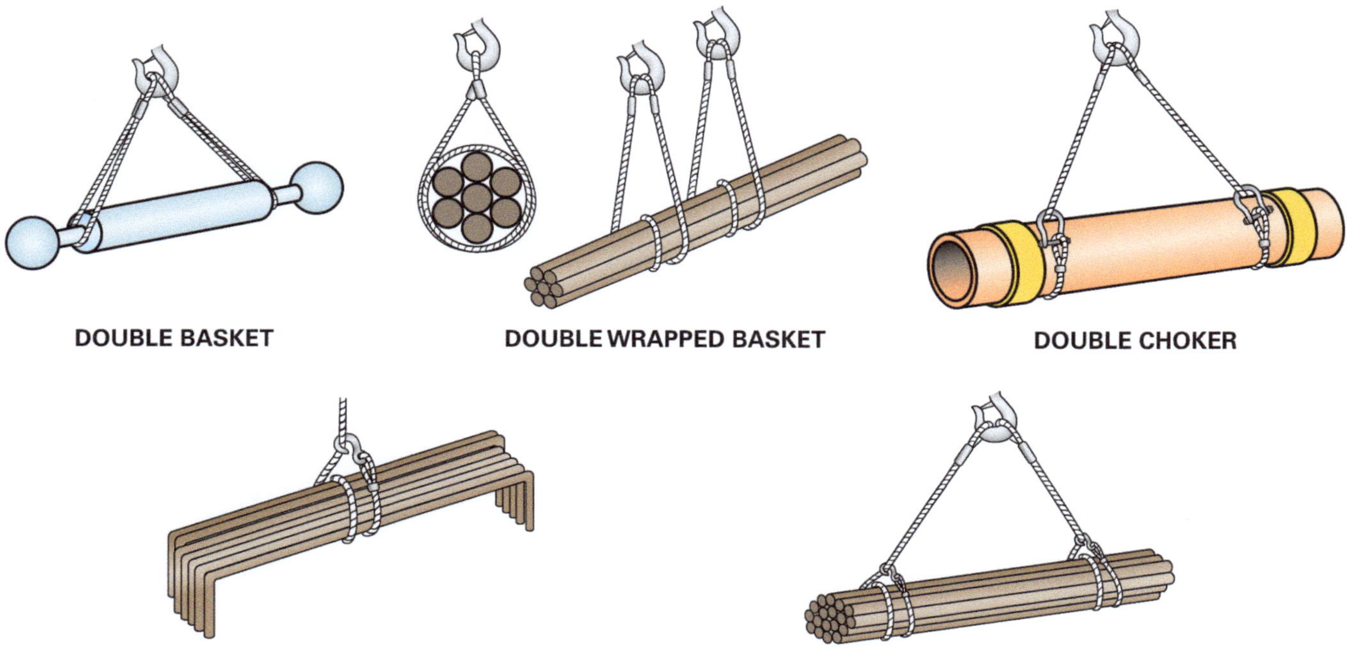

Figure 15 Examples of wire-rope sling applications.

Wire-rope slings must be properly tagged with specific information. This is required by both ASME and OSHA standards. *Figure 18* shows a typical wire-rope tag. A missing or illegible identification tag or data plate is a cause for sling disposal. Information that must be on the tag includes the following:

- Name or trademark of manufacturer or repair organization
- Rated load for at least one hitch and the angle upon which it is based

- Size/diameter
- Number of legs, if more than one

Wire-rope slings should also be removed from service if any of the following conditions are discovered; refer to *Figure 19* for examples:

- Localized abrasion or scraping that reduces the diameter of the sling by more than 5 percent of its original size
- Rope distortion, which includes kinking, crushing, and **bird caging**

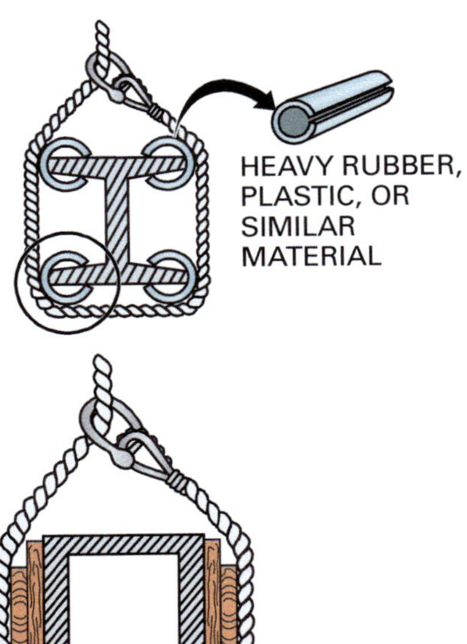

Figure 16 Use of sling protection.

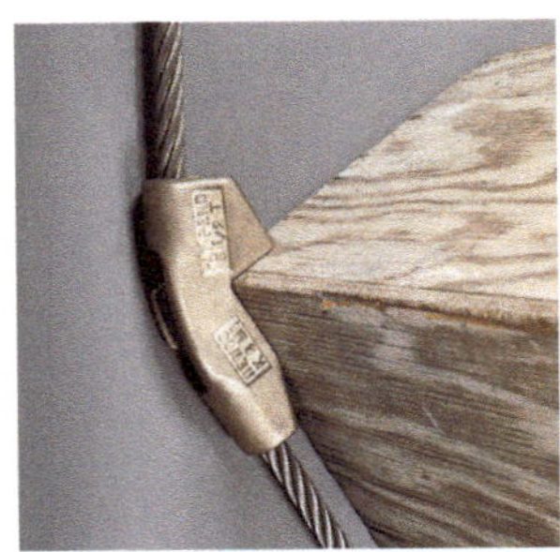

(A) UPPER AND LOWER WIRE-ROPE SADDLES

- Evidence of heat damage, usually indicated by discoloration
- Damaged end fittings, such as cracks, deformation, and excessive wear
- Severe corrosion
- Any other type of damage that results in doubt about the integrity of the sling
- Broken wires

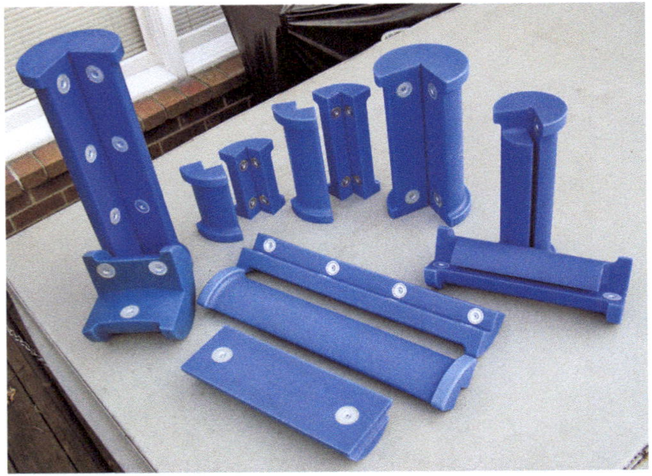

(B) PLASTIC SLING PROTECTORS FOR ALL SLING TYPES

Figure 17 Sling protection products.

ASME Standard B30.9 does allow a specific number of broken wires depending on the rope design; consult the standard for details.

Sling Maintenance

Slings are required to have periodic inspections. *ASME Standard B30.9* requires that documentation of the most recent periodic inspection be maintained. Someone is responsible for maintaining that documentation as well as the slings themselves.

Most sling manufacturers now offer radio-frequency identification (RFID) chips permanently attached to new slings. Passive RFID chips called transponders are scanned by devices that receive the signal and synchronize important data with a database based on the embedded serial number. RFID tags are extremely durable and use Bluetooth technology; the tag can be located and information downloaded from a reasonable distance and without a direct line of sight to the receiver. The technology is not only used for slings, but for many other rigging components such as hooks and shackles.

Figure Credit: Lift-All Company, Inc.

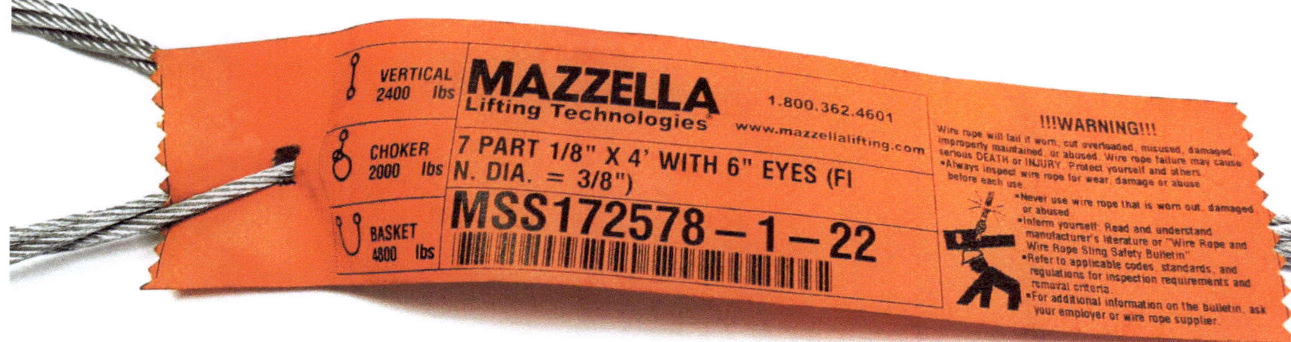

Figure 18 Wire-rope sling tag.

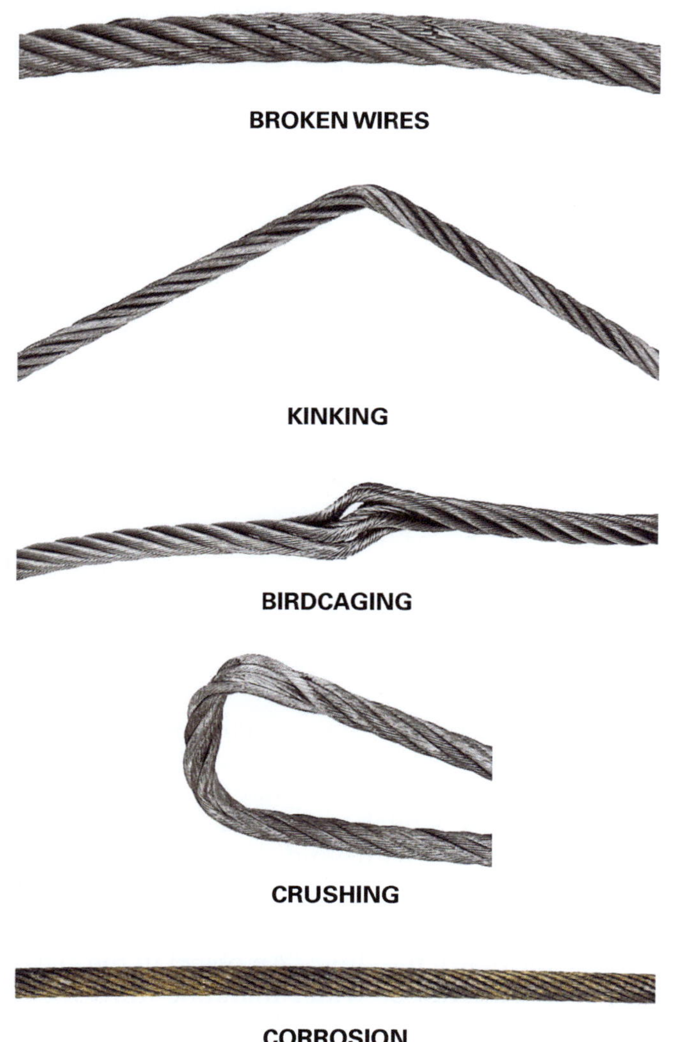

BROKEN WIRES

KINKING

BIRDCAGING

CRUSHING

CORROSION

Figure 19 Common types of wire-rope damage.

2.2.0 Synthetic Web Slings

Synthetic slings are widely used to lift loads, and they are especially suitable for easily damaged ones. Common types of synthetic slings, as well as guidelines for their storage and inspection, are outlined in the following sections.

2.2.1 Characteristics and Applications

Synthetic web slings commonly used for construction rigging are made of polyester, nylon, or other high-performance synthetic materials. While synthetic slings are useful in many situations, there are some applications and environments that are likely to damage them.

Synthetic web slings are available in a number of configurations, with common types shown in *Figure 20*. These include the following:

- *Endless slings* – Endless slings are also referred to as *continuous loop slings* or *grommet slings*. The ends of a piece of webbing are overlapped and sewn together to form an endless loop. They can be used for a **vertical hitch**, bridle hitch, choker hitch, or basket hitch.
- *Standard eye-and-eye slings* – Webbing in these slings is sewn to form a flat body with an eye on each end. The eye is in the same plane as the sling body. They are used for the same purposes as wire-rope slings of similar design.
- *Round slings* – Round slings are available in endless and eye-and-eye styles. They are generally made from a continuous length of polyester filament yarn covered by a woven sleeve. The eye-and-eye style of round sling simply has a sleeve wrapped around the two sides, forming eyes on the end.
- *Twisted eye-and-eye slings* – The eyes in twisted eye-and-eye slings are sewn at right angles to the plane of the sling body.

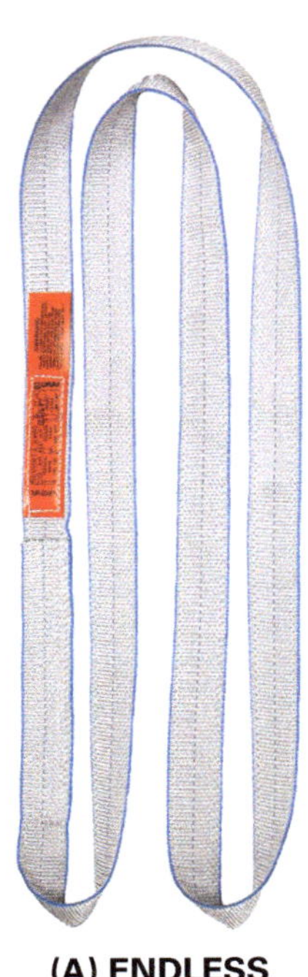

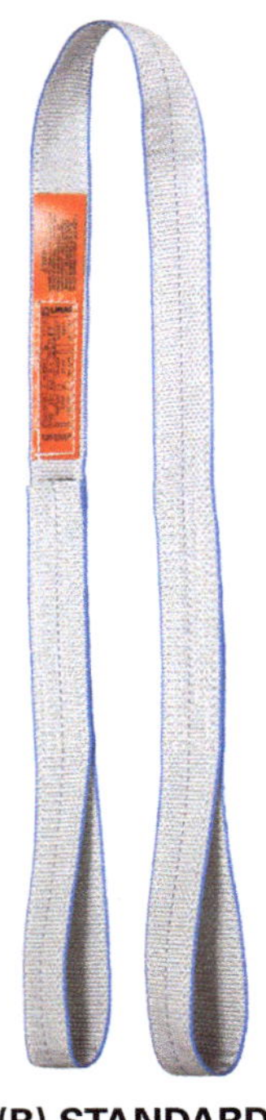

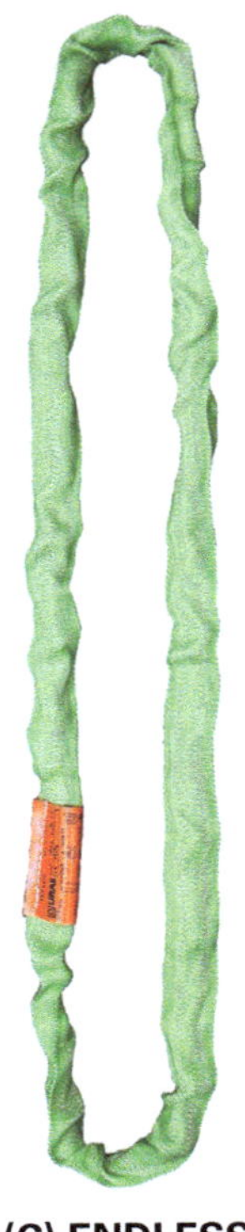

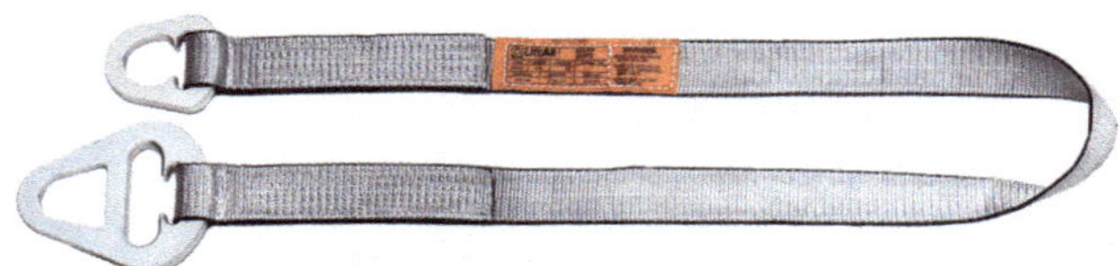

Figure 20 Examples of synthetic web slings.

- *Slings with attached hardware* – Web slings are also available with triangular and choker hardware end fittings.

Synthetic web slings can be cut by sharp edges and corners or damaged by excessive sling angles. A razor sharp edge is not required for damage to be done; a sling stressed with weight against a seemingly dull edge can result in failure. The same type of protection used with wire-rope slings must also be applied to web slings. Advantages of synthetic web slings include the following:

- Their texture and width reduce marring or scratching of polished, finished, or soft surfaces.
- They are less likely to crush fragile surfaces.
- They mold themselves to the shape of the load.
- Moisture and a variety of chemicals do not damage them.
- As long as they remain clean and dry, they are nonsparking and nonconductive. (Excessive soil can also make a sling conductive.)
- They minimize twisting of the load during lifting.
- They do not stain ornamental materials.
- They are lightweight and soft, making them easier and safer to handle.
- They stretch somewhat under load, allowing them to withstand shock.

ASME Standard B30.9, Section 5.10.4(p) provides the following guidance for fitting hardware into the eye of the sling: "An object in the eye of a sling should not be wider than one-third the length of the eye." For example, if the eye of a synthetic sling is 9" (23 cm) long, the maximum width of a hook or other object placed through the sling eye is 3" (7.5 cm). This is smaller than the maximum width allowed for wire-rope sling eyes, which is half the length of the eye.

2.2.2 Storage and Inspection

Nylon and polyester web slings should not be stored or allowed to contact objects at temperatures over 194°F (90°C) or below −40°F (−40°C). This temperature range is a requirement of *ASME Standard B30.9*; note that manufacturers may specify a more limited range for their products that would supersede the standard. It is unlikely that storage facilities would reach these temperatures, but the surface of a rigged load could in some cases. It is also important to limit long-term exposure to sunlight or ultraviolet (UV) light. Both can affect the strength of synthetic web slings. Sling manufacturers can assist with information related to sunlight and UV light exposure.

The strength of synthetic webbing slings can also be degraded by certain chemicals. This includes exposure in the form of solids, liquids, vapors or fumes. Consult the sling manufacturer for a specific list of the chemicals and their variations that can damage synthetic slings.

Like other types of slings, synthetic web slings should not be laid on the ground where they can be damaged or run over by heavy equipment. They also must be properly labeled with specific information. An unreadable or missing label is cause for removing a sling from service. However, a damaged label can be repaired by the manufacturer or other qualified party. When repairs are made, the repair organization must record their information on the label. Labels must include the following information:

- Name or trademark of manufacturer, or if repaired, the entity performing repairs
- Manufacturer's code or stock number
- Rated load for at least one hitch type and the angle upon which it is based
- Type of synthetic web material
- Number of legs, if more than one

Per *ASME Standard B30.9*, synthetic web slings should be removed from service if any of the following conditions are found during an inspection; refer to *Figure 21* for examples:

- Missing or illegible sling identification and rated load information
- Acid or caustic burns
- Melting or charring of any part of the sling
- Holes, tears, cuts, or snags
- Broken or worn stitching
- Excessive abrasion
- Knots in any part of the sling
- Discoloration and brittle or stiff areas on any part of the sling, which may indicate chemical or ultraviolet/sunlight damage has occurred
- For attached hooks, removal criteria as stated in *ASME Standard B30.10*
- For other attached rigging hardware, removal criteria as stated in *ASME Standard B30.26*
- Other conditions, including visible damage, that create doubt as to the continued use of the sling

2.3.0 Chain and Metal-Mesh Slings

Chain slings and metal-mesh slings are often used for lifts in high heat or rugged conditions. They are versatile because they can be adjusted over the center of gravity, and they are also very durable. However, because chain and metal-mesh are very heavy, they can be harder to inspect than other types.

2.3.1 Chain Slings

For some lifts, chains slings are more appropriate than wire-rope or web slings. For example, the use of chain slings is recommended when lifting rough castings that would quickly destroy wire or synthetic web slings. They are also used in high-heat applications or where wire-rope chokers are not suitable, and for dredging and other marine work because they withstand abrasion and corrosion better than wire rope. *Figure 22* shows some common configurations of chain slings and hooks.

Synthetic Slings and Rigging Incidents

Industrial Training International (ITI) compiled a webinar entitled "Rigging & Sling Failures: Case Studies & Solutions" in 2013. During ITI's research for the webinar, the results of which were supported by a poll taken of attendees during the webinar, it was found that over 80 percent of rigging accidents were related to synthetic slings. The incidents were generally due to cutting or severe abrasion resulting from a lack of sling protection used during lifting. As many rigging professionals have stated over the years, if you have a synthetic sling in your right hand, you should have sling protection in your left. All slings, regardless of their materials of construction, should be protected in use.

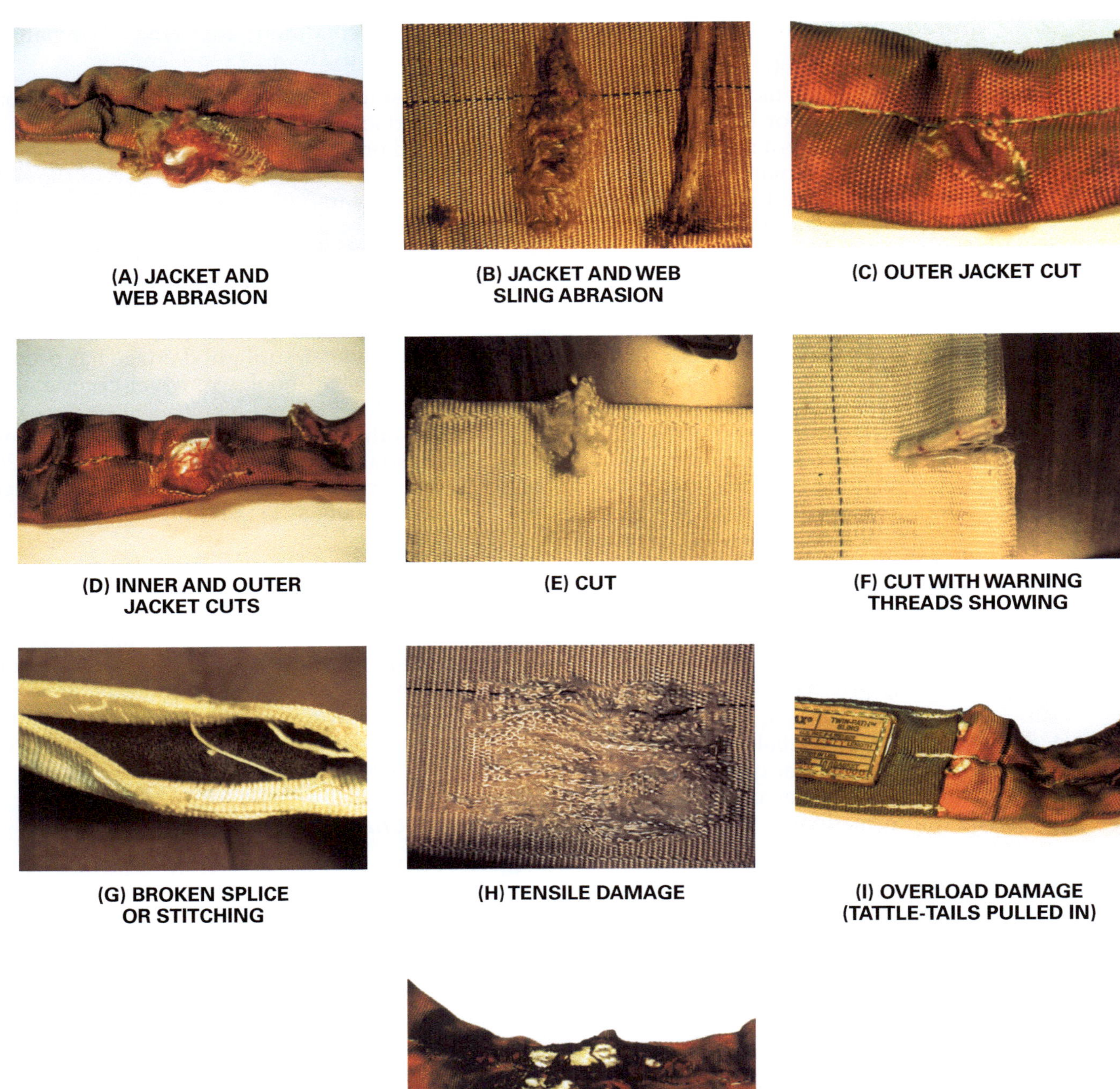

Figure 21 Examples of synthetic-web sling damage.

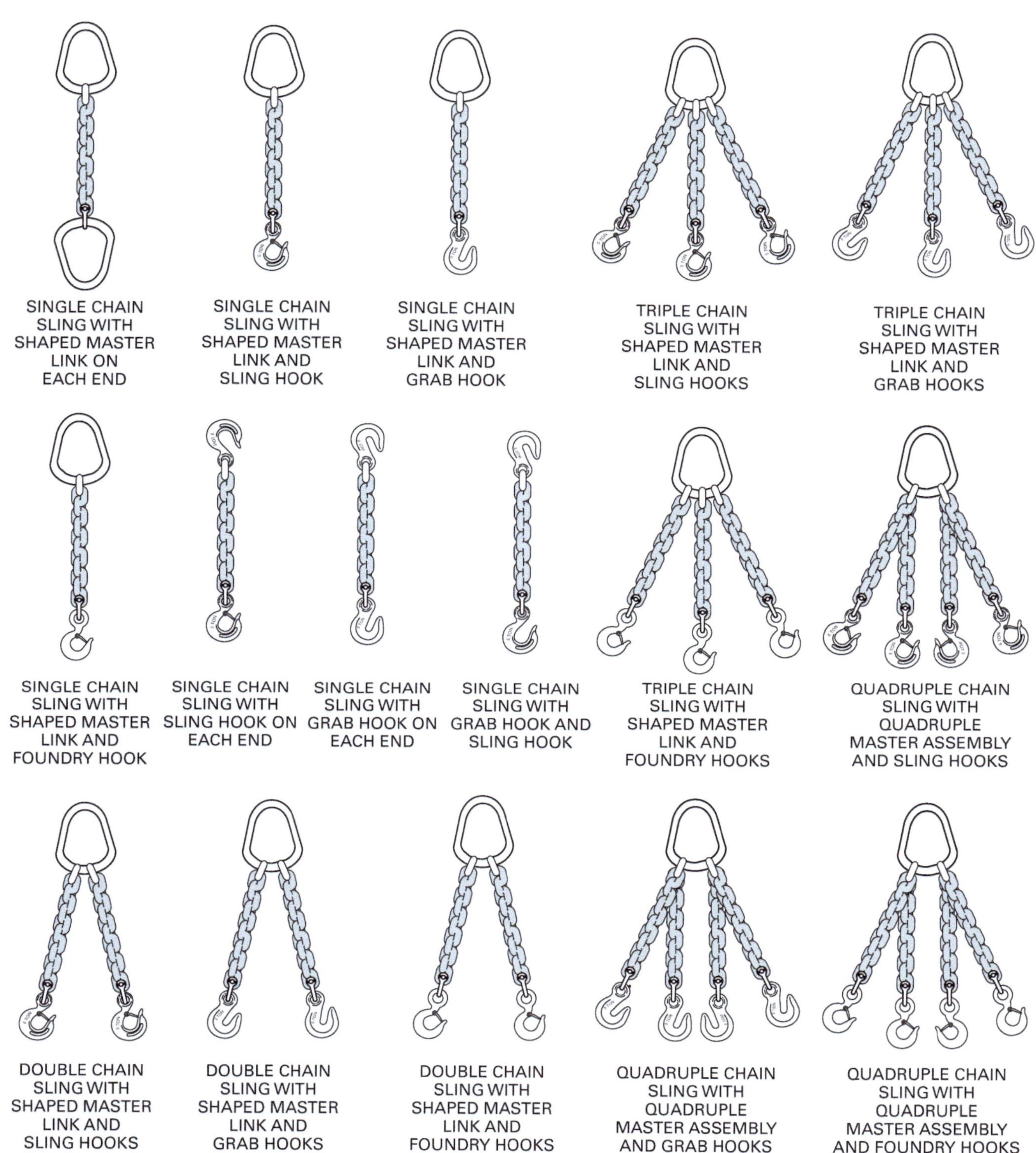

Figure 22 Common chain slings and hooks.

Chain links have two sides. Failure of either side causes the link to open and drop the load. Wire rope is frequently composed of as many as 114 individual wires, all of which must fail before the rope finally breaks. In other words, wire rope is more likely to experience a progressive failure than chain. Chains have less reserve strength and are more likely to fail quickly once the process begins.

Chains will stretch under excessive loading. This causes elongating and narrowing of the links until they bind on each other, giving visible warning. If overloading is severe, the chain will fail with less warning than a wire rope. When a chain link breaks, there is little or no warning.

2.3.2 Metal-Mesh Slings

Metal-mesh slings (*Figure 23*) are typically made of wire or chain mesh. They are similar in appearance and flexibility to web slings and are suited for some situations where other slings do not perform well. Metal-mesh slings have the following advantages:

- Resist abrasion and cutting
- Grip the load firmly without stretching
- Conform to irregular shapes
- Do not kink or tangle
- Can withstand high temperatures

ASME Standard B30.9 requires that the use of metal-mesh slings at temperatures above 550°F (288°C) requires manufacturer approval. These slings are available in several mesh sizes and can be coated with a variety of substances, such as rubber or plastic, to help protect the load. When used in high-temperature applications though, slings with coatings are not typically approved.

2.3.3 Storage and Inspection

Chain and metal-mesh slings must be stored inside a building or vehicle and hung on racks to reduce deterioration due to weather-related rust or corrosion. Never let chain slings lie on the ground in areas where heavy machinery can run over them.

Some manufacturers suggest lubrication of alloy chains while in use. However, slippery chains increase handling hazards. Chains coated with oil or grease also attract dirt and grit that may cause abrasive wear. This is especially true of metal-mesh slings that are far more difficult to clean. Chain slings to be stored in exposed areas should be coated with a film of oil or grease for rust and corrosion protection.

Like all other slings, chain slings should be visually inspected before every lift. They should be removed from service if any of the following conditions are found during inspection:

- Missing or unreadable identification and/or rated load information
- Cracks or breaks
- Nicks, gouges, and excess wear
- Stretched, bent, twisted, or deformed links or end fittings;*ASME Standard B30.9*, Table 9-1.9.5.1 provides the minimum allowable thickness of any point on a link, based on the nominal size of the link.
- Evidence of overheating
- Excessive pitting or corrosion
- Lack of ability of chain or fittings to hinge freely
- Weld splatter
- Any other condition, including visible damage, that causes doubt about the continued use of the sling

Although metal-mesh slings are similar to chain slings, they are constructed in a very different way. Metal-mesh slings should be removed from service if any of the following conditions are found during inspection:

- Missing or unreadable identification tag
- A broken weld or brazed joint along the sling edge
- A broken wire in any part of the mesh
- A reduction in wire diameter of 25 percent due to abrasion, or 15 percent as the result of corrosion
- Lack of flexibility due to distortion of the mesh
- Distortion of the choker fitting so the depth of the slot is increased by 10 percent
- Distortion of either end fitting so the width of the eye opening is decreased by more than 10 percent
- Fittings that are pitted, corroded, cracked, bent, twisted, gouged, or broken
- A 15 percent reduction in the original cross-sectional area of metal at any point around
- Slings with individual spirals that are locked in place
- Any other condition, including visible damage, that causes doubt about the continued use of the sling

Figure 23 Metal-mesh sling.

NCCER – *Mobile Crane Operations*

2.4.0 Sling Angles and Basic Hitches

Every sling, regardless of type and manufacturer, has a specified rated load capacity that should never be exceeded. Less obvious is that the load on a sling changes dramatically as the sling angle changes. What appears to be a light load for a sling can quickly become a very stressful load due to the additional stress placed on the sling as the sling angle changes. Sling angles and other significant factors in sling selection and use are explored in this section.

Since there are an infinite number of loads and load configurations, slings must be used in different ways to connect the load to the hoist line. An understanding of basic hitches and their performance characteristics help the crane operator understand how a reliable and properly made hitch should look.

2.4.1 Sling Capacity

Sling capacity depends on the sling material, sling construction, hitch configuration, number of slings, and the angle of the sling in the hitch used. This type of information, along with other relevant rigging information, is available from rigging equipment manufacturers and trade organizations in the form of easy-to-use pocket guides like the one shown in *Figure 24*. Sling capacity information is also provided in *ASME Standard B30.9* for various types of slings. *Table 1* is an example of a manufacturer's capacity table for wire-rope slings. Note that the capacity of a sling used in a simple basket hitch with a vertical pull is double the capacity for a straight vertical hitch. This is common, since two legs are being used to suspend the load instead of one. However, as the angle between the legs of the hitch widens, the rated load capacity of the sling is reduced. Remember that most anything you do with a sling that varies from a straight vertical pull is likely to negatively affect its rated load capacity. Choking a sling, for example, even for a vertical pull, significantly reduces the capacity of the sling.

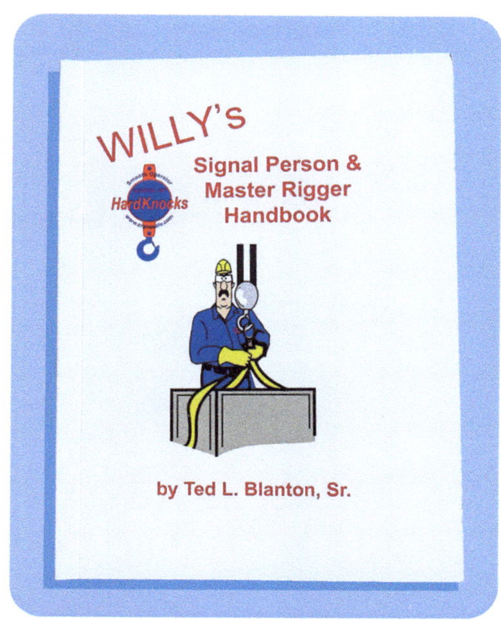

Figure 24 A rigging pocket guide.

2.4.2 Sling Angles

The angle formed by the legs of a sling, with respect to the horizontal plane, when tension is placed on the load is called the *sling angle*. The sling angle directly affects the tension applied to each sling. For this reason, maintaining acceptable sling angles is crucial to safe rigging. In addition, since the sling angle affects its rated load capacity, the rigger must ensure that even a sling at an acceptable angle remains within its rated load capacity.

To determine the effective weight placed on a sling at an angle, a sling angle factor is needed. The factor is applied to the actual load weight per sling to determine the tension on the sling. The proper way to calculate the sling angle factor is through some simple field measurements and math.

Figure 25 shows a load suspended from two slings in a vertical hitch. "L" represents the distance between the points of sling contact with the hook and the load. "H" represents the distance from where the sling contacts the hook down to

Convenient Rigging Information

Over the years, there have been quite a few rigging handbooks developed and marketed. Many are pocket-sized for the convenience of the rigger, allowing quick access to the needed information. As you might expect, rigging information has also gone electronic. Apps are now available from various manufacturers and trade organizations that are compatible with all popular smartphones, and more are sure to reach the market in the future. As OSHA and/or ASME standards change, it is much easier to update an app than it is to throw away and replace books.

CLASS	SIZE (IN)	RATED CAPACITY - LBS*						EYE DIMENSIONS (APPROXIMATE)	
		VERTICAL	CHOKER**	BASKET HITCH		30	60	90	WIDTH (IN)	LENGTH (IN)
6 × 19 IWRC	1/4	1,120	820	2,200	2,200	1,940	1,580	2	4
	5/16	1,740	1,280	3,400	3,400	3,000	2,400	2 1/2	5
	3/8	2,400	1,840	4,800	4,600	4,200	3,400	3	6
	7/16	3,400	2,400	6,800	6,600	5,800	4,800	3 1/2	7
	1/2	4,400	3,200	8,800	8,600	7,600	6,200	4	8
	9/16	5,600	4,000	11,200	10,800	9,600	8,000	4 1/2	9
	5/8	6,800	5,000	13,600	13,200	11,800	9,600	5	10
	3/4	9,800	7,200	19,600	19,000	17,000	13,800	6	12
	7/8	13,200	9,600	26,000	26,000	22,000	18,600	7	14
	1	17,000	12,600	34,000	32,000	30,000	24,000	8	16
	1 1/8	20,000	15,800	40,000	38,000	34,000	28,000	9	18
6 × 37 IWRC	1 1/4	26,000	19,400	52,000	50,000	46,000	36,000	10	20
	1 3/8	30,000	24,000	60,000	58,000	52,000	42,000	11	22
	1 1/2	36,000	28,000	72,000	70,000	62,000	50,000	12	24
	1 5/8	42,000	32,000	84,000	82,000	72,000	60,000	13	26
	1 3/4	50,000	38,000	100,000	96,000	86,000	70,000	14	28
	2	64,000	48,000	128,000	124,000	110,000	90,000	16	32
	2 1/4	78,000	60,000	156,000	150,000	136,000	110,000	18	36
	2 1/2	94,000	74,000	188,000	182,000	162,000	132,000	20	40

* Rated capacities for unprotected eyes apply only when attachment is made over An object narrower than the natural width of the eye and apply for basket hitches only when the d/d ratio is 20 or greater, where d=diameter of curvature around which the body of the sling is bent, and d=nominal diameter of the rope.

** See choker hitch rated capacity adjustment chart.

an imaginary line that connects the two points of sling contact with the load. To determine the sling angle factor, the length (L) is divided by the height (H). The result is the factor to be applied to the weight of the load on each sling.

In the example shown in *Figure 25*, the height (H) is 74" (188 cm), and the length (L) is 86" (218 cm). (The length will always be longer than the height when the sling is at an angle.) The sling angle factor is calculated as follows:

US measure:
> **Sling angle factor = L ÷ H**
> **Sling angle factor = 86" ÷ 74"**
> **Sling angle factor = 1.162**

Metric:
> **Sling angle factor = L ÷ H**
> **Sling angle factor = 218.4 cm ÷ 187.9 cm**
> **Sling angle factor = 1.162**

The total weight of the load in this example is 2,000 lbs (907 kg). Since there are two slings, each sling must lift 1,000 lbs (454 kg). (2,000 lbs ÷ 2 slings = 1,000 lbs per sling.) Now apply the sling angle factor to determine the actual tension placed on each sling:

US measure:
> **Sling tension = load per sling × sling factor**
> **Sling tension = 1,000 lbs × 1.162**
> **Sling tension = 1,162 lbs**

Metric:
> **Sling tension = load per sling × sling factor**
> **Sling tension = 453.5 kg × 1.162**
> **Sling tension = 527 kg**

Figure 26 shows the effect of various sling angles on sling loading when 1,000 pounds (454 kg) of load are applied to each sling. Note that the tension on the slings is much higher when the legs are positioned at an angle of 30 degrees relative to the horizontal plane than when the legs are at an angle of 60 degrees. Optimum sling angles fall between 60 and 45 degrees to the horizontal plane. Angles of 30 degrees are occasionally required, but lesser angles are generally considered hazardous and unnecessary. At 30 degrees, the sling angle factor is 2.0, meaning that the tension on the sling is already double the actual load weight. If a lift plan leads to sling angles less than 30 degrees, changes in the rigging approach are likely needed.

These calculations for finding sling angle factor and tension help to determine whether slings are applied within their rated load capacity, and what that load actually is. However, they do not tell the rigger the sling angle. Tables have been developed that equate the sling angle factor to the angle. These tables can be used to determine the sling angle if the factor is known, or to determine the sling angle factor if the angle is known. *Table 2* is an example of such a table. Using the sling angle factor of 1.162 calculated in the previous example, it can be determined from the table that the sling angle is between 55 and 60 degrees. This is a very acceptable and safe sling angle for lifting.

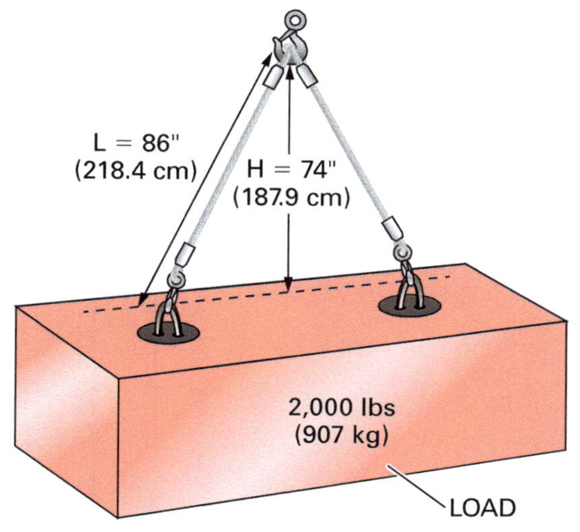

US MEASURE	METRIC
Sling angle factor = L ÷ H	Sling angle factor = L ÷ H
Sling angle factor = 86" ÷ 74"	Sling angle factor = 218.4 ÷ 187.9 cm
Sling angle factor = 1.162	Sling angle factor = 1.162

Figure 25 Determining the sling angle factor.

2.4.3 Common Hitches

The way a sling is arranged to hold the load is referred to as a *hitch*. Hitches can be made using just the sling or by combining slings with connecting hardware. There are three basic types of hitches: vertical hitches, choker hitches, and basket hitches.

One of the most important parts of a rigger's job is making sure that the load is held securely. The type of hitch used depends on the nature of the load. For example, different hitches are used to secure a load of pipes, a concrete slab, or heavy machinery. Controlling the movement of the load once the lift is in progress is another extremely important part of the rigger's job. Therefore, the rigger must also consider the intended movement of the load when choosing a hitch. For example, some loads are lifted straight up and then lowered down to the same spot. Other loads may be lifted, turned 180 degrees in midair, and then set down in a completely different place. The

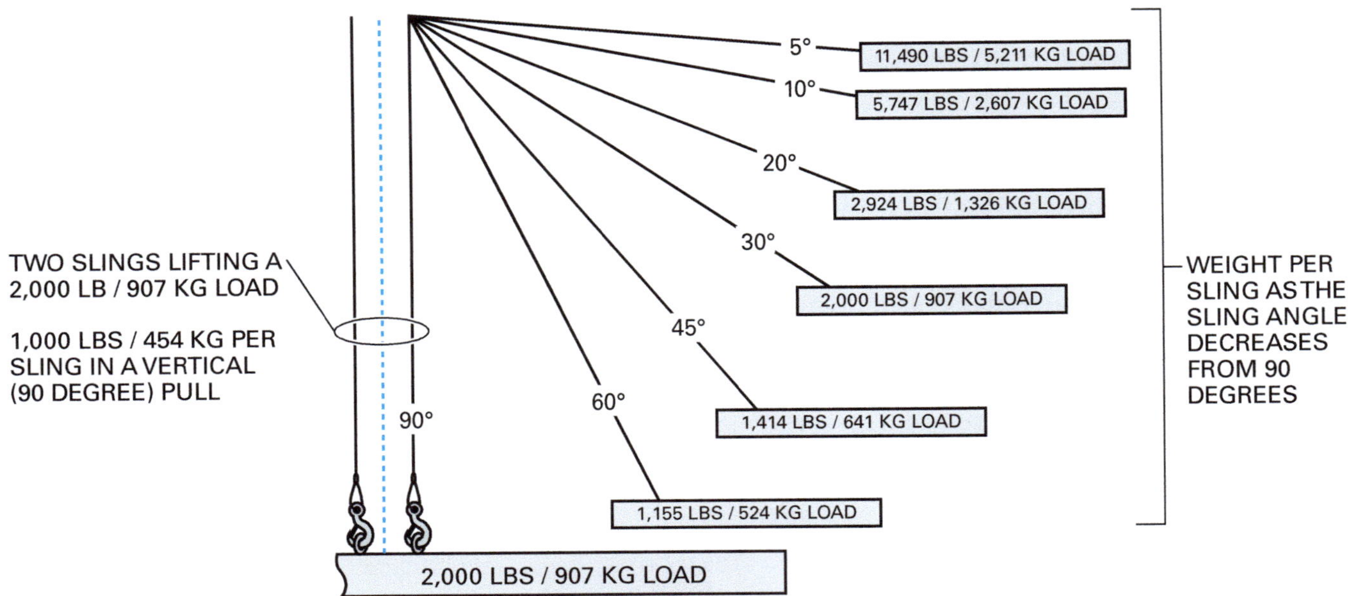

Figure 26 Effect of various sling angles.

Table 2 Sling Angle and Sling Angle Factor

Sling Angle	Sling Angle Factor
5	11.490
10	5.747
15	3.861
20	2.924
25	2.364
30	2.000
35	1.742
40	1.555
45	1.414
50	1.305
55	1.221
60	1.155
65	1.104
70	1.064
75	1.035
80	1.015
85	1.004
90	1.000

Figure 27 A vertical hitch using a plate clamp.

construction of the three basic types of hitches will be reviewed here.

The single vertical hitch (*Figure 27*) is used to lift a load straight up. With this hitch, some type of attachment hardware is needed to connect the sling to the load. The single vertical hitch allows the load to rotate freely. If you do not want the load to rotate freely, some method of load control must be used, such as a tagline.

Another version of the vertical hitch is the bridle hitch (*Figure 28*). The bridle hitch consists of two or more vertical hitches attached to the same hook, master link, or ring. This hitch allows the slings to be connected to the same load without the use of such devices such as spreader beams. Multiple-leg bridle hitches provide increased stability and balance for the load being lifted.

The Ten-Inch Rule

A alternate way to determine the sling angle factor in the field is done using a tape measure. First, measure the distance up from a horizontal line between the connection points, and find the point where the sling is exactly 10" above the line (as shown in the image). Then measure the distance from that point down the angled sling to the connection point. That line will be longer, since it travels at an angle. In the example shown here, that distance measures 14". Now simply place a decimal point between the 1 and the 4 to arrive at a sling angle factor of 1.4. If the factor is 2.0 or greater, the sling angle is excessive and applies too much stress to the slings. Change the rigging to reduce the sling angle.

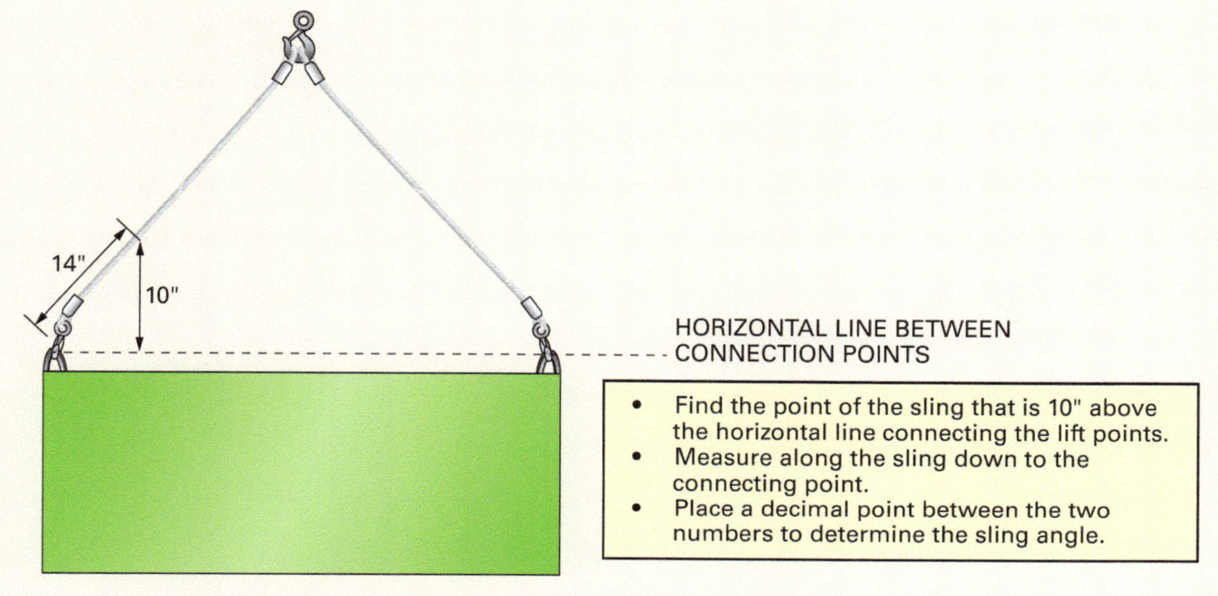

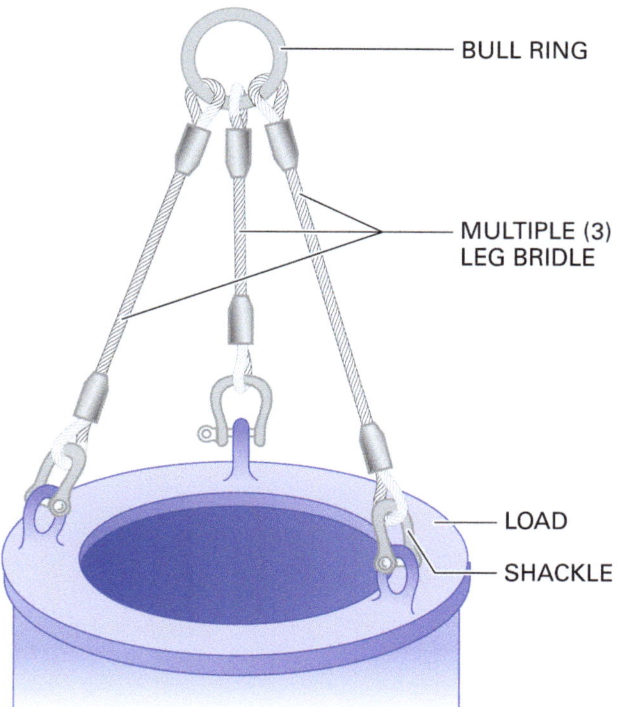

Figure 28 A multi-leg bridle hitch.

However, it is important to note that a bridle hitch results in slings that are at an angle other than 90 degrees to the horizontal plane. Therefore, the stress applied to the sling is increased and must be accounted for in sling selection.

A choker hitch is often used when a load has no attachment points or when the attachment points are not practical for lifting (*Figure 29*). The hitch is made by wrapping the sling around the

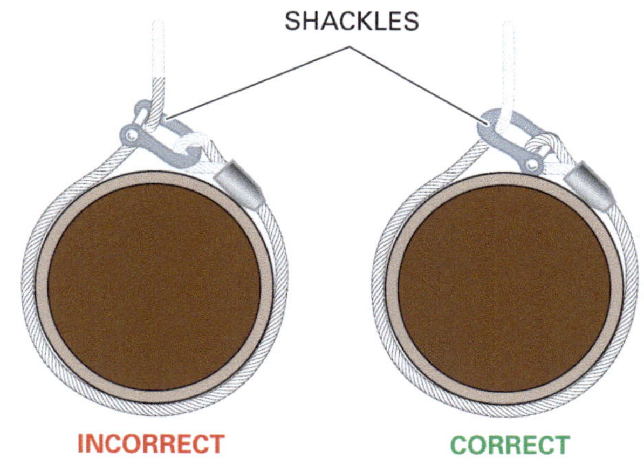

Figure 29 A choker hitch.

load and passing one eye of the sling through a shackle to form a constricting loop around the load. It is important that the shackle used in a choker hitch be oriented properly, as shown in *Figure 29*. It is also important to place a single choker hitch at the load's CG. Otherwise, the load will be unbalanced when lifted and will slip out of the hitch. The choker hitch affects the capacity of the sling, reducing it by a minimum of 25 percent. This reduction must be considered when choosing the proper sling.

A choker hitch does not grip the load as securely as the name implies. It is not recommended for loose bundles of materials because it tends to push loose items up and out of the choker. Many riggers use the choker hitch for loose bundles, mistakenly believing that forcing the choke down provides a tight grip. This actually increases the stress on the choked leg of the sling.

Instead, to gain gripping power, use a double-wrap choker hitch (*Figure 30*). The double-wrap choker uses the load weight to provide the constricting force, so there is no need to try and force the sling into a tighter choke. A double-wrap choker hitch is ideal for lifting bundles of items, such as pipes and structural steel. It will also keep the load in a certain position, which makes it ideal for equipment installation lifts.

When an item more than 12' (3.7 m) long is being rigged, the general rule is to use two choker hitches spaced far enough apart to provide the stability needed to transport the load. When two hitches are used, the hoist line should be positioned over the load's CG. To lift a bundle of loose items, or to maintain the load in a certain position during transport, remember to use the double-wrap choker hitch instead. Loads that are long enough to cause the sling angle to be too great should be rigged using a spreader beam.

Basket hitches (*Figure 31*) are very versatile and can be used to lift a variety of loads. A basket hitch is formed by passing the sling around the load and placing both eyes in the hook. Placing a sling into a basket hitch effectively doubles

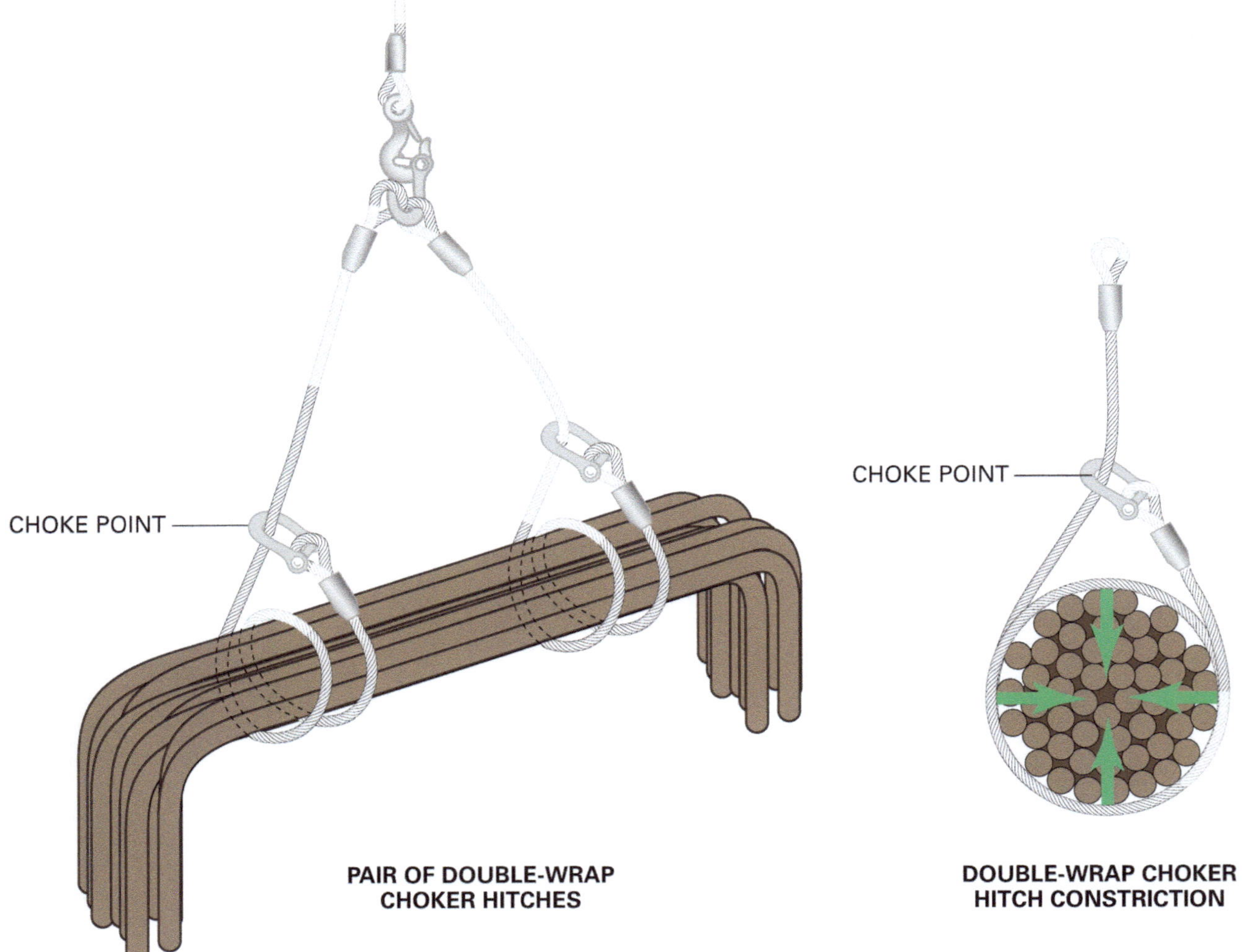

Figure 30 A double-wrap choker hitch.

NCCER – *Mobile Crane Operations*

the capacity of the sling. This is because the basket hitch creates two sling legs from one sling. However, this does not provide secure control of the load.

The double-wrap basket hitch (*Figure 32*) combines the constricting power of the double-wrap choker hitch with the capacity advantages of a basket hitch. This means it is able to hold a larger load more tightly. The double-wrap basket hitch requires a considerably longer sling length than a double-wrap choker hitch, since both sling eyes must be connected to the crane hook. If it is necessary to join two or more slings together, the load must be in contact with the sling body only, not with the hardware used to join the slings. The double-wrap basket hitch provides support around the load. Just as with the double-wrap choker hitch, the load weight provides the constricting force for the hitch.

2.4.4 *Finding the Load's Center of Gravity*

The center of gravity (CG) of an object is the point around which the weight of the object is concentrated. The CG must be known when using any hitch in order to properly and safely position the load line over the load.

The CG of an object is the point where the weight times the length of the object on one side is equal to the weight times the length of the other side; the point in between these two sides is the CG. This is illustrated in *Figure 33*. When the two calculations are equal, the CG has been located. The symbol to identify the CG, also shown in *Figure 33*, will be seen on drawings for lifted components and equipment and sometimes on the load itself.

Figure 31 A simple basket hitch.

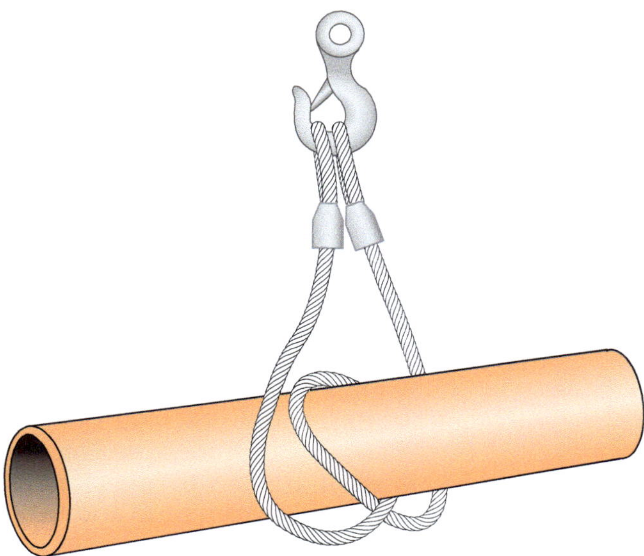

Figure 32 A double-wrap basket hitch.

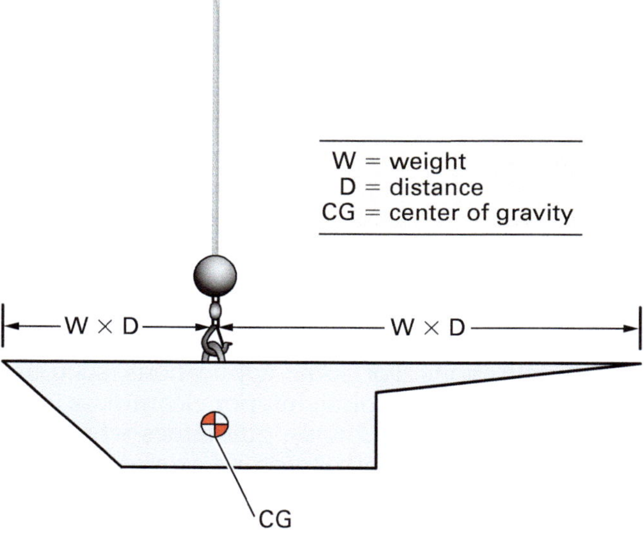

Figure 33 Calculating the center of gravity.

The simplest way of determining the CG of many objects is to experiment by lifting the object a few inches off the ground by a single point. If the object tilts, the single point by which the object is hanging is not directly over the CG. That means the lifting point needs to move in the same direction the load is tilting. When the object hangs level, the CG is vertically below the point from which the object is hung.

2.5.0 Rigging and Handling Piping Materials and Rebar

Reinforcing bars, piping, and similarly shaped construction materials are ordered from a supplier (usually the lowest bidder meeting the specification). Delivery of these materials is scheduled to coincide with the construction schedule as precisely as possible. Long-term storage of construction materials at the jobsite is not generally desirable.

These materials usually arrive at the job site on flatbed trucks or tractor trailers. If a **spur track** is available at the job site, shipments may be received on flatbed or gondola-style railroad cars. Where delivery of steel is scheduled to meet daily placement requirements, truckloads are delivered to the points of placement. Workers should be aware of safety factors that are necessary to achieve safe, efficient handling, storage, and hoisting of pipe and structural materials.

Trucks should be promptly unloaded to minimize jobsite traffic and potential safety hazards. Structural materials that are stored on the ground must be placed on timbers or other suitable blocking to keep them free from mud and to allow safe and easy handling.

Piping materials, rebar, and structural stock are normally stored by type and size first, then by length within each size if necessary. Materials such as pipe are usually of the same standard length unless sections or structural components have been precut or prefabricated before shipping. Tags are kept at the same end of each piece or bundle for easy identification. When a bundle is opened and part of the stock removed, the unit with the tag should remain with the bundle, and a new number reflecting the remaining quantity should be recorded. All mud and debris generally must be washed off before placement. The interior of piping materials should be kept clean and dry for most applications. For some applications, such as HVACR refrigerant piping, interior cleanliness is essential.

Use the following guidelines when rigging and handling a load of pipe, rebar, and similar products:

- Pipe of the same length should be lifted at the same time. If significantly different lengths of pipe must be lifted, lift each different length separately.
- Ensure that the open part of the hook is facing away from center when using slings with fixed or sliding hooks.
- Do not use carbon steel slings on stainless steel pipe to prevent cross-contamination of the metals. Carbon steel particles embedded in the stainless steel can rust and not only ruin the outer appearance, but also contaminate welds to the point of rejection. Use synthetic web slings on stainless steel and painted pipe.
- Lay the pipe on blocks so the slings can be worked on and off the load. Blocks should be made from hardwood and be thick enough to prevent the pipe from contacting uneven ground. Adequate ground clearance is also needed to slip a sling around the bundle. Pipe placed on blocks requires wedges to keep it from rolling off the blocking. Large diameter pipe requires larger chocks to keep it from rolling. In general, there should be 1" (2.5 cm) of chock per 1' (3 m) of diameter. For instance, a 42" (1,050 DN) pipe should have a chock that is about $3\frac{1}{2}$" thick (9 cm) to prevent rolling. Note that a chock that is too thick can be pushed away by a rolling load.
- Use the hook and load line of the crane as the centerline with which to line up the load.
- As soon as the load clears the ground, check its orientation. If it is not level, the signal person should signal the operator to stop lifting the load, and guide the load back onto the blocks to adjust the position of the slings.
- Never stand underneath the load.
- Control all loads with one or more tag lines.
- Keep the load as low to the ground as possible.
- Store all pipe on level ground whenever possible.
- Stand to one side of the load, and guide it onto the blocks. Move to the end of the pipe before the rigging is released. Remember that a bundle of pipe can easily roll when the sling tension is released. Avoid pinch points when removing the sling, and do not stand in a position where the pipe may roll on you.
- Lay pipe side by side. Do not stack pipe if it can be avoided.

To select slings and properly rig a load, the weight of the load must be known. It is therefore beneficial to understand how to determine the weight of these materials.

2.5.1 Determining the Weight of Pipe

Every foot of pipe weighs a given amount, depending on the wall thickness and nominal size of the pipe. Therefore, if you know the weight per foot and the length, the total weight of the pipe can be determined.

Table 3 is an example of a weight chart for carbon steel pipe. Always be sure you consult a chart for the proper material as well as the correct size and weight. There are also separate charts for metric pipe sizes, which will report the weight as kilograms per meter rather than pounds per foot. Tables like the one shown here are available from a variety of sources online, including pipe distributors.

The numbers in the leftmost column represent the nominal pipe size in inches. The designations and numbers across the top represent wall thickness. The wall thickness of pipe is often designated by a schedule number; the numbers 10 through 160 across the top are schedule numbers. To use the table, find the point at which the nominal size and wall thickness, or schedule, of the pipe meet. The number in this block is the weight of one foot of pipe. To find the total weight, multiply the total number of feet by the weight per foot. Note that there is a significant difference in the weight of pipe made from other materials, such as PVC, copper, and stainless steel.

For example, to find the weight of five 10-foot lengths of 20" Schedule 40 carbon steel pipe, read right from the 20.0 in the Nominal Pipe Size column until reaching the Schedule 40 column. The number found there is 123.1. This is the weight, in pounds, of one foot of pipe. There is a total of 50' of pipe (5 lengths × 10' = 50'). Multiply the weight per foot from the table by the total length in feet to find the total weight of the pipe load (123.1 × 50' = 6,155 lbs).

Table 3 Carbon Steel Pipe Weights

Nominal Pipe Size (inches)	Wall Thickness								
	STD	XS	XXS	10	40	60	80	120	160
	Weight Per Foot in Pounds								
2.0	3.65	5.02	9.03	–	3.65	–	5.02	–	7.06
2.5	5.79	7.66	13.7	–	5.79	–	7.66	–	10.01
3.0	7.58	10.25	18.58	–	7.58	–	10.25	–	14.31
3.5	9.11	12.51	22.85	–	9.11	–	12.51	–	–
4.0	10.79	14.98	27.54	–	10.79	–	14.98	18.98	22.52
6.0	18.97	28.57	53.16	–	18.97	–	28.57	36.42	45.34
8.0	28.55	43.39	72.42	–	28.55	35.66	43.39	60.69	74.71
10.0	40.48	54.74	104.1	–	40.48	54.74	64.40	89.27	115.7
12.0	49.56	65.42	125.5	–	53.56	73.22	88.57	125.5	160.3
14.0	54.57	72.09	–	36.71	63.37	85.01	106.1	150.8	189.2
16.0	62.58	82.77	–	42.05	82.77	107.5	136.6	192.4	245.2
18.0	70.59	93.45	–	47.39	104.8	138.2	170.8	244.1	308.6
20.0	78.60	104.1	–	52.73	123.1	166.5	208.9	296.4	379.1
22.0	86.61	114.8	–	58.07	–	197.4	250.8	353.6	451.1
24.0	94.62	125.5	–	63.41	171.2	238.3	296.5	429.5	542.1
26.0	102.6	136.2	–	85.73	–	–	–	–	–
28.0	110.6	146.9	–	92.41	–	–	–	–	–
30.0	118.7	157.5	–	99.08	–	–	–	–	–
32.0	126.7	168.2	–	105.8	229.9	–	–	–	–
34.0	134.7	178.9	–	112.4	244.9	–	–	–	–
36.0	142.7	189.6	–	119.1	282.4	–	–	–	–
42.0	166.7	221.6	–	–	330.4	–	–	–	–

2.5.2 Determining the Weight of Rebar

The weight of rebar is determined the same way as the weight of pipe. *Table 4* shows a chart of rebar weight by bar size. This particular example shows the bar size and other characteristics in both imperial and metric units. Virtually all rebar is made from carbon steel, so it would be rare to encounter rebar made from another material. If that is the case however, the supplier would need to be contacted for the weight information.

Once the size and total length of rebar in a bundle is identified, the weight per foot (or per meter) is multiplied by the appropriate length to determine the weight. Always remember to work within the same system of units—kilograms per meter, or pounds per foot.

2.5.3 Rigging Valves

Rigging a valve correctly involves knowing where to place the sling, how to place the sling, what kind of sling to use, and what kind of valve is being lifted. Manufacturers of large valves often provide drawings to show how the part is best rigged for lifting. Very large valves and valve components may be equipped with factory-installed lifting lugs.

Synthetic slings are usually best for rigging all types of valves, since many are also painted and easily scarred by metal slings. Never use a carbon steel sling to rig a stainless steel valve to avoid surface contamination of the stainless steel.

To rig a valve like the one shown in *Figure 34*, place a synthetic sling on each side of the valve body between the bonnet and the flanges. If the handwheel remains attached, bring the slings up

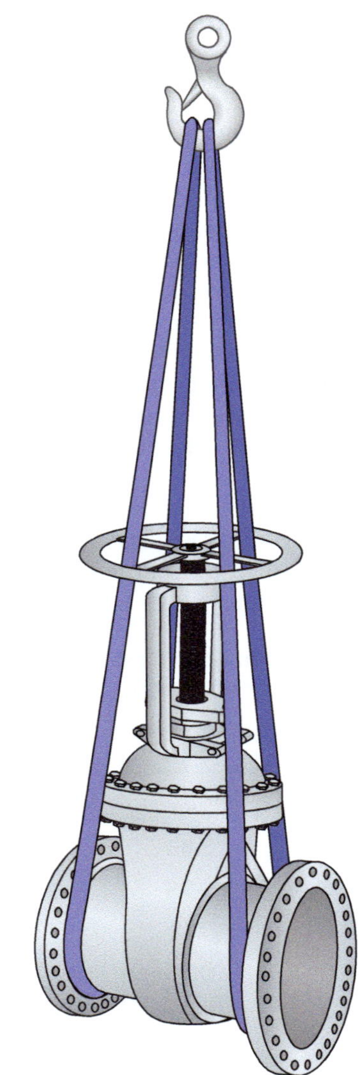

Figure 34 Rigging a valve.

Table 4 ASTM Standard Metric and US Measure Reinforcing Bar

Bar Size		Nominal Characteristics*					
		Diameter		Cross-Sectional Area		Weight	
Metric	[in-lb]	mm	[in]	mm	[in]	kg/m	[lbs/ft]
#10	[#3]	9.5	[0.375]	71	[0.11]	0.560	[0.376]
#13	[#4]	12.7	[0.500]	129	[0.20]	0.944	[0.668]
#16	[#5]	15.9	[0.625]	199	[0.31]	1.552	[1.043]
#19	[#6]	19.1	[0.750]	284	[0.44]	2.235	[1.502]
#22	[#7]	22.2	[0.875]	387	[0.60]	3.042	[2.044]
#25	[#8]	25.4	[1.000]	510	[0.79]	3.973	[2.670]
#29	[#9]	28.7	[1.128]	645	[1.00]	5.060	[3.400]
#32	[#10]	32.3	[1.270]	819	[1.27]	6.404	[4.303]
#36	[#11]	35.8	[1.410]	1006	[1.56]	7.907	[5.313]
#43	[#14]	43.0	[1.693]	1452	[2.25]	11.38	[7.65]
#57	[#18]	57.3	[2.257]	2581	[4.00]	20.24	[13.60]

*The equivalent nominal characteristics of inch-pound bars are the values enclosed within the brackets.

NCCER – *Mobile Crane Operations*

through the handwheel spokes so that the valve cannot tilt from front to back. Do not place a sling around the handwheel or through the valve bore. The handwheel is not built to support the weight of the valve. A valve rigged around the handwheel is unsafe, even if the valve is going to be moved a short distance. Placing a sling through the bore of the valve can destroy the inner workings, even if soft synthetic slings are used.

2.6.0 Taglines and Knots

Taglines are natural fiber or synthetic ropes used to control the load (*Figure 35*). The absence or improper use of taglines can turn a simple hoisting operation into a hazardous situation. Taglines are used to maintain lateral control and prevent spinning of a suspended load as the crane and/or boom move. Loads of all shapes and sizes are subject to some level of dynamic and wind forces once in the air.

When selecting a tagline, several factors need to be considered. Natural fiber, or manila, is notably weaker than synthetic fibers such as nylon, polyester, polypropylene, or polyethylene. Although any rope that is wet becomes an electrical conductor, natural fiber rope absorbs water readily and may remain wet enough to conduct electricity for a long time. Most (but not all) synthetic ropes do not absorb moisture. A nonconductive tagline should be used when working in the vicinity of power lines.

Synthetic rope is lighter than natural fiber and has a high strength-to-weight ratio. Its resistance to water and reduced electrical conductivity do give it a distinct safety advantage. However, synthetic ropes are more easily damaged by heat. Significant contact with surfaces at 150°F (66°C) can result in a loss of strength, and many synthetic rope materials begin to melt at 300°F (149°C).

The diameter of a tagline should be large enough so that it can be gripped well when wearing gloves. Rope with a diameter of ½" (12 mm) is common, but ¾" (19 mm) and 1" (25 mm) diameter rope is sometimes used on heavy loads or where the tagline must be extremely long. Taglines should never have knots or loops tied in them. Terminating hardware, such as snaps and carabiners (*Figure 36*), may be added to make an easy connection to some loads. This hardware does not need to be designed and rated for lifting purposes, but must be substantial enough to handle the stress.

Always wear gloves when using a tagline to control a load. The momentum of a moving load can cause the rope to slide through the hands unexpectedly, causing severe rope burns. Never wrap a tagline around an arm or leg in an attempt to stop a load's swing. Never place yourself between a fixed object and a suspended load. Manning a tagline is a significant responsibility that requires careful thought and attention to execute the task safely and prevent serious injury to personnel and/or damage to property.

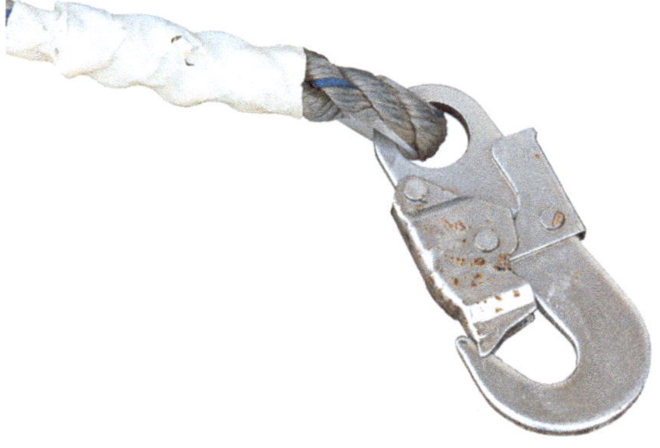

(A) HEAVY-DUTY SNAP

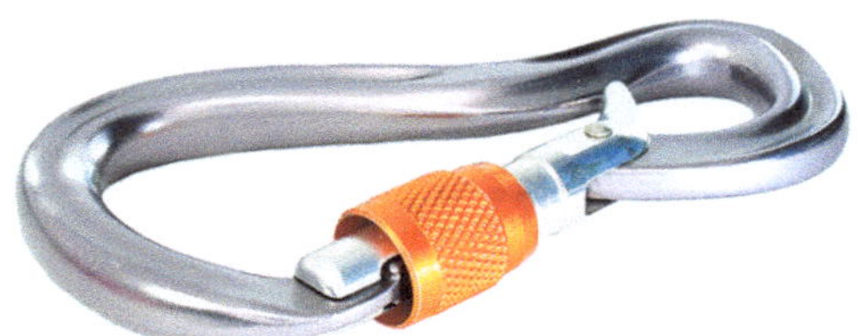

(B) CARABINER WITH LOCK

Figure 36 Rope hardware for quick connections.

Figure 35 Taglines in control of a load.

Taglines should be of sufficient length to allow control of the load from its original lift location until it is safely placed or until load control is transferred to other team members. Special consideration should be given to situations where a long tagline could interfere with the safe handling of loads, such as steel erection projects. Think through the lift and the material movement and consider where the tagline(s) will be and what obstructions might be encountered.

When working near power lines, there is always the hazard of electric shock or electrocution. An insulating link (*Figure 37*) can be added to taglines to help protect the user if the load or tagline contacts a power source.

Taglines should be attached to loads at a location that provides the best physical advantage in maintaining control. Long loads, for example, should have taglines attached as close to the ends as possible. The tagline should also be located in a place that allows personnel to access it for removal after the load is placed.

> Avoid overcompensating during load control through exaggerated movements of the tagline; do not jerk the line. Taglines that are too long can be caught on objects or drag unnecessarily. Avoid tying knots in the line, as the potential for being caught on objects as it moves along the ground are significantly increased. Never tie the tagline to the load hook.

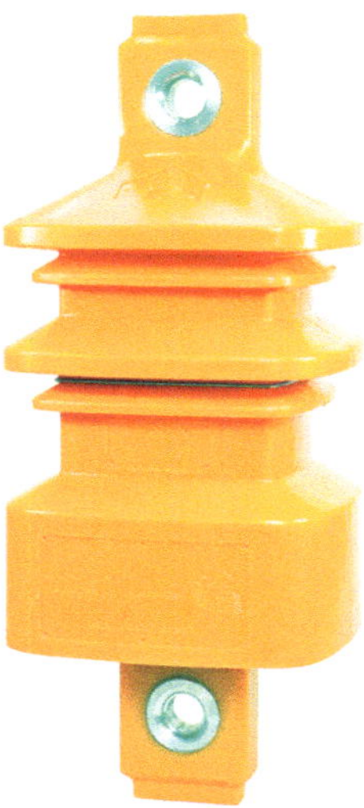

Figure 37 Insulating link.

To properly handle a tagline, you must determine the physical advantage intended. Consider the lift and which direction the load may be most likely to swing or rotate. With a long load, for example, try to maintain a position that is 90 degrees to the length of the object (*Figure 38*).

Whenever possible, keep yourself and the tagline in view of the crane operator. Stay alert. Do not become complacent during the lift. Be aware of the location of any excess rope and do not allow it to become fouled or entangled around your legs or on nearby objects.

Taglines should not to be used to pull or yank a load away from its natural vertical suspension. They also cannot be used in any way that results in them supporting or carrying any portion of the load. Large loads often require the use of multiple taglines. In these cases, tagline personnel must work as a team and coordinate their actions.

2.6.1 Knots

Knots used to attach taglines to loads should be tied properly to prevent slipping or accidental loosening, but they also must be easy to untie after the load is placed. Whenever possible, taglines should be of one continuous length and free of splices. If joining two taglines together becomes necessary, it is best done using the short-splice method. The short-splice method results in a knot diameter roughly twice that of the rope itself, and retains more rope strength than other methods. Larger knots tied in the middle of the taglines can sometimes create difficulties. However, the short splice can be a challenging knot to create, as it involves weaving of the individual rope strands. The short-splice method can be learned from numerous websites and internet videos if desired. In general, it is always best to avoid knots of any kind in a tagline and use a continuous length of rope instead.

Some recommended knots for rigging are the bowline and the clove hitch. The bowline (*Figure 39*) is used to form a secure loop in the end of a rope. It is sometimes called a *rescue knot* or the *king of knots* because it is reliable enough to be used for rescue work. It can be backed up with a second knot for extra security or tied twice, into a double bowline, for extra strength. A bowline does not slip or bind when under load, but it is easy to untie when there is no load.

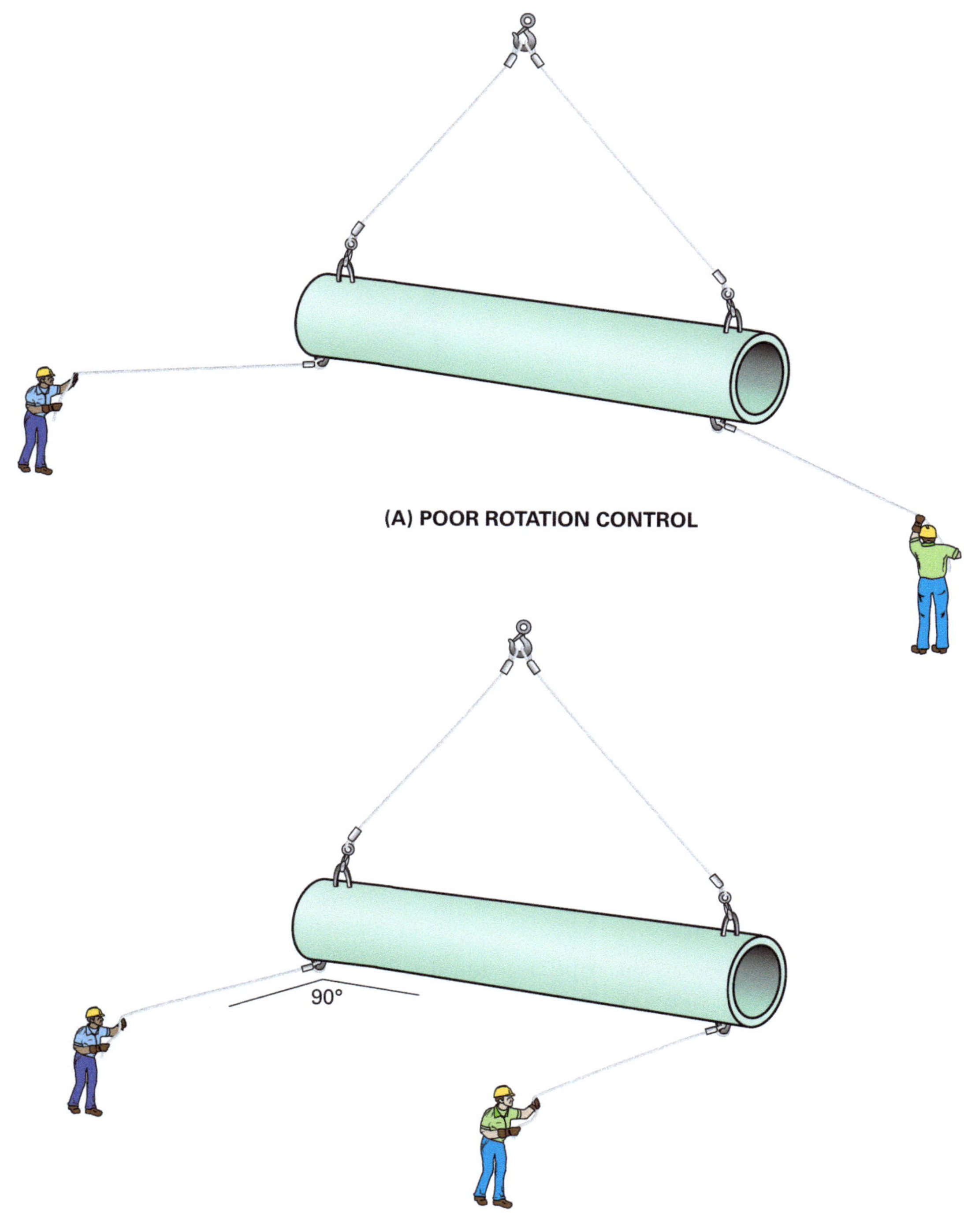

Figure 38 Best position for controlling load rotation.

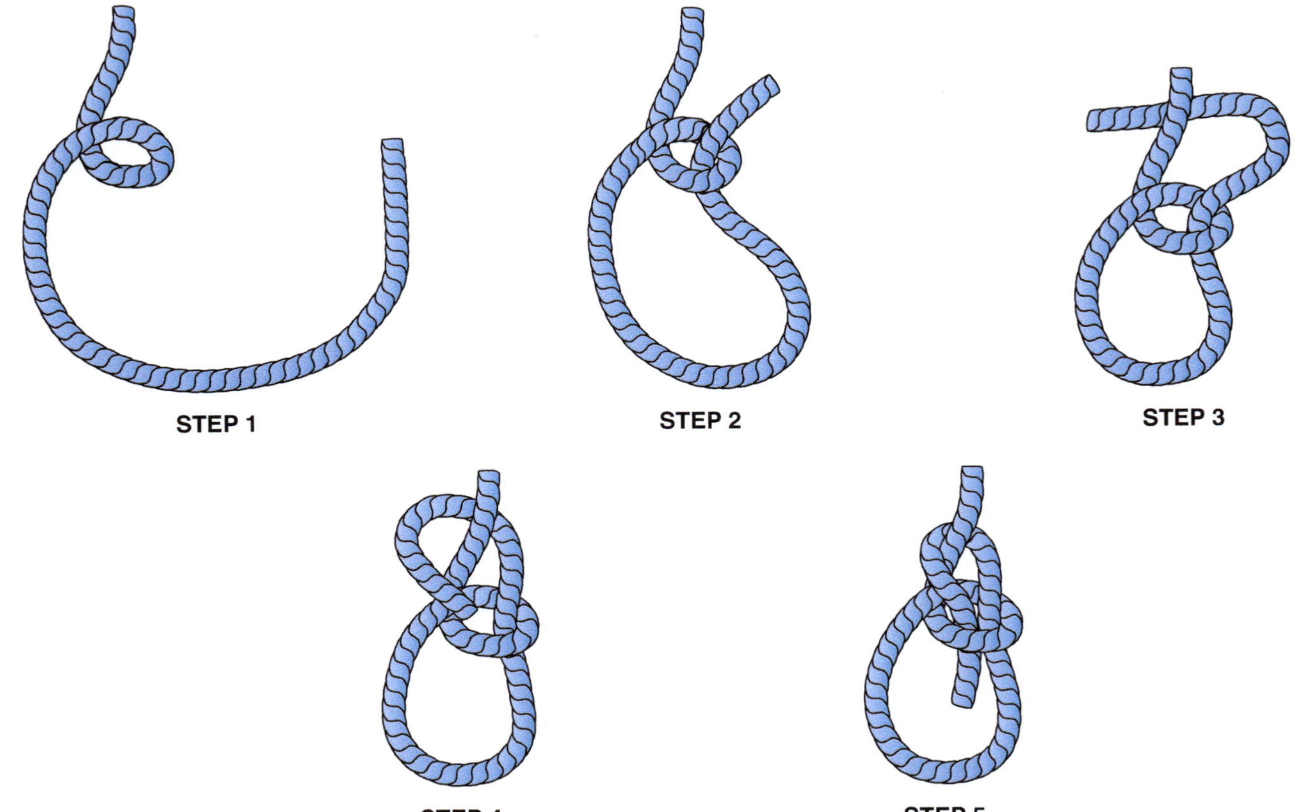

Figure 39 Tying a bowline.

The bowline reduces the strength of the rope that it is tied in by as much as 40 percent. Bowline knots also require a long tail for reliability.

The following steps here, shown in *Figure 39*, can be used to tie a bowline:

Step 1 Form a small loop in the rope, leaving a long enough tail to make the desired size of main loop.

Step 2 Pass the tail of the rope through the small loop.

Step 3 Pass the tail under the standing end.

Step 4 Pass the tail back down through the small loop.

Step 5 Adjust the main loop to the required size and tighten the knot, making sure to maintain a sufficiently long tail.

> **NOTE**
> Some people use this saying to help them remember how to tie a bowline: "The rabbit comes out of his hole, around a tree, and back into the hole."

A half hitch (*Figure 40*) can be used to tie a rope around an object such as a rail, bar, post, or ring. It is commonly used for tasks such as suspending items from overhead beams and carrying light loads that have to be removed easily. The half hitch is also widely used in making other knots, such as the clove hitch. Because the half hitch is not very stable by itself, it is not suitable for heavy loads or tasks in which safety is paramount. It is often used to back up and secure another knot that has already been tied.

The following steps, shown in *Figure 40*, can be used to tie a half hitch:

Step 1 Form a loop around the object.

Step 2 Pass the tail of the rope around the standing part and through the loop.

Step 3 Tighten the hitch by pulling on the working end and the standing part of the rope simultaneously.

A clove hitch (*Figure 41*) is one of the most widely used general hitches. It is typically used to make a quick and secure tension knot on a fixed object that serves as an anchor, such as a post, pole, or beam. A clove hitch can also be used as the first knot when lashing items together.

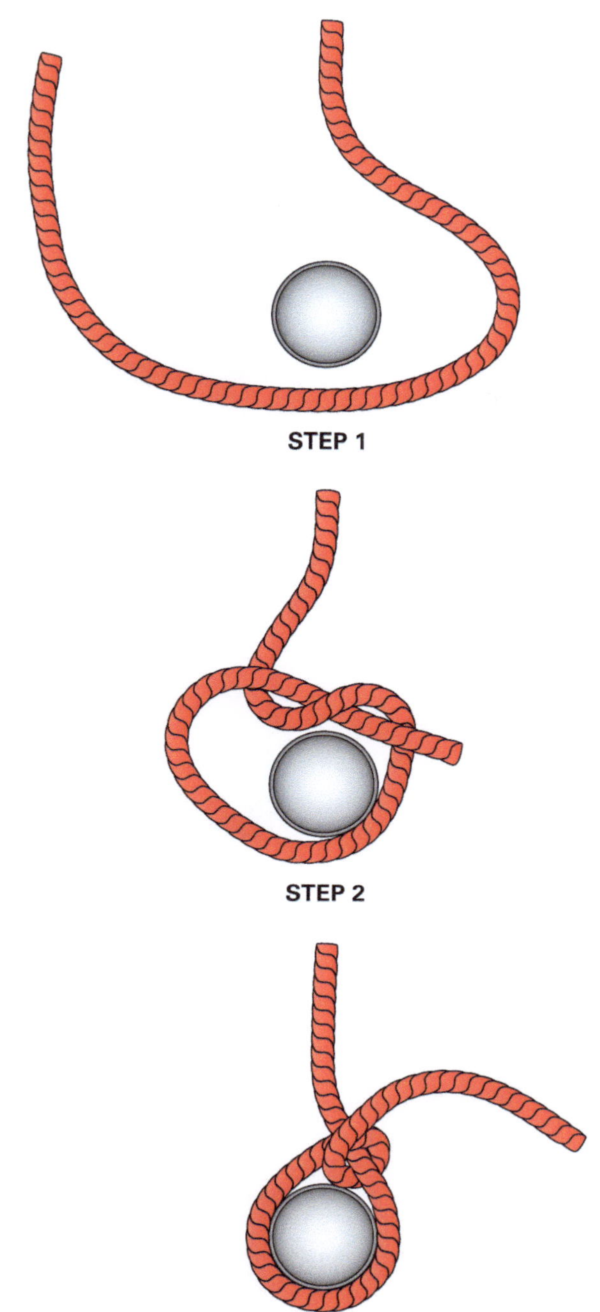

STEP 1

STEP 2

STEP 3

Figure 40 Tying a half hitch.

Because it is a tension knot, a clove hitch loosens when tension is removed from the rope. For these reasons, a clove hitch should not always be used as-is. Additional half hitches or an overhand safety knot should be added to make a clove hitch more secure, unless loosening when tension is released is intentional.

There are alternate techniques for tying a clove hitch. Regardless of the technique, the structure of the resulting knot consists of two half hitches made in opposite directions.

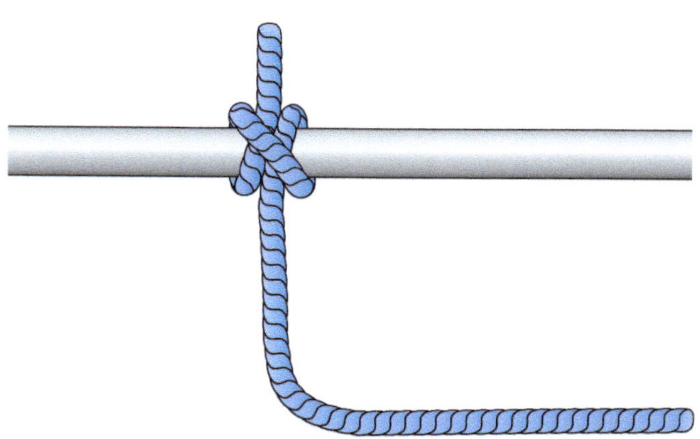

Figure 41 Clove hitch.

One common technique for tying a clove hitch that is used to attach a rope to a ring or upright structural component is to thread the end. This is called the *threading-the-end technique.* Use the following steps, shown in *Figure 42*, to tie a clove hitch using the threading-the-end technique:

Step 1   Pass the working end of the rope over the object.

Step 2   Pass the working end back over the standing part and then over the object. This forms a half hitch.

Step 3   Thread the working end back under itself and up through the loop to form a second half hitch.

Step 4   Pull both the working and standing ends evenly to tighten the hitch.

Another method for tying a clove hitch, the stacked-loops technique, allows the rope to be dropped quickly over a standing object, instead of tying the knot around the object. It also allows the hitch to be tied at any point in a rope, not just at an end. Use the following steps, shown in *Figure 43*, to tie a clove hitch the stacked-loops technique:

Step 1   Form two identical loops in a rope, one in the right hand and one in the left.

Step 2   Cross the loops one above the other to form a knot.

Step 3   Place the knot over the stake or post.

Step 4   Tighten the knot by pulling simultaneously on both ends of the rope. Note that the completed clove hitch consists of two half hitches stacked on each other.

Remember that an additional half hitch is often added for increased security.

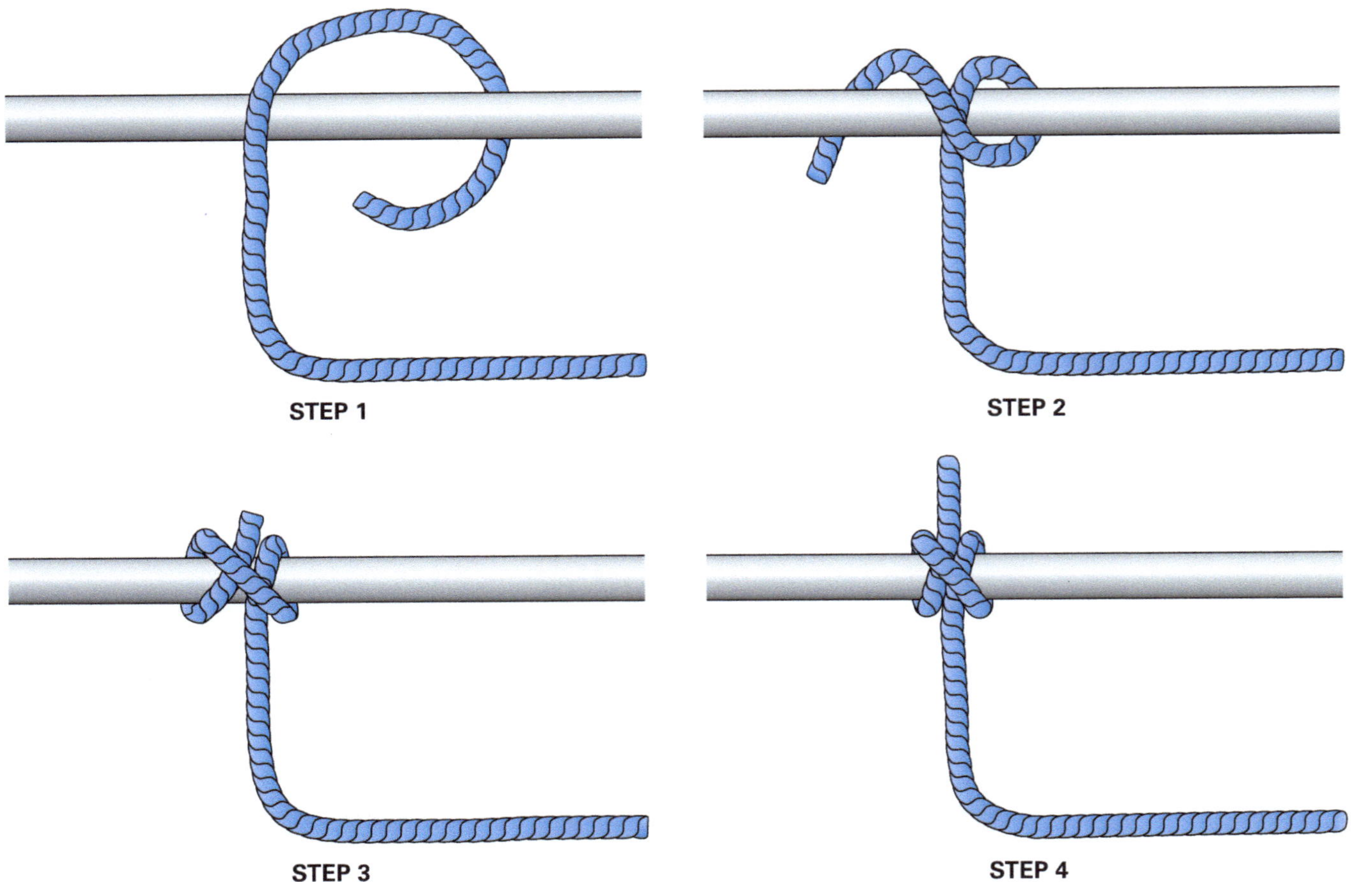

Figure 42 Threading-the-end technique for tying a clove hitch.

2.7.0 Rigging Safety Precautions

Workers on the job are responsible for their own safety and the safety of their fellow workers. Project and corporate management, in turn, has a responsibility to each worker. The responsibility of management and supervisors is to ensure that the workers who prepare and use the equipment, and those who work with or around it, are well trained in operating procedures and safety practices. Each worker is expected to put that training to good use. This section describes safety guidelines that are related to rigging.

NOTE

Safety guidelines pertaining to mobile cranes are covered in detail in NCCER Module 21106, "Crane Safety and Emergency Procedures" from *Mobile Crane Operations Level One*.

Did You Know?

The Ashley Book of Knots

Although *The Ashley Book of Knots* by Clifford W. Ashley is no longer in print, sailors, climbers, campers, macramé artists—anyone with more than a passing interest in knots—continue to find this book cited as the definitive work on the subject of knot tying. Originally published by Doubleday & Company in 1944, it contains the histories of and instructions for tying over 3,900 types of knots, accompanied by 7,000 pen-and-ink drawings.

Clifford Ashley was born in 1881, in New Bedford, Massachusetts. By trade he was an author and artist, but while sailing on many types of boats and researching knots he performed a wide variety of jobs. He spent 11 years writing and drawing the illustrations for his book. He died three years after the book's publication. Ashley continues to be regarded as a fine marine painter as well as one of the world's leading authorities on knot tying.

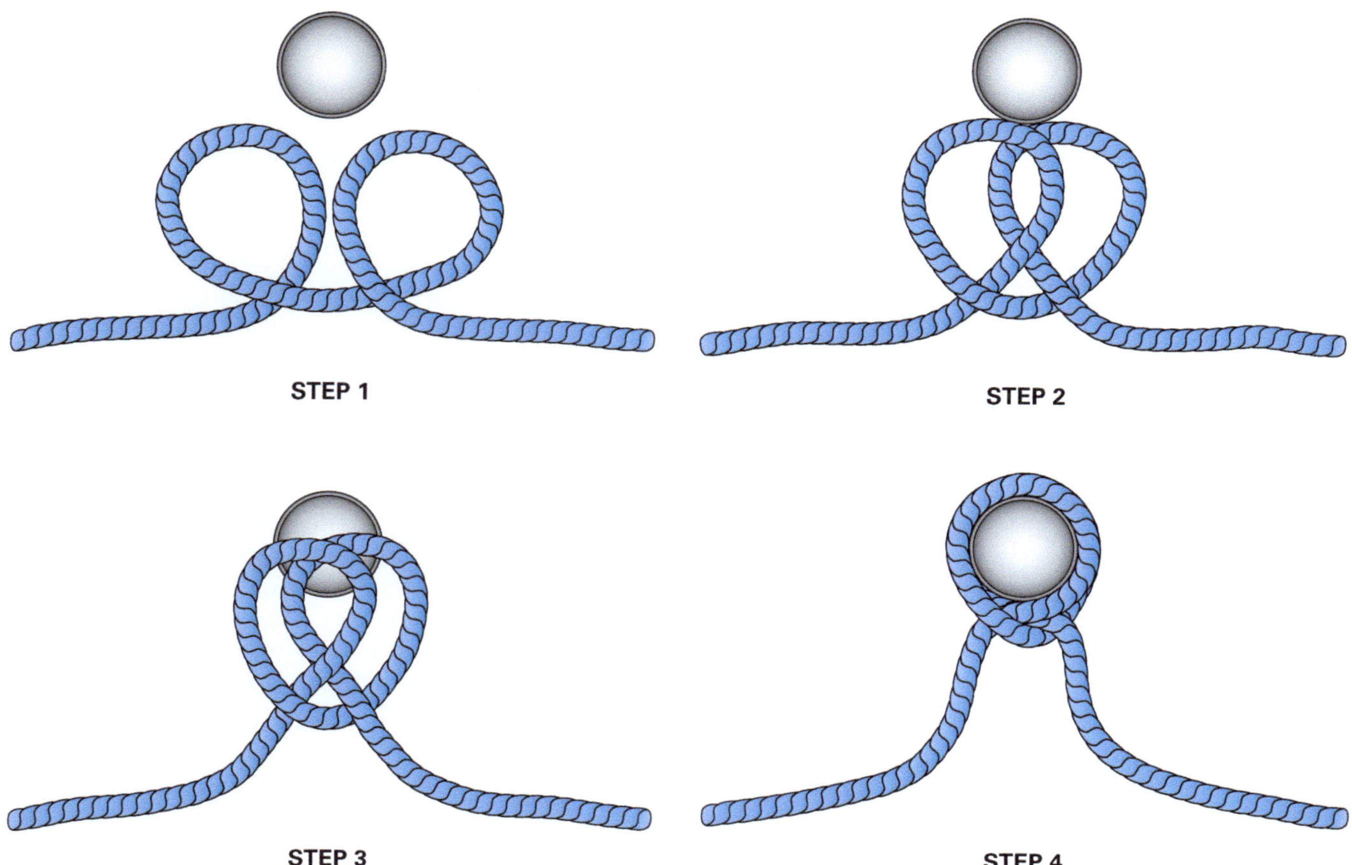

Figure 43 Stacked-loops technique for tying a clove hitch.

One very important rigging precaution is to determine the weight of every load before attempting to lift it. The weight of the load must obviously be within the capacity of the crane, but this is not the responsibility of the rigger. However, the load weight must be considered in the selection of the rigging components before the lift ever begins.

Unfortunately, accurately determining the weight is not always possible. This issue is addressed in 29 *CFR* 1926.1417, which requires that the crane operator verify the weight of the load in one of the following two ways:

- The weight can be determined from an industry-recognized source such as the manufacturer of the product(s), or by an industry-recognized calculation method. This latter method could be used, for example, to calculate the weight of an I-beam based on the weight per foot of such products.
- If the weight cannot be verified through the above methods, the operator may begin the lift using a load-weighing device, load-moment indicator, rated-capacity indicator, or rated-capacity limiter. This exercise is used to

determine if the lifted load is more than 75 percent of the crane's maximum rated capacity at the longest operating radius that the lift will require. If it is greater than 75 percent, the lift must be aborted until the actual weight can be verified through the first method.

It is equally important to rig the load so that it is stable and its center of gravity is below the hook. The personal safety of riggers and hoisting operators depends a lot on common sense. The following safety practices should be observed:

- Following the OSHA procedures from 29 *CFR* 1926.1417, determine the weight of loads before rigging whenever possible.
- Coordinate the rigging process as necessary with other personnel on the job site to minimize foot and vehicle traffic in the area.
- Know the rated load capacity of the equipment and tackle, and never exceed it. Remember that the rated load capacity of all hoisting and rigging equipment is based on ideal conditions. Normal working conditions are rarely ideal, so it is important to recognize factors that can reduce equipment capacity or increase the load imposed, such as sling angles.

- Examine all hardware, tackle, slings, and equipment before use. Remove defective components from service for evaluation and possible repair by others.
- Always wear gloves when handling slings, taglines, and similar rigging equipment.
- Never wrap hoist ropes around a load. Use only slings or other appropriate lifting devices.
- Ensure that all slings are made from the same material when using two or more slings on a single load.
- Never lift loads with one or two legs of a multi-leg sling until the unused slings are secured.
- Only use lifting beams for the purpose for which they were designed.
- All personnel must stand clear and not underneath loads being lifted and lowered, and when slings are being withdrawn from beneath the load.
- Never point-load a hook unless it has been specially designed for that task.
- Keep hands and feet away from pinch points as the slack is taken up by the crane.
- Use one or more taglines as necessary to keep the load under control.
- Immediately report defective equipment or rigging issues to the rigging supervisor or lift director, who must issue orders to proceed after safe conditions have been ensured.
- Prepare blocking at the load landing zone before the load arrives, rather than after it is suspended above. Allow the load to safely land before removing the slings.

Weather conditions can be a major factor in lifting operations. Stop hoisting and/or rigging operations when weather conditions such as the following present a hazard to property, workers, or bystanders:

- Winds exceeding crane manufacturer recommendations
- Lightning or thunder
- Poor visibility due to conditions such as darkness, dust, fog, rain, or snow, impairing view of rigger or hoist crew
- Temperature low enough to cause crane structures to fracture upon shock or impact

Crane Collapse in Manhattan

In early 2016, a lattice-boom crawler crane was attempting to lower its 565-foot (172-meter) boom and secure the crane due to high winds. Unfortunately, the wind gusts intensified and the boom came crashing into the streets of Manhattan. One bystander was killed and several others were injured. The crew had already begun clearing the streets of traffic to prepare to receive and secure the boom, minimizing the loss of life and injuries.

Figure Credit: © a katz/Shutterstock.com

Additional Resources

ASME Standard B30.5, Mobile and Locomotive Cranes. Current edition. New York, NY: American Society of Mechanical Engineers.

ASME Standard B30.9, Slings. Current edition. New York, NY: American Society of Mechanical Engineers.

ASME Standard B30.20, Below-The-Hook Lifting Devices. Current edition. New York, NY: American Society of Mechanical Engineers.

ASME Standard BTH-1, Design of Below-The-Hook Lifting Devices. Current edition, New York, NY: American Society of Mechanical Engineers.

29 *CFR* 1926, Subpart CC, **www.ecfr.gov**

29 *CFR* 1926.251, **www.ecfr.gov**

29 *CFR* 1926.753, **www.ecfr.gov**

Mobile Crane Operations Level One, NCCER. Third Edition. 2018. New York, NY: Pearson Education, Inc.

NCCER Module 00106, *Introduction to Basic Rigging.*

Mobile Crane Safety Manual (AEM MC-1407). 2014. Milwaukee, WI: Association of Equipment Manufacturers.

Willy's Signal Person and Master Rigger Handbook, Ted L. Blanton, Sr. Current edition. Altamonte Springs, FL: NorAm Productions, Inc.

Knots: The Complete Visual Guide, Des Pawson. First American Edition. 2012. New York, NY: DK Publishing.

North American Crane Bureau, Inc. website offers resources for products and training, **www.cranesafe.com**

2.0.0 Section Review

1. A sling with a wide, flat bearing surface that is especially well-suited for basket hitches where sharp bends are encountered is the ______.

    a. strand-core wire rope sling
    b. fiber-core wire rope sling
    c. braided-belt sling
    d. chain sling

2. If a sling is needed to lift a load with a polished surface, the best choice would be a ______.

    a. synthetic web sling
    b. chain sling
    c. metal-mesh sling
    d. wire-rope sling

3. Metal-mesh slings must have the approval of the manufacturer if they are to be used at temperatures above ______.

    a. 195°F (91°C)
    b. 375°F (191°C)
    c. 550°F (288°C)
    d. 750°F (399°C)

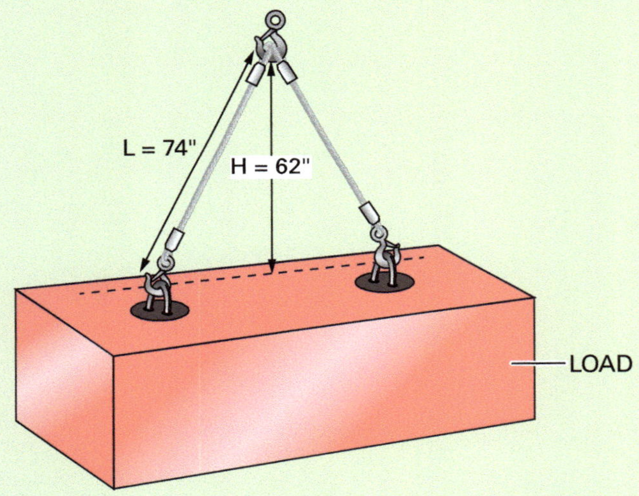

Figure SR01

4. Refer to *Figure SR01*. Based on the measurements shown, what is the sling angle factor that should be applied to the weight of the lifted load on each sling?

    a. 0.841
    b. 1.194
    c. 1.213
    d. 1.414

5. Refer to *Table 2* in Section 2.4.2. Based on the generally accepted safe range of sling angles, is the sling angle associated with a sling angle factor of 1.194 within the acceptable range?

    a. Yes
    b. No

6. To prevent 24" pipe material (600 DN) from rolling, the chock needs to be ______.

    a. 1" thick
    b. 2" thick
    c. 3½" thick
    d. 5" thick

7. One disadvantage of synthetic rope compared to natural fiber is that synthetic rope ______.

    a. is significantly heavier
    b. readily conducts electricity
    c. has a low strength-to-weight ratio
    d. is more easily damaged by heat

8. The procedures for determining the weight of a load to be lifted are covered in ______.

    a. *29 CFR* 1910.140
    b. *ASME Standard B30.9*
    c. *29 CFR* 1926.1417
    d. *ASME Standard BTH-1*

3.0.0 HOISTING EQUIPMENT AND JACKS

Objective

Identify and describe how to use various types of hoisting and jacking equipment.

a. Identify and describe how to use manual and powered hoisting equipment.
b. Identify and describe how to use jacks.

Performance Tasks

8. Select, inspect, and demonstrate the safe use of the following rigging equipment:
   - Block and tackle
   - Chain hoist
   - Ratchet-lever hoist
   - One or more types of jacks

Trade Terms

Gantry: A framed overhead structure supported by legs on each end, used to cross over obstructions. Gantries can be portable or permanent, providing support for hoisting equipment or raising and supporting lighting, cameras, and similar equipment.

Hauling line: The portion of a rope or chain on hoisting equipment that the operator uses to raise or lower the load. Also known as a hauling part.

Parts of line: The resulting number of lines that are supporting the load block when a line is reeved more than once.

Rigging isn't only about connecting loads to cranes. Riggers are also involved in equipment movement and placement inside of buildings and other locations where a crane is either unnecessary or impractical. This section presents various types of hoisting and jacking equipment.

3.1.0 Manual and Powered Hoisting Equipment

Both manual and powered hoisting equipment are used when it is necessary to lift components into position, or raise one component to insert something beneath it. Common hoisting equipment used includes block and tackle rigs, chain hoists, and ratchet-lever hoists.

3.1.1 Block and Tackle

The block and tackle is the most basic lifting device. It is used to lift or pull light loads. A block consists of one or more sheaves (pulleys) fitted into a wood or metal frame with a hook attached to the top. The tackle is the line or rope and end attachments connected to the block. Some block and tackle rigs have a brake that holds the load once it is lifted and others do not. The types that do not have a brake require continuous pull on the hauling line, or the hauling line must be tied off to hold the load.

There are two types of block and tackle rigs: simple and compound. A simple block and tackle consists of one sheave and a single line (*Figure 44*). It is used to lift or pull very light loads. The hook is attached to the load, and the load is lifted by pulling the line. The load capacity of this type of block and tackle is equal to the capacity of the load line. The block must be attached to a building structure or other support by a method that provides adequate load capacity to support the load and the tackle. Adequate capacity and stability of the supporting structure is crucial and it must be evaluated carefully by qualified personnel.

A compound block and tackle (*Figure 45*) uses more than one block. It has an upper, fixed block that is attached to the building structure or other support and a lower, traveling block that is attached to the load. Each block may have one or more sheaves. The more sheaves the blocks have, the more parts of line the block and tackle has, and the higher the lifting capacity. The compound block and tackle multiplies the power applied to the rope, so a worker can lift a much heavier load than is possible with a simple block and tackle.

3.1.2 Chain Hoists

A chain hoist, also called a *chain fall*, is a very useful and commonly used by a number of crafts. Chain hoists should be used for straight, vertical lifts only—just like a crane. They may be damaged if used for angled lifts or horizontal pulls. Although chain hoists are more popular, generally more durable, and capable of lifting heavier loads, there are electric and pneumatic cable hoists as well.

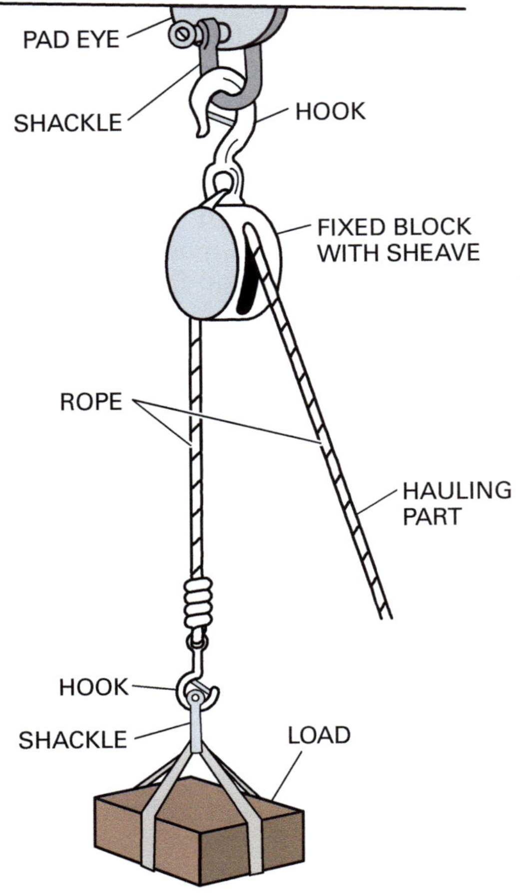

Figure 44 Simple block and tackle.

Figure 45 Compound block and tackle.

Chain hoists are standard equipment in many fabrication shops and rigging departments because they are dependable, portable, and easy to use. Some common types of chain hoists are the spur-geared (manual) chain and the electric (powered) chain.

The spur-geared chain hoist has two chains. An endless chain, the hand chain, drives a single pocketed sheave, which in turn drives a gear-reduction unit. The load chain is fitted to the gear-reduction unit and has a hook that attaches to the load. The gear-reduction unit provides the mechanical advantage, just as it does in many motor-driven conveyors and similar industrial systems. The spur-geared chain hoist is commonly used for significant loads and is convenient since it does not require a power source. Two common types are shown in *Figure 46*. The trolley-style chain hoist is designed to run on the rails of an I-beam, like the beam of a gantry.

A gantry (*Figure 47*) is often used indoors to hoist and move components. Gantries have maximum-load limitations just like chain hoists, but very large gantries with enormous capacities

Around the World

Block and Tackle History

Archimedes, a Greek mathematician, physicist, engineer, inventor, and astronomer, is credited with the development of the block and tackle around 250 BC. However, Archytas of Tarentum is thought to have developed the concept of a pulley—an essential component of the assembly to reduce friction. Block and tackle has been used in the construction of many structures that continue to fascinate us, including Stonehenge in the United Kingdom. Block and tackle systems are also considered to be the source of inspiration for modern cranes.

NCCER – *Mobile Crane Operations*

are used in shipping harbors and similar locations. These smaller versions are portable and very handy. A trolley-style chain hoist can be placed on the horizontal beam, or a separate trolley assembly can be purchased and a hook-type chain hoist can be attached to the trolley.

Electric chain hoists (*Figure 48*) work much like manual chain hoists, except that they have an electric motor instead of a pull chain to raise and lower the load. Electric chain hoists are faster and more efficient than manual chain hoists. Most are controlled by a pendant with pushbuttons, suspended from the hoist. There are a few battery-powered hoists on the market, but with limited capacity. As battery technology continues to advance, battery-operated hoists with more capacity may eventually be available. Most electric hoists operate on 120 and/or 240 VAC. Note that air-powered (pneumatic) chain hoists are also available.

Special care should be taken when raising a load with an electric chain hoist, since it is hard to tell how much force the electric motor is exerting. Motor sound may provide a clue. If the load gets caught on something immoveable, the chain hoist could be overloaded and damaged. A good electric chain hoist has reliable built-in overload protection that will open the motor circuit to prevent damage.

The load capacity of any chain hoist should be clearly marked on the data plate. The capacity should never be exceeded. Like a crane, it is necessary to know the weight of the load before beginning. Always use a sling to rig the load for a chain hoist if there is no single, dedicated lifting eye on the load. The chain of the hoist should never be wrapped around an object and used as a sling.

Like other lifting devices, chain hoists must be carefully inspected before each use. The chain must be checked to ensure that it has no significant defects. Before using a hoist, read the manufacturer's instructions and be aware of its limitations.

(A) HOOK-STYLE CHAIN HOIST

(B) TROLLEY-STYLE CHAIN HOIST

Figure 46 Manual chain hoists.

Figure 47 Portable gantry.

Figure 48 Electric chain hoist.

Follow these basic steps to select, inspect, use, and maintain a chain hoist:

Step 1 Select a chain hoist of adequate capacity to handle the load. Ensure that the overhead structure used to suspend the hoist also has adequate capacity.

Step 2 Inspect the load chain and hook to ensure that they are not excessively worn, bent, or deformed in any way.

Step 3 Inspect the sheaves to ensure that they are not bent or excessively worn.

Step 4 Ensure that the chain hoist has proper lubrication. Maintain per manufacturer's instructions.

Step 5 Hang the chain hoist from a suitable structure using the proper rigging. Provide an adequate power source for electric and pneumatic models, ensuring that it is routed to the hoist neatly and out of the way. The power cord or air hose should not be allowed to dangle straight down where it can interfere with the task.

Step 6 Position the hoist directly over the load. Lower the load hook, and connect it to the load using the proper rigging.

Step 7 Raise and place the load. To raise the load using a manual chain hoist, pull the hand chain. To raise the load using an electric chain hoist, press and hold the Up pushbutton on the handheld control.

Step 8 Disconnect the load hook from the load once it is in place.

Step 9 Remove the chain hoist from its support.

Step 10 Coil the chain so that it will not get tangled.

Step 11 Store the chain hoist in its proper place.

For detailed information about chain hoist inspection and operation, consult *ASME Standard B30.16, Overhead Hoists (Underhung).*

3.1.3 Ratchet-Lever Hoists and Come-Alongs

Ratchet-lever hoists and come-alongs (*Figure 49*) are used for short pulls on heavy loads. You can see by the length of the chain in the figure that the ratchet-lever hoist cannot accommodate a long pull. The term come-along is widely used to identify both tools, but they are not the same thing. A come-along uses a cable, whereas a ratchet-lever hoist uses a chain. Linemen often use ratchet-lever hoists with a flat synthetic strap. Ratchet-lever hoists are often designed and rated for vertical lifts, but cable-type come-alongs can only be used to pull horizontally. Do not use a come-along or ratchet-lever hoist for lifting unless you are certain it is rated by the manufacturer to do so.

Never use a come-along for vertical lifts. They do not have the same safety-braking mechanisms as ratchet-lever hoists to prevent the load from slipping.

Come-alongs and ratchet-lever hoists are portable, easy to use, and available in varying capacities. Always read and follow the manufacturer's instructions. Follow these steps to select, inspect, use, and maintain a ratchet-lever hoist:

Step 1 Select a hoist of adequate capacity to handle the load. Ensure that it is rated for vertical lifting.

Step 2 Inspect the chain and hooks to ensure that they are not excessively worn, bent, or deformed in any way. Inspect the device overall for significant damage or signs of being overloaded.

Step 3 Hang the device from a suitable structure, and ensure that structure is adequate for the load.

Step 4 Turn the ratchet release to the mid position.

NCCER – *Mobile Crane Operations*

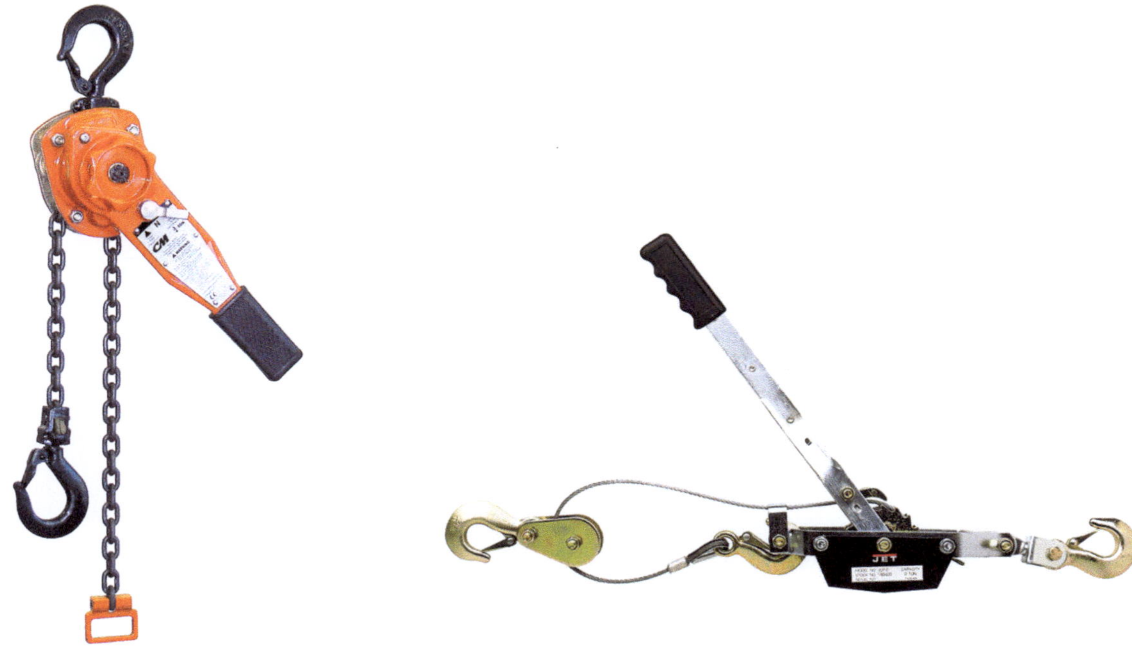

Figure 49 A ratchet-lever hoist and a come-along.

Step 5 Position the hoist directly over the load. Pull the chain out enough to attach it to the load.

Step 6 Attach the load hook to the load, using the proper rigging.

Step 7 Turn the fast-wind handle to take the slack out of the chain.

Step 8 Turn the ratchet control to the Up position.

Step 9 Pump the ratchet handle to raise the load.

Step 10 Turn the ratchet control to the Down position.

Step 11 Pump the ratchet handle to lower the load until there is slack in the chain.

Step 12 Disconnect the hook from the load.

Step 13 Dismount and store the hoist in its proper place.

3.2.0 Jacks

A jack is a device used to raise or lower equipment. Jacks are also used to move heavy loads a short distance, with good control over the movement. The following are the three basic types of jacks:

- Ratchet
- Screw
- Hydraulic

3.2.1 Ratchet Jacks

The ratchet jack (*Figure 50*), also called a *railroad jack*, is used to raise loads under 25 tons. It uses the lever-and-fulcrum principle. The downward stroke of the lever raises the rack bar one notch at a time. A latching mechanism, called a pawl, automatically springs into position, holding the load and releasing the lever for the next lifting stroke. They can lift full jack capacity on the toe or on the cap. Ratchet jacks are rated by lifting capacity and the length of their stroke, or lifting distance.

3.2.2 Screw Jacks

Screw jacks (*Figure 51*) are used to lift heavier loads than ratchet jacks, but more slowly. A simple screw jack uses the screw-and-nut principle to lift the stem. A simple lever is placed into the hole and turned to raise or lower the threaded stem. For heavier loads and jacks, a gear-reduction unit is used to reduce the amount of power required to turn the nut.

3.2.3 Hydraulic Jacks

Hydraulic jacks (*Figure 52*) are operated by the pressure of pumped fluid. These jacks can lift a surprising amount of weight. Simple bottle jacks have a hand-operated lever to pump the jack and raise it. A bypass valve is opened to allow fluid to flow out of the cylinder and lower the jack.

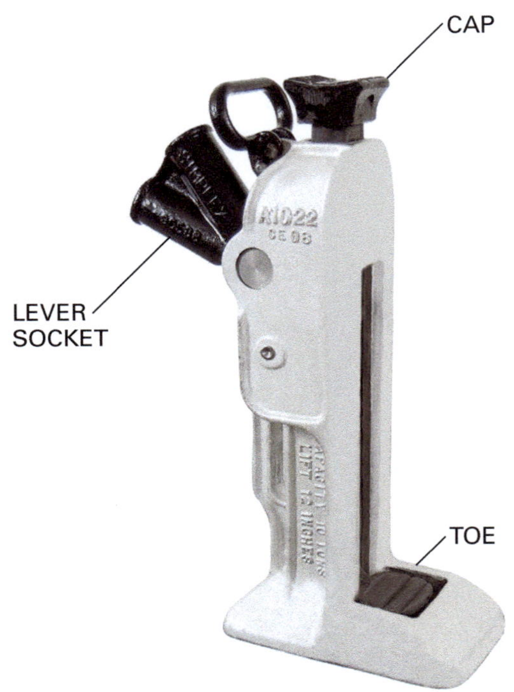

Figure 50 Ratchet jack.

Figure 51 Screw jack.

Toe jacks are a different version of bottle jacks that provide a lifting toe at the base. The toe can be pushed under a load that is much closer to the floor, while a bottle jack requires a lot of clearance to fit under the load.

Jacks with higher lifting capacities use an external pump. The pump is connected to the jack by a hose and a quick-disconnect coupling. The pump can be hand-operated, foot-operated, electric, air-operated, or even gasoline-engine driven. It is not uncommon to see portable hydraulic jacks that are rated for 1,000 tons.

3.2.4 Jack Use and Maintenance

In time, a jack can become damaged or worn and fail under a load. To avoid such failures, all jacks should be carefully inspected before each use. Apply the following general guidelines when using jacks:

- Inspect jacks before using them to ensure that they are not damaged in any way. For jacks that use an external pump, also inspect the pump, hoses, and couplings carefully. The connecting hose is the most fragile part of the system and is subject to cuts and deep scarring during normal use.

- Thoroughly clean all hydraulic hose connectors before connecting them.
- Never exceed the load capacity of the jack.
- Use wood softeners when jacking against metal.
- Never place jacks directly on earth when lifting; provide a solid footing.
- Position jacks so the direction of force is perpendicular to the base and the surface of the load. Then raise the load evenly to prevent the load from shifting or falling.
- Use the proper jack handle, and remove it from the jack when it is not in use. Do not use extensions to the jack handles. If the added power of an extension is necessary, the wrong jack is being used.
- Never step on a jack handle to create additional force; use a foot-operated pump and an appropriate jack.
- Always apply blocking or cribbing under a raised load when jacking. Never leave a jack under a load without having the load blocked up so that it will not fall if the jack fails suddenly.
- Brace loads to prevent the jacks from tipping.
- Lash or block jacks when using them in a horizontal position to move an object.
- Never jack against any kind of roller or wheel.
- Match multiple jacks for uniform lifting of a single object.

NCCER – *Mobile Crane Operations*

Figure 52 Hydraulic jacks.

Additional Resources

29 *CFR* 1926.251, **www.ecfr.gov**

Willy's Signal Person and Master Rigger Handbook. Current edition. Ted L. Blanton, Sr. Altamonte Springs, FL: NorAm Productions, Inc.

North American Crane Bureau, Inc. website offers resources for products and training, **www.cranesafe.com**

3.0.0 Section Review

1. The mechanical advantage gained by using a spur-geared chain hoist is provided by _____.

   a. multiple parts of chain
   b. a trolley
   c. an operating lever
   d. the gear-reduction unit

2. Which of the following is a true statement about the use of jacks?

   a. If a load is difficult to lift, you can use your foot to operate the jack handle.
   b. Always apply blocking or cribbing under a raised load.
   c. The capacity of a hydraulic jack is only limited by the hydraulic pressure applied.
   d. A toe jack is just a different version of a screw jack.

Summary

Selecting and setting up hoisting equipment, hooking cables to the load to be lifted or moved, and helping guide the load into position are all part of the rigging process. Performing this process safely and efficiently requires the rigger to properly select equipment, use it in the correct way, understand safety hazards, and know how to prevent accidents while working efficiently.

Riggers must have great respect for the hardware and equipment of their trade because their lives may depend on them functioning correctly.

Although they are not required to be rigging professionals, crane operators must also understand rigging practices and be able to distinguish between proper and improper methods for the safety of all jobsite personnel.

1. Threaded blind holes used with eyebolts should have a minimum depth of ______.

   a. 1¼ times the bolt diameter
   b. 1½ times the bolt diameter
   c. 1¾ times the bolt diameter
   d. 2 times the bolt diameter

2. A device used as a connection point for a load that has no reduction in rated load when an angular pull is applied is a ______.

   a. shouldered eyebolt
   b. swivel hoist ring
   c. rigging hook
   d. shoulderless eyebolt

3. Custom-fabricated spreader and equalizer beams are tested at ______.

   a. 125 percent of their rated load
   b. 200 percent of their rated load
   c. 250 percent of their rated load
   d. 350 percent of their rated load

4. To keep them in place while the sling is positioned, manufactured protective materials for slings may be equipped with ______.

   a. adhesive strips
   b. magnets
   c. hook-and-loop fasteners
   d. wooden edges

5. A wire rope sling should be removed from service if localized abrasion and scraping has reduced the diameter more than ______.

   a. 25 percent of the original rope diameter
   b. 15 percent of the original rope diameter
   c. 10 percent of the original rope diameter
   d. 5 percent of the original rope diameter

6. When the eyes of a synthetic web sling are sewn at right angles to the plane of the sling body, it is called a(n) ______.

   a. round sling
   b. endless sling
   c. standard eye-and-eye sling
   d. twisted eye-and-eye sling

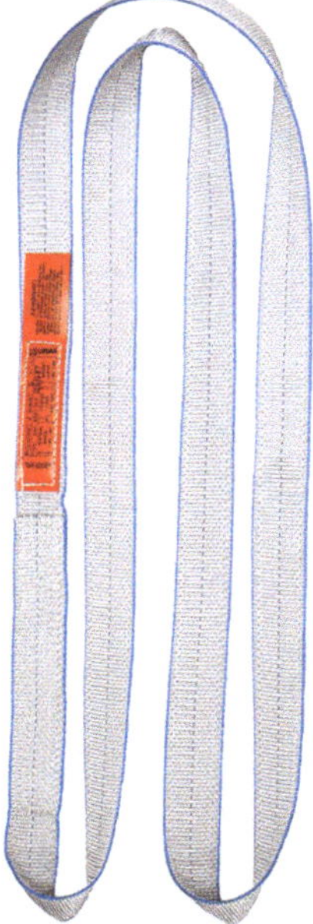

Figure RQ01

7. The type of sling shown in *Figure RQ01* is a(n) ______.

   a. standard eye-and-eye
   b. twisted eye-and-eye
   c. endless
   d. round

8. Which of the following is a disadvantage of chain slings when compared to wire-rope slings?

   a. Chain slings do not handle abrasion and corrosion as well as wire rope.
   b. Chain slings cannot be used in high-heat applications.
   c. Chain slings have less reserve strength and are more likely to fail without warning.
   d. Chain slings are more easily damaged by sharp corners than wire rope.

9. Which of the following slings is best suited for situations where the load is abrasive, hot, and/or tends to cut?

   a. Wire-rope sling
   b. Metal-mesh sling
   c. Synthetic web sling
   d. Endless grommet sling

10. Which of the following types of damage to a metal-mesh sling is allowed to be more extensive than the other before it is removed from service?

a. Abrasion
b. Corrosion

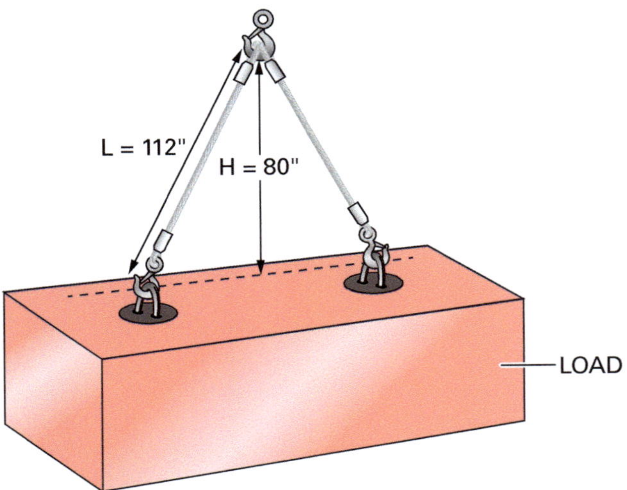

Figure RQ02

11. Refer to *Figure RQ02*. Based on the measurements shown, what is the sling angle factor that should be applied to the weight of the lifted load on each sling?

a. 0.902
b. 1.003
c. 1.213
d. 1.400

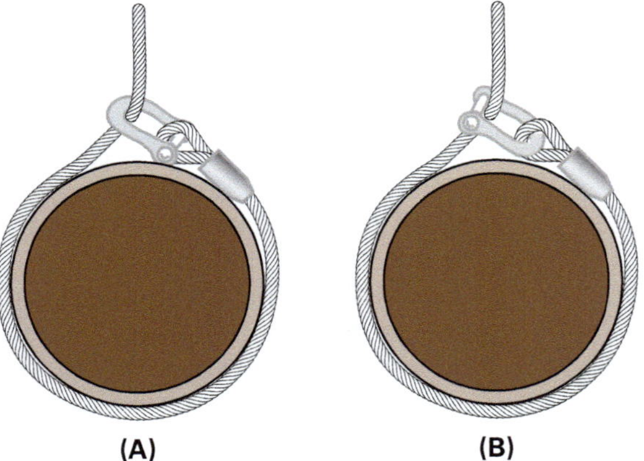

Figure RQ03

12. Refer to *Figure RQ03*. Which of the two drawings, A or B, represent the proper way to orient the shackle?

a. Drawing A
b. Drawing B

13. Regardless of how a clove hitch is actually tied, the result is the same as ______.

a. two bowlines tied one after another
b. the combination of a bowline and a half hitch
c. two half hitches made in the same direction
d. two half hitches made in opposite directions

14. When the weight of a load cannot be determined in other ways, it can be lifted slightly while monitoring the weight using instruments in the crane. The lift can proceed if the load is less than ______.

a. 40 percent of the crane's capacity at the longest required operating radius
b. 75 percent of the crane's capacity at the longest required operating radius
c. 85 percent of the crane's capacity at the longest required operating radius
d. 100 percent of the crane's capacity at the longest required operating radius

15. Which of the following is a correct statement?

a. All come-alongs are rated and best used for vertical lifts.
b. A ratchet-lever hoist can accommodate any length of chain.
c. A come-along uses a cable while a ratchet-lever hoist uses a chain.
d. Come-alongs and ratchet-lever hoists use the same type of braking mechanisms.

NCCER – *Mobile Crane Operations*

Trade Terms Quiz

Fill in the blank with the correct term that you learned from your study of this module.

1. Damage that occurs to wire rope identified by the separation of strands that then balloon outward like vertical bars is called __________.

2. The type of wire rope construction considered to be the most durable for rigging applications is the __________ type.

3. A(n) __________ is attached to a lifted load for the purpose of controlling load spinning and swinging.

4. Rigging devices used to distribute the weight of a load on multi-crane (tandem) lifts are called __________.

5. If a hole does not fully penetrate the material, resulting in a hole with a bottom, it is referred to as a(n) __________.

6. The portion of a hook directly below the center of the lifting eye is called the __________.

7. A common hitch made by passing a sling through a load or connection and attaching both sling eyes to the hoist line is called a(n) __________.

8. The __________ identifies the amount of stress required to bring a rigging component to its breaking point.

9. The terms *safe working load, working load limit,* and *rated capacity* are all synonyms for __________.

10. A simple hitch that uses one end of a sling to connect to a point on the load and the opposite end to connect to the hoist line is the __________.

11. The angle formed by the legs of a sling with respect to the horizontal plane when tension is placed on the rigging is referred to as the __________.

12. A(n) __________ is made by passing a sling around the load and then passing one eye of the sling through the other.

13. A(n) __________ is comprised of 2 or more single-leg hitches that is used for lifting objects equipped with lifting lugs or similar points of connection.

14. Rigging devices that are often used when the object being lifted is too long or large to be lifted from a single point, or when the use of slings around the load may crush the sides, are __________.

15. __________ have three or more holes in them and are used to level loads when sling lengths are not equal.

16. Plates with two holes, referred to as __________, are used as termination hardware to connect rigging to specific lift points.

17. A hook should always be positioned over a load's __________.

18. A major load of pipe or similar materials might be delivered to a site on a __________.

19. A chain hoist or ratchet lever hoist used for lifting can be suspended under a __________.

20. The __________ have a lot to do with the lifting assistance provided by a block and tackle.

21. To raise a load using a block and tackle, the user must pull on the __________.

Trade Terms

Basket hitch	Equalizer beams	Minimum breaking strength (MBS)	Sling angle
Bird caging	Equalizer plates	Parts of line	Spreader beams
Blind hole	Gantry	Rated load	Spur track
Bridle hitch	Hauling line	Rigging links	Tagline
Center of gravity (CG)	Independent wire rope core (IWRC)	Saddle	Vertical hitch
Choker hitch			

Harold "Ed" Burke

Rigging Training Specialist,
Mammoet USA South

Please give a brief synopsis of your construction career and your current position.

I began working as a rigger with Mammoet in 1997. I had never been around cranes before, so I was green, for sure. I was fortunate, though, to learn and grow under the guidance of some of the best riggers in the industry—people like Dean and Mark Bell, Jeff Bland, and Jason Marks, just to name a few of those I respect and owe a debt of gratitude. I've been a part of some great projects across the United States and abroad, such as The Netherlands, Belgium, Italy, Taiwan, and Chile. In October 2013, I was approached by our safety director and asked if I would consider serving as a trainer for Mammoet USA. I accepted and I'm proud to still be a part of it.

How did you get started in the construction industry?

I had been doing general industry work since high school, with some time working in residential construction. I was fortunate to get an opportunity with Mammoet through a connection there. When I was hired, the same person told me, "I got you the job—you'll have to keep it."

Who or what inspired you to enter the construction industry?

Honestly, it was exactly what I needed at that point in my life. I was looking to better myself. When I started with Mammoet, I made more money for the remaining eight months of the year than in any full year before that. The earning potential was a key factor, but I really needed some positive changes in my life, and the construction industry surely provided that.

How has training in construction impacted your life and career? What types of training have you completed?

When I got into crane and rigging in the late 1990s, there was not so much emphasis on formal training. Most training happened through mentoring on the job. The industry has changed dramatically in the 20 years since I became involved. I really believe quality training contributes to a safer and more efficient work force. Knowledge truly is power; I say that all the time.

Why do you think credentials are important in the construction industry?

I believe valid credentials provide benchmarks, showing where an individual's skill is in relation to an accepted standard of competency. Credentials provide a glimpse of their knowledge foundation on a given topic or craft.

What do you enjoy most about your career?

The people, without a doubt. I have had the pleasure of working with some of the best people across the world from my days in the field. I continue to work with some of the best in my training environment today. I wouldn't be an effective educator without the people who have been a part of my career and growth as a craft professional. If you surround yourself with great professionals, you have a much better chance of becoming one yourself.

Would you recommend construction as a career to others? Why?

Absolutely! Why, you ask? Not everyone is suited for the college experience or for a career in an office. The earning potential of young men and women graduating from high school who are willing to work hard, learn, and be an asset to their chosen craft is really fantastic. They can start to build real wealth immediately instead of taking on crippling college-loan debt.

What advice would you give to someone who is new to the construction industry?

I see new hires all the time, and those that are also new to the industry. What I tell them is this: Show up early, stay late, do more than is expected, be the first to volunteer, learn all you can by keeping your ears open and your mouth closed, and do a good enough job to convince your supervisor he can't do the next job without you. You must build a reputation for yourself, and it must be a good one.

How do you define craftsmanship?

Many would say that it is skill in a particular craft, and I agree with that. But there is a noticeable level of pride and passion evident in the work and attitude of a true craftsman. You really have to love what you do to be happy and productive, and you pour yourself into it to be a craftsman. That's my take on it.

Trade Terms Introduced in This Module

Basket hitch: A common hitch made by passing a sling around a load or through a connection and attaching both sling eyes to the hoist line.

Bird caging: A deformation of wire rope that causes the strands or lays to separate and balloon outward like the vertical bars of a bird cage.

Blind hole: A hole that does not penetrate the material completely, leaving a hole with a bottom.

Bridle hitch: A type of hitch comprised of 2 or more single-leg hitches, used for lifting objects equipped with lifting lugs or other points of connection.

Center of gravity (CG): The point at which the entire weight of an object is considered to be concentrated, such that supporting the object at this specific point would result in its remaining balanced in position.

Choker hitch: A hitch made by passing a sling around the load, and then passing one eye of the sling through the other. The one eye is then connected to the hoist line, creating a choke-hold on the load.

Equalizer beams: Beams used to distribute the load weight on multi-crane lifts. The beam attaches to the load below, with two or more cranes attached to lifting eyes on the top.

Equalizer plates: A type of rigging plate that has three or more holes, used to level loads when sling lengths are unequal.

Gantry: A framed overhead structure supported by legs on each end, used to cross over obstructions. Gantries can be portable or permanent, providing support for hoisting equipment or raising and supporting lighting, cameras, and similar equipment.

Independent wire rope core (IWRC): Wire rope with a core consisting of wire rope, as opposed to a fiber or single-stranded core; considered to be the most durable for rigging applications.

Hauling line: The portion of a rope or chain on hoisting equipment that the operator uses to raise or lower the load. Also known as a *hauling part*.

Minimum breaking strength (MBS): The amount of stress required to bring a rigging component to its breaking point. The MBS is a factor in determining a components' rated load capacity.

Parts of line: The resulting number of lines that are supporting the load block when a line is reeved more than once.

Rated load: The maximum working load permitted by a component manufacturer under a specific set of conditions. Alternate names for rated load include *working load limit (WLL)*, *rated capacity*, and *safe working load (SWL)*.

Rigging links: Links or plates with two holes used as termination hardware to appropriate lifting points.

Saddle: The portion of a hook directly below the center of the lifting eye.

Sling angle: The angle formed by the legs of a sling with respect to the horizontal plane when tension is placed on the rigging.

Spreader beams: Beams or bars used to distribute the load of a lift across more than one point to increase stability. Spreader beams are often used when the object being lifted is too long or large to be lifted from a single point, or when the use of slings around the load may crush the sides.

Spur track: A relatively short branch leading from a primary railroad track to a destination for loading or unloading. A spur is typically connected to the main at its origin only (a dead end).

Tagline: A rope attached to a lifted load for the purpose of controlling load spinning and swinging, or used to stabilize and control suspended attachments.

Vertical hitch: A simple hitch that uses one end of a sling to connect to a point on the load and the opposite end to connect to the hoist line. Also known as a *straight-line hitch*.

Additional Resources

This module presents thorough resources for task training. The following reference material is recommended for further study.

ASME Standard B30.5, Mobile and Locomotive Cranes. Current edition. New York, NY: American Society of Mechanical Engineers.

ASME Standard B30.9, Slings. Current edition. New York, NY: American Society of Mechanical Engineers.

ASME Standard B30.10, Hooks. Current edition. New York, NY: American Society of Mechanical Engineers.

ASME Standard B30.16, Overhead Hoists (Underhung). Current edition. New York, NY: American Society of Mechanical Engineers.

ASME Standard B30.20, Below-The-Hook Lifting Devices. Current edition. New York, NY: American Society of Mechanical Engineers.

ASME Standard BTH-1, Design of Below-The-Hook Lifting Devices. Current edition, New York, NY: American Society of Mechanical Engineers.

29 CFR 1926, Subpart CC, **www.ecfr.gov**

29 CFR 1926.251, **www.ecfr.gov**

29 CFR 1926.753, **www.ecfr.gov**

Mobile Crane Operations Level One, NCCER. Third Edition. 2018. New York, NY: Pearson Education, Inc.

NCCER Module 00106-15, *Introduction to Basic Rigging*.

Mobile Crane Safety Manual (AEM MC-1407). 2014. Milwaukee, WI: Association of Equipment Manufacturers.

Willy's Signal Person and Master Rigger Handbook, Ted L. Blanton, Sr. Current edition. Altamonte Springs, FL: NorAm Productions, Inc.

Knots: The Complete Visual Guide, Des Pawson. First American Edition. 2012. New York, NY: DK Publishing.

The following websites offer resources for products and training:

Occupational Safety and Health Administration (OSHA), **www.osha.gov**

Electronic Code of Federal Regulations, **www.ecfr.gov**

North American Crane Bureau, Inc. website offers resources for products and training, **www.cranesafe.com**

Figure Credits

Link-Belt Construction Equipment Company, Module opener

Columbus McKinnon Corporation, Figures 1, 4A, 4C, 46, 48, 49A

Konecranes Americas, Inc., SA01

Courtesy of The Crosby Group LLC, Figures 4B, 27

J. C. Renfroe & Sons, Figure 9

Vaculift™, Inc. d.b.a. Vacuworx®, SA02

Lift-All Company, Inc., Figures 14, SA03, 19, 20, 23, Review Question Figure 1, Exam Figure 1

Lift-It Manufacturing Co., Inc., Figure 17A

Linton Rigging Gear Supplies LLC - www.lrgsupplies.com, Figure 17B

Mazzella Companies, Figure 18

Ed Gloninger, Figure 21

© Donvictorio/Dreamstime.com, Figure 35

© iStock.com/krungchingpixs, Figure 36A

© iStock.com/Cajarima01, Figure 36B

Insulatus Company, Inc., Figure 37

© a katz/Shutterstock.com, SA05

© Steve Norman/Shutterstock.com, Figure 45

Vestil Manufacturing, Figure 47

Walter Meier Manufacturing Americas, Figures 49B, 51, 52A, 52B, Exam Figure 3

Photos courtesy of Enerpac, Figures 50, 52C, 52D

Answer	Section Reference	Objective
Section One		
1. a	1.1.2	1a
2. a	1.2.4	1b
Section Two		
1. c	2.1.1	2a
2. a	2.2.1	2b
3. c	2.3.2	2c
4. b	2.4.2	2d
5. a	2.4.2	2d
6. b	2.5.0	2e
7. d	2.6.0	2f
8. c	2.7.0	2g
Section Three		
1. d	3.1.2	3a
2. b	3.2.4	3b

Section Review Calculations

2.0.0 SECTION REVIEW

Question 4

Divide the length (L) by the height (H) to determine the sling angle factor:

Sling angle factor = L ÷ H
Sling angle factor = 74" ÷ 62"
Sling angle factor = 1.194

The sling angle factor is **1.194**.

NCCER CURRICULA — USER UPDATE

NCCER makes every effort to keep its textbooks up-to-date and free of technical errors. We appreciate your help in this process. If you find an error, a typographical mistake, or an inaccuracy in NCCER's curricula, please fill out this form (or a photocopy), or complete the online form at **www.nccer.org/olf**. Be sure to include the exact module ID number, page number, a detailed description, and your recommended correction. Your input will be brought to the attention of the Authoring Team. Thank you for your assistance.

Instructors – If you have an idea for improving this textbook, or have found that additional materials were necessary to teach this module effectively, please let us know so that we may present your suggestions to the Authoring Team.

NCCER Product Development and Revision

13614 Progress Blvd., Alachua, FL 32615

Email: curriculum@nccer.org
Online: www.nccer.org/olf

❏ Trainee Guide   ❏ Lesson Plans   ❏ Exam   ❏ PowerPoints   Other ___________________

Craft / Level: ____   Copyright Date: ______________

Module ID Number / Title: ____

Section Number(s): ____

Description: ____

____

____

____

Recommended Correction: ____

____

____

____

____

Your Name: ____

Address: ____

____

Email: ____   Phone: ____________________

Crane Communications

OVERVIEW

This module focuses on the methods and modes of communication required in crane operations. General information about the communication process is also presented to help workers better understand the mechanics of communication in all environments. Signal persons are relied upon to properly communicate both verbally and nonverbally with crane operators, and crane operators must learn how to interpret verbal messages and hand signals provided by a signal person.

Module 53101

CRANE COMMUNICATIONS

Objectives

When you have completed this module, you will be able to do the following:

1. Describe the communication process and identify barriers to effective communication.
   a. Describe the basic communication process.
   b. Identify common barriers to effective communication.
2. Identify and interpret the OSHA regulations related to crane communications and explain how to communicate with crane operators verbally and nonverbally.
   a. Identify and interpret construction-related OSHA regulations associated with crane communications and signaling.
   b. Describe the equipment used for verbal communications and how to communicate with and direct a crane operator verbally.
   c. Explain how to communicate with and direct a crane operator nonverbally.

Performance Tasks

Under the supervision of your instructor, you should be able to do the following:

1. Demonstrate proper crane-communication techniques using a handheld radio or another acceptable verbal-signaling device.
2. Demonstrate each standard hand signal depicted in 29 *CFR* 1926.1400, Subpart CC, Appendix A.
3. Direct an operator to move and place a load using the appropriate hand signals.
4. Direct an operator to move and place a load using voice communication.

Trade Terms

Abstraction
Blind lift
Bridge
Consensus standard
Dedicated spotter
Diver tender

Line of sight
Nonverbal communication
Open mike
Paraphrasing
Trucks

Industry Recognized Credentials

If you are training through an NCCER-accredited sponsor, you may be eligible for credentials from NCCER's Registry. The ID number for this module is 53101. Note that this module may have been used in other NCCER curricula and may apply to other level completions. Contact NCCER's Registry at 888.622.3720 or go to **www.nccer.org** for more information.

Contents

Figures

SECTION ONE

1.0.0 THE COMMUNICATION PROCESS

Objective

Describe the communication process and identify barriers to effective communication.

a. Describe the basic communication process.
b. Identify common barriers to effective communication.

Trade Terms

Abstraction: Any form of verbal, graphical, or written communication representing a generalized and nonspecific idea or quality of a thing, action, or event.

Nonverbal communication: All communication that does not use words. This includes appearance, personal environment, use of time, and body language.

Paraphrasing: Expressing the perceived meaning of something read or heard in one's own words, generally to ensure clarity. Paraphrasing is an important component of active listening.

The ability of workers to communicate effectively is essential to the safe operation of cranes. The clarity of the information being exchanged is very important, and can be difficult to maintain due to distractions and noise that accompany crane operation and other construction activities. The techniques presented in this module can assist anyone that provides or responds to crane operating signals in strengthening their ability to communicate effectively.

Before presenting the detailed requirements of crane signaling and communication, it is beneficial to understand the general communication process. This knowledge helps workers understand how and why information may not be accurately transferred and take the necessary steps to avoid problems. The sections that follow provide general information about communication that are valuable both on and off the job.

1.1.0 Exchanging a Message

The communication process (*Figure 1*) consists of a message being sent and, ideally, received and interpreted precisely as intended. A message may be sent verbally or through nonverbal communication.

The challenge for those involved in operating cranes is not only to communicate the right information to co-workers, but to do so effectively.

The communication process consists of the following three general components:

- Sending the message
- Receiving the message
- Feedback

1.1.1 Sending the Message

There are four elements involved in sending a message. First, the sender prepares the message that is intended for communication. Next, the sender considers possible internal barriers that may affect the message. This includes the sender's own experiences, the terms used, and even the sender's feelings toward the receiver. External barriers, such as noise, must also be considered. Third, the sender encodes the message; that is, the sender puts the message into spoken words (verbal), written words (nonverbal), or gestures (nonverbal) that he or she wants to use. Finally, the sender sends the message through the chosen method.

1.1.2 Receiving the Message

There are also four elements involved in receiving a message. The receiver will first hear and/or see the message that was sent. Second, the influence of any active external or internal barriers takes effect. Possible internal barriers, for example, may include the receiver's experience level, the receiver's understanding of the terms used by the sender, the receiver's attitude toward the job, or even the way the receiver feels about the sender. Third, the receiver often decodes the

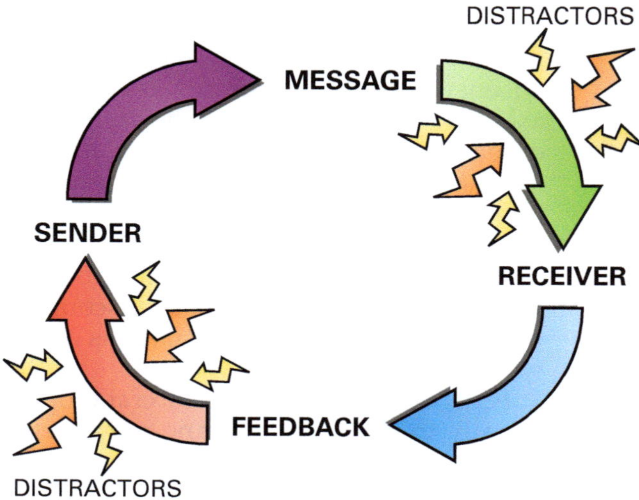

Figure 1 The communication process.

---

message through a rapid set of mental images. For instance, when a sender speaks the term *two-block*, a receiver does not generally imagine the letters that form the word. Instead, a mental image represented by the words appears. Many different mental images might arise among those who hear this word—perhaps a crane boom, a wire rope, or a load block might be briefly pictured. Fourth, the message is interpreted from the receiver's perspective. At this point in the process, there is no way to determine that the message was received as intended. To determine this, the receiver must provide feedback to the sender.

1.1.3 Feedback

Feedback, which may take several forms in general communication, provides essential information about the success in communicating a message. To get feedback, ask or look for a response. A simple yes or no may be all that is needed for simpler messages. Other times, a question may be asked in response to the message, indicating a lack of understanding or a need to verify the information. Nonverbal behaviors can also provide important clues about the clear reception of a message. Facial expressions and body language often indicate when the receiver clearly understands the message or is unsure of its meaning. Even a nonverbal response such as a nod of the head or a thumbs-up signal can be considered feedback.

Paraphrasing is a communication tool that often helps ensure clear communication between a sender and a receiver. It can be used when the message is not clear, or when verification is desired to avoid a mistake. Paraphrasing means to express the perceived meaning of something read or heard back to the sender using different words. For example, another worker may say, "I don't understand that guy. Some days he is on fire, but he's asleep at the wheel on other days." To ensure understanding, the other party may ask, "So you're not sure if he is consistently focused on his job?" This provides an opportunity for the receiver to find out if the message was interpreted correctly, and provides feedback to the first worker about how the message was interpreted. Paraphrasing is a valuable tool that can be used to reduce or eliminate miscommunication in many cases, including situations outside of the workplace.

Feedback to a message in crane operations is usually in the form of crane movement. Ideally, the crane operator repeats a signal back to the sender to ensure accuracy. The incorrect response to a signal may be uneventful, but it can also be

disastrous. For this reason, crane signaling goes further to formally close the communication loop with a three-step process:

- The signal person signals the crane operator.
- The operator repeats the signal back to the signal person.
- The signal person confirms that the crane operator has interpreted the signal correctly.

The terms and signals used to communicate with crane operators must be learned and practiced until they are deeply embedded in memory and are as natural to use as any other part of speech.

1.2.0 Communication Barriers

The purpose of effective communication in any environment is to ensure that the receiver accurately understands the material or information provided by the sender. The existence of certain barriers or filters increases the potential for poor communication. A barrier is something that can halt communication altogether. An example would be a jackhammer starting up nearby when you are speaking to someone. A filter is an internal, personal screening mechanism that receives a message and alters its meaning in some way. A filter can be a person's knowledge or prior experience, or even emotions. For example, if a signal person has sent the same signal incorrectly to a crane operator seven times in a row, the operator most likely will develop a filter that renders future uses of the same signal untrustworthy. By being aware of barriers and filters, lift teams can avoid them and ensure clear understanding of the information they are attempting to communicate. The following sections describe some of the barriers and filters that can damage effective communications, including those related to crane operations.

1.2.1 Lack of Common Experience

One potential barrier to effective communication is the lack of common experience. Operators are likely to find that their co-workers have many different backgrounds. Some individuals may have worked exclusively in one type of construction, such as bridge/road, structural steel, or residential, and are unable to immediately relate to another area of construction. To prevent misunderstandings, the sender and receiver first need to determine their experience level. Then, the sender should develop comparisons between the different types of work. For example, an operator who is accustomed to working on a building construction site might not understand some of the terminology used on bridge projects. Indeed, the word *bridge* itself has a completely different meaning in the world of overhead cranes than it does in bridge construction.

1.2.2 Language

Today's construction workplace is a reflection of the global marketplace. Some employees may not be fluent in the principal language of the jobsite or organization. A language barrier can certainly complicate the process of communication in any environment. The challenge of clear communication between individuals that do not share a native language is very common today, and a challenge that will likely continue to grow. Even when a non-native speaker has advanced knowledge of the vocabulary in a language, their pronunciation of some words may be difficult for a native speaker to understand.

One option is for all workers to become fluent in all the languages spoken on the job. Of course, that is an impractical solution. However, it is beneficial to understand some of the basic concepts and words of the language(s) you hear on the job regularly. A few words spoken in the language of the listener may be all it takes to clarify the instruction or statement.

Signal persons and crane operators must be fluent in a standardized language that is commonly understood by both. Using the standardized crane hand signals as outlined in 29 *CFR* 1926, Subpart CC, Appendix A is a reliable method of bridging language barriers in crane operations.

1.2.3 Overuse of Abstractions

An **abstraction** is a generalized, nonspecific concept or idea. To avoid confusion, speak in common, accepted, standardized terms, and be specific. Many abstractions are too general in nature, and therefore do not provide enough detail to the receiver. Be aware of the abilities of your co-workers and use appropriate terms and examples that suit their communication needs. Remember that signal persons are also responsible for the safe operation of the crane, so it is crucial that all communications between the operator and signal person are clearly understood. For example, if a signal person tells an operator to move a load "a little to the left"—an example of an abstraction—one operator may move it a few inches while another may move it several feet. The correct verbal instruction would include how far the load needs to be moved to the left. This removes the abstraction.

1.2.4 Fear

Fear may be one of the greatest barriers to effective communication. The fear of showing ignorance, the fear of disapproval, the fear of losing status, and the fear of judgment are all common barriers. Co-workers may have hidden anxieties or fears about their own abilities. They may lack confidence or be afraid of appearing ignorant, thus avoiding communication when they should speak out.

Around the World

Hand Signals for Everyone

More than ever before, the world is a global economy supported by a global workforce. Workers from China operate in the United States, workers from the United States operate in Brazil, and workers from Brazil might operate in Italy. This is true for crane operators as well. If someone has a taste for travel, an opportunity to serve as a crane operator can be found in almost any country.

Although it is certainly helpful, an English-speaking crane operator is not expected to master Portuguese, French, or any other language to function on the job. The most important language to know as a crane operator in a foreign country, regardless of your native language, is the language of hand signals.

The International Organization for Standardization (ISO) develops many different standards that help to ensure that products and services are safe and reliable, regardless of their origin. ISO is engaged in the world of crane operations through *ISO Standard 16715, Cranes—Hand Signals Used with Cranes*. The goal is to create a standard of hand signals for jobsites with an international workforce. If you plan to work outside of the Unites States, this is an important standard to be familiar with. In the United States, 29 *CFR* 1926, *Subpart CC, Appendix A* is the US standard for crane hand signals.

Provide a threat-free environment by being encouraging and nonjudgmental. It is important that communication flow smoothly. During meetings of all kinds, including pre-lift meetings, encourage everyone to ask questions without fear of judgment and ridicule.

Fear can affect communication in crane operations as well. When confidence is lacking, a worker may avoid displaying the Stop or Emergency Stop signal on the job when a hazard clearly exists or is developing. A signal person may also hesitate to display a common signal because the worker is not confident it is the correct signal. This is yet another reason why crane signals must be learned and practiced until they become second nature. Signal persons must be prepared to stop a lift at any time if things are not progressing as expected.

1.2.5 Environmental Factors

Environmental factors such as noise and weather often interfere with the communication process on the job. Pile-driving operations (*Figure 2*) are a perfect example because the noise involved can be a barrier to hearing what is being said. In inclement weather, it may be difficult to see clearly or listen attentively. In good weather, environmental factors such as the sun in your eyes can pose a problem. However, subtle factors may also affect the communication process. The background color of a wall or structure, the position of the signal person, and the location of the load in relation to the signal person can all interfere with effective communication.

Both signal persons and crane operators should constantly strive to identify and eliminate interference from the communication process. As stated earlier, clarity in communication and adhering to standard signal sets is a crucial part of crane operations, since the feedback is usually in the form of crane movement. As a result, miscommunication can lead to equipment damage and/or personal injury.

Figure 2 Pile-driving operations.

NCCER – *Mobile Crane Operations*

Interplay: The Process of Interpersonal Communication, Ronald Adler, Lawrence Rosenfeld, and Russell Proctor. 13th Edition. New York, NY: Oxford University Press.

NCCER Module 00107, *Basic Communication Skills*.

1.0.0 Section Review

1. Which of the following is an example of an internal barrier that may influence the receipt of a message?

   a. Noise on a jobsite
   b. The receiver's attitude toward the job
   c. Glare on the windshield of the crane
   d. The weather

2. Which of the following is a true statement?

   a. A barrier may affect how a message is interpreted, but a filter keeps it from being heard.
   b. The experience level of an individual does not represent any sort of communication barrier.
   c. A non-native speaker of a language may have problems with accurate pronunciation, even with an extensive knowledge of the vocabulary.
   d. Learning all the languages spoken by non-native speakers on a jobsite is a practical solution to a language barrier.

2.0.0 CRANE COMMUNICATIONS

Objective

Identify and interpret the OSHA regulations related to crane communications and explain how to communicate with crane operators verbally and nonverbally.

a. Identify and interpret construction-related OSHA regulations associated with crane communications and signaling.
b. Describe the equipment used for verbal communications and how to communicate with and direct a crane operator verbally.
c. Explain how to communicate with and direct a crane operator nonverbally.

Performance Tasks

1. Demonstrate proper crane-communication techniques using a handheld radio or another acceptable verbal-signaling device.
2. Demonstrate each standard hand signal depicted in 29 *CFR* 1926.1400, Subpart CC, Appendix A.
3. Direct an operator to move and place a load using the appropriate hand signals.
4. Direct an operator to move and place a load using voice communication.

Trade Terms

Blind lift: Any lift involving a load that is out of the direct view of the operator. Blind lifts are generally always categorized as critical lifts.

Bridge: In relation to overhead cranes, the part of an overhead crane consisting of one or more girders or beams and the supporting trucks. The bridge is the overhead, weight-bearing structure along which the trolley(s) and load block assembly travels.

Consensus standard: A set of proprietary guidelines published and agreed to by a consensus (representative majority) of members of a given industry. While not legally binding, they are often cited in governmental regulations, such as OSHA standards.

Dedicated spotter: An individual qualified as a signal person who is charged with monitoring the separation between power lines and the equipment, load line, and load, so that the minimum approach distance is not compromised per OSHA standards.

Diver tender: One or more individuals assigned to attend to a diver's needs, including providing assistance in equipment preparation and managing the diver's cables and hoses.

Line of sight: The straight-line path between an observer's eyes and the thing being observed.

Open mike: In electronic communications, the condition where a radio's Transmit button or switch is held continuously without releasing it, even during pauses in speaking.

Trucks: In relation to overhead cranes, a mechanical assembly consisting of a frame, wheels, bearings, and axles that support the bridge of an overhead crane and provide the ability for it to move along a set of parallel tracks.

The methods and modes of communication vary widely in mobile crane operations. The method of communication refers to whether the communication is verbal or nonverbal. The mode is defined by the means of performing the communication. Modes may include bullhorns, radios, hand signals, flags, etc.

2.1.0 OSHA Standards and Requirements

29 *CFR* 1926, Subpart CC, *Cranes and Derricks in Construction* is the primary safety standard related to crane operations in the construction environment. The directives and guidelines it contains are enforceable as federal laws. They are not simply suggestions or examples of best practices. The specific sections that are dedicated to crane communications and the role of signal persons are 29 *CFR* 1926.1419, *Signals—General Requirements* through 1926.1422, *Signals—Hand-Signal Chart*. In addition, 29 *CFR* 1926.1428, *Signal Person Qualifications* provides the information indicated by its title.

There is also one very important communication-related statement found in 29 *CFR* 1926.1417(y), as follows: "The operator must obey a stop (or emergency stop) signal, irrespective of who gives it." This is a very important provision, meant to ensure that any potentially unsafe condition can be recognized and the operation stopped by anyone that recognizes it. All members of a lift team have the authority to display the Stop or Emergency Stop signal, and the crane operator is required to obey it. Even a bystander can display the signal if a hazardous condition becomes apparent, and the crane operator must obey. This is rare, however, as those directly involved with a properly manned lift operation are more likely to recognize a hazard before a casual observer.

The remainder of this section presents detailed information related to crane communications from the relevant OSHA standards. In the case of crane communications, *ASME Standard B30.5, Mobile and Locomotive Cranes,* has not been incorporated by reference into the OSHA standards. As a result, the primary focus of this section is on the enforceable OSHA standards. However, the hand signal set and most of the signaling-related provisions of *ASME Standard B30.5* parallel the OSHA standard.

2.1.1 29 CFR 1926.1419, Signals—General Requirements

Per 29 *CFR* 1926.1419, there are situations in which a signal person must be provided. These situations are outlined as follows:

- When any part of the path of a load being moved, including its origin and destination, is outside of the view of the crane operator
- When the equipment is traveling but the driver's view in the direction of travel is blocked
- When either the crane operator or the personnel handling the load determine that it is necessary

A dedicated spotter is required under certain conditions when a crane is working in the vicinity of power lines. Although they are referred to as *spotters* in 29 *CFR* 1926.1411, *Power Line Safety,* and are assigned to focus their attention on the crane's proximity to the power lines, they must stay in continuous contact with the crane operator. To a certain extent, then, the spotter is serving as an auxiliary signal person, and must meet the signal person qualifications of 29 *CFR* 1926.1428 as stated in 29 *CFR* 1926.1401, *Definitions.* Verbal communication is best for these situations, as the information that a spotter may need to relay to the crane operator may not have any kind of standard hand signal assigned to it.

The standard allows for signals between the signal person(s) and crane operator to be by hand, by voice, or by an audible means other than voice. OSHA simply directs that the chosen method of communication be appropriate for the site conditions. New signals, other than hand, voice, and audible signals, may be used in some conditions if they meet specific OSHA requirements. Note that signal persons are required to provide the signals from the crane operator's perspective of direction. Signal persons must signal in the direction of required crane movement.

Hand signals are designed to be universal and do not require any special equipment. When hand signals are used, they must be those found in 29 *CFR* 1926, Subpart CC, Appendix A. This is referred to as the *Standard Method.* There are two exceptions to the rule of using the OSHA Standard Method:

- When the use of the Standard Method is determined to be infeasible
- When an attachment or required crane movement is not covered by the Standard Method

When an attachment or required crane movement is not covered by the Standard Method, a nonstandard hand signal can be identified and used. OSHA requires that the signal person, crane operator, and lift director discuss any nonstandard hand signals before the operation begins and agree to the signals and their use. Note that there is a difference between the terms *nonstandard signals* and *new signals.* New signals refer to a new method or mode of communicating, while a nonstandard signal refers only to a hand signal that is not part of the Standard Method.

Regardless of the method or mode used, the crane operator must bring the operation to a safe stop if the line of communication is interrupted. For example, if another construction vehicle suddenly blocks the operator's view of the hand signals, the operation must be safely halted until the line of sight is restored. The crane operator can also stop the operation temporarily when any unsafe condition is encountered and discuss the issue with the signal person. The operation resumes only after both parties agree that the issue has been satisfactorily resolved.

29 *CFR* 1926.1419(j) is related to the responsibility of others to display the Stop or Emergency Stop signal. It also refers the reader back to 29 *CFR* 1926.1417(y), which outlines the crane operator's responsibility to obey the signal, regardless of its source. This standard states that "anyone who becomes aware of a safety problem must alert the operator or signal person by giving the Stop or Emergency Stop signal."

A signal person can be in contact with and provide signals to more than one crane or derrick at a time. A system to identify which crane is being signaled must be developed and used. This is generally satisfied by a specific signal chosen and agreed upon by both parties to identify each crane before the functional signal is communicated.

2.1.2 29 CFR 1926.1420, Signals—Radio, Telephone, or Other Signals

OSHA allows a variety of devices to be used to communicate electronically in crane operations. In some cases, hand signals are not practical or appropriate. Any electronic means of communication must be tested at the site to ensure the equipment is fully functional and reliable before the operation begins.

As a general rule, electronic communication must take place on a dedicated channel between the signal person and the operator, with the following exceptions allowed:

- When multiple cranes are using the same signal person, or when multiple signal persons are in use, they may share the channel.
- When a crane is being operated on or near railroad tracks and the crane's movement needs to be coordinated with trains or other equipment moving on the tracks, the channel can be shared.

The final provision of this section is that the crane operator's method of receiving signals electronically must be hands-free. It is acceptable for the operator to use a push-to-talk system, where a button must be pressed to speak, as long as reception is hands-free.

In some cases, the signal person keeps the radio microphone keyed to continuously communicate with the crane operator. An **open mike** is often used when the signal person is speaking progressive information, such as distance to touchdown, during the lift. However, an open mike can prevent receipt of a Stop signal from others on the channel, depending on the equipment involved. This practice may differ from company to company and project to project. Open-mike procedures should be discussed and clarified in the pre-lift meeting. Some radios have an Emergency button on them, which allows Stop signals to be transmitted even when another mike is keyed.

Everyone involved in a lift must exercise some radio discipline. Perhaps you have been using a different radio channel to speak to another craft or your supervisor. Before entering or reentering the crane operations channel, take a few moments to monitor the conversations occurring before speaking. This practice will prevent talking over or interrupting an important communication in progress.

2.1.3 29 CFR 1926.1421, Signals—Voice Signals; Additional Requirements

The OSHA standard in 29 *CFR* 1926.1421 requires crane operators, signal persons, and lift directors to discuss and agree on the verbal signals that will be used. The team will only need to meet and discuss the matter again if a different worker is added or substituted, there seems to be confusion about a signal, or a signal needs to be changed or added. This standard also outlines the structure of voice signals. Each voice signal must contain these components that are spoken in the following order:

- Function and direction, such as "Hoist Up" or "Hoist Down"
- Distance and/or speed, such as "10 feet and slowly"
- Function, followed by the Stop command, such as "Swing Stop" or "Load Stop"

Providing distance and/or speed information is required. It is typically modified and repeated as the function progresses. A typical voice command from a signal person sounds something like this: "Swing right, 30 feet… 20 feet… 10 feet… 5 feet… swing stop." Another example is "Load down slow… slow… load stop." Note that a verbal instruction such as "Swing left… swing left… swing left… swing stop" does not meet the requirements since no speed or distance information is provided. It is also important to remember to supply the distance as the amount remaining before the load reaches the desired point, rather than as the distance it has already moved. Crane operators typically prefer that the signal person paint a real-time picture of progress and movement. Give the operator as much information as you think the operator needs—if you are providing too much, the operator will tell you.

This section of the OSHA standard ends by requiring that the crane operator, signal person(s), and lift director be able to effectively communicate in the language being used.

2.1.4 29 CFR 1926.1422, Signals—Hand-Signal Chart

Per 29 *CFR* 1926.1422, hand-signal charts must be posted on the equipment in use, typically the crane itself, or in a conspicuous location in the area of the hoisting operation. Most mobile cranes have a hand-signal chart posted on or near the cab of the crane in an easily visible location to satisfy this requirement. Where overhead or tower cranes are in use, the hand-signal chart should be posted in a more accessible location than the equipment itself generally provides.

Organizations that employ signal persons have a responsibility to ensure that they are properly qualified as outlined in 29 *CFR* 1926.1428. Employers can choose one of two options to satisfy this OSHA standard:

- The signal person already possesses or obtains documentation from a third-party qualified evaluator (such as NCCER) showing that the individual has demonstrated the necessary skills and meets the qualification requirements.
- The employer provides a qualified evaluator from within the organization that ensures the signal person can demonstrate the necessary skills and then provides the supporting documentation.

This standard also requires employers to maintain the required documentation of signal-persons' skills at the site where they are employed. The documents must indicate which type of signals, such as hand signals or radio communications, that each individual is qualified to perform.

Employers also have a responsibility to ensure that a signal person's performance continues to be sound and accurate. If the performance of a previously-qualified signal person indicates that the individual is not properly qualified, the employer must not allow the worker to continue serving as a signal person until retraining has been completed and the skills are reevaluated through one of the two options outlined above.

The standard provides the following qualification requirements for signal persons:

- Knowledge and understanding of the type of signals being used. For hand signals, the individual must know the Standard Method signals provided in *Appendix A* of 29 *CFR* 1926,*Subpart CC.*
- Competence in the application of the signals used.
- A basic understanding of equipment operation and its limitations. This includes the crane dynamics related to swinging and stopping the movement of loads, and the boom deflection that results from hoisting a load.
- Knowledge and understanding of the requirements found in 29 *CFR* 1926.1419 through 1926.1422, as well as 1926.1428. The requirements of each of these standards have been covered in this section.
- Documentation that these requirements have been met through the successful completion of either an oral or a written test, and a practical

test. These tests are administered by the employer's qualified evaluator or a third-party organization such as NCCER.

2.1.6 Miscellaneous Requirements

There are several additional sections of the OSHA standard that mention the role or actions of signal persons. One such section is 29 *CFR* 1926.1431,*Hoisting Personnel.* An occupant of a hoisted personnel platform can serve as the signal person. If the signal person, when needed, is operating from a location other than the personnel platform, the occupants of the platform must remain in direct contact with the signal person at all times. This same standard also requires a pre-lift meeting to be held when personnel are to be hoisted, regardless of the environment. If a signal person will be used during the lift, he or she must attend the pre-lift meeting. If any workers assigned to the operation are later replaced, including the signal person, another meeting is required.

Another requirement is related to cranes that are supporting one or more divers in the water, found in 29 *CFR* 1926.1437(j). Note, however, that this standard clearly applies to floating cranes and derricks, or to land cranes positioned on a floating vessel. When a crane is devoted to diver entry and exit, it cannot be used for any other purpose until the diver(s) is/are safely out of the water. Divers require a **diver tender** who monitors them at all times to ensure safety. Their primary responsibility is to manage the bundle of cables and hoses, known as the diver's umbilical. The diver tender generally serves as the crane signal person when the crane is directly supporting the diver, and must be a qualified signal person to do so. The standard permits hand signals between the diver tender and the crane operator as long as a clear line of sight is maintained. Otherwise, signals must be transmitted electronically. If the diver is not directly connected to the crane and is instead swimming freely, a dive supervisor is likely in direct contact with both the diver and crane operator.

When cranes are working near power lines, a dedicated spotter is often required to ensure that no part of the crane or the load enters the safe zone established around the lines. The dedicated spotter is an individual separate from that of the primary signal person. Per OSHA's definition of a dedicated spotter, the individual must meet the same requirements as a signal person. These requirements are found in 29 *CFR* 1926.1428. In 29 *CFR* 1926.1410(d)(2)(i), dedicated spotters are

Know Your Craft

In the late 1990s, a young 24-year-old man was killed on his first day on the job, working at Shoreham Docks in the United Kingdom. He was assigned to assist with unloading bags of aggregate from the hold of a ship. The load block of a crane was descending through the opening in the hold from above when it struck the young man in the head, killing him instantly.

There were a number of errors that lead up to this tragedy. Although he had never done such work before, he received no training before taking his place on the crew. He was not even issued a hard hat by his employer, which could very well have saved his life that day. However, there was already an accident waiting to happen: the crane operator could not see into the hold, and the signal person calling the shots did not speak any English and was not familiar with standard hand signals. The language barrier prevented the signal person and crane operator from even considering a conversation related to signals.

Although many lifts may seem very simple and the idea of a serious accident seems remote, there is always the potential for a tragedy to occur. Being an effective signal person goes beyond just knowing what each hand signal means. It also means having the ability to evaluate and re-evaluate what is happening moment by moment, looking forward into the lift process without losing focus on the here and now. Learning to do that is a process. But the process does begin with knowing how to properly signal a crane without having to dig deep into your memory to find the correct signal. Hand signals should be practiced extensively. Practicing the signals in the mirror can help you determine if the hand and arm positions are appropriate. Practice by looking at the signals and identifying them to yourself, and practice making the signals while someone else calls them out rapidly. Practice in class and practice at home. Whatever you do, don't serve in the role of a signal person until you know the signals. You want to be thinking about what to direct the crane to do next, and not how to make the signal.

further required to be in continuous contact with the crane operator, rather than communicating through the signal person. The spotter's responsibility is to monitor the separation of the crane, load line, and load from the power lines. No communication is generally necessary unless one of the three components comes too close to, or enters, the safe zone around the power lines. There are other significant requirements related to the use of a dedicated spotter, but they are not related to signaling or communication.

There are occasions when more than one crane is required to manage and position a single load. When more than one crane is involved, clear communication becomes even more critical. According to 29 *CFR* 1926.1432, one designated person who meets the OSHA criteria as both a qualified person and a competent person shall serve as the lift director. Alternatively, a competent person can be assisted by one or more qualified people. In any case, when multiple cranes are used, a pre-lift planning meeting must be held beforehand with all workers participating in the lift. One signal person is generally assigned to control both cranes, but complex situations may lead the lift director to use more than one. A nonstandard signal must be developed and agreed upon to identify which crane is being signaled, since this type of signal is not a part of OSHA's Standard Method.

Whenever multiple signal persons are used, regardless of the reason, it is important for all members of the lift team to know and understand the plan and how it will be implemented. Lifts involving multiple signal persons increase the complexity of the process. A thorough pre-lift meeting must be held to discuss the responsibilities. A blind lift is a good example of when it may be necessary to use multiple signal persons. If one signal person cannot observe both the complete load path and the crane, additional signal person or persons may be required to relay information. Note that relay signaling is not typically preferred and provides an additional opportunity for errors. Lifting operations should therefore move more slowly. When one signal person hands off responsibility to a second, it should be clear that the responsibility for the load has been transferred. The method of hand-off should be clearly discussed and agreed upon during the pre-lift meeting.

> **NOTE**
>
> Relay signaling automatically creates some lag time in signaling that can create problems and must be taken into consideration.

2.2.0 Verbal Crane Communications and Equipment

Verbal modes of communication vary depending on the requirements of the situation. Some of the most common devices used are portable radios, often referred to as *walkie-talkies* (*Figure 3*). Compact, low-power, inexpensive pairs of units can enable a crane operator and signal person to communicate verbally in some environments. They do meet OSHA standards since the operator can hear the signals hands-free, but they usually require a hand to press a button when speaking.

There are some disadvantages to using low-power and inexpensive equipment in an industrial setting. One disadvantage is interference. With basic, low-quality units, the frequency used to carry the signal may have many other users. A crowded channel can cause signal disruptions and lead to accidents. Another disadvantage is the effect of background noise. When attempting to transmit in a noisy area, the person may transmit unintended noise, resulting in a garbled, unintelligible signal for the receiver. On the receiving end, the individual may not be able to hear the transmission clearly due to a high level of background noise.

> **NOTE**
>
> Wind and other external factors can create noise, distortion, and feedback in low-fidelity radio communications microphones. You can reduce these effects by cupping your hand around the microphone.

To overcome the shortcomings associated with low-power, handheld units, more expensive units with the ability to use specific, dedicated frequencies and transmit at a higher power level may be needed. An electronic communication standard referred to as DECT (Digital Enhanced Cordless Telecommunications), developed in the 1990s, provides improved range as well as encrypted and secure communications. Only DECT-compatible devices can operate on the specific frequencies allocated to them by the US Federal Communications Commission (FCC), and the frequencies are far enough away from those of other devices to eliminate cross-frequency interference issues. DECT-compatible headsets (*Figure 4*) offer high-quality communications for a large number of users on a single dedicated system. Pairing, or connecting, a group of headsets to a local communications hub, also shown in *Figure 4*, establishes a private audio network. Each headset can be configured for full-duplex operation, meaning that the user can both listen to others and transmit at will, completely hands-free. They can also be configured for listening only (broadcast mode), but returned quickly to full-duplex operation when necessary. Note that the over-ear headsets also provide hearing protection.

Cell phones may also qualify as an electronic device that can be used for crane communications. With a compatible headset of good quality to make them hands-free, preferably with functional noise-cancelling technology, they can provide good service. However, it is important to note that signal strength on some networks may be weak or even non-existent in some work areas. As a result, it is difficult to rely on them exclusively. DECT-compatible wireless communication networks can be established just about anywhere, but even this equipment has limitations related to line of sight, elevation, and distance between hub stations and headsets.

Hardwired communications systems are also still available. These units overcome some of the

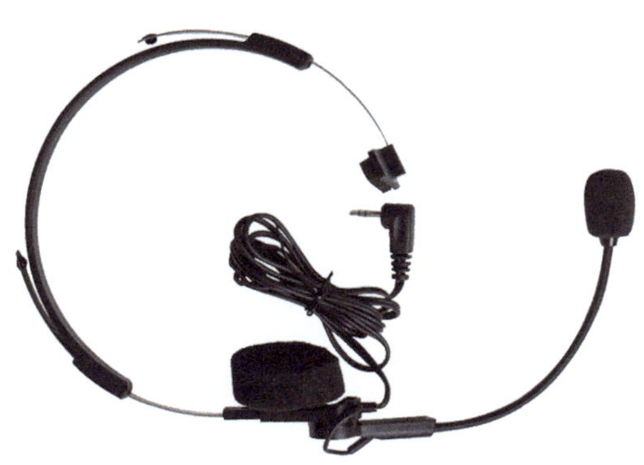

Figure 3 Walkie-talkies with an accessory headset.

(A) HEADSET AND MICROPHONE

(B) HARDHAT-MOUNTED HEADSET

(C) WIRELESS COMMUNICATIONS HUB

Figure 4 Examples of DECT-compatible wireless communications gear.

disadvantages of radio use. When using this type of system, interference from another unit is unlikely because this system does not use a radio frequency to transmit information. Like a telephone system, occasional interference may be encountered if the wiring is not properly shielded from very strong radio transmissions or other electromagnetic interference. Hardwire systems, however, are not very portable or practical, especially when the crane must be moved often. Very long wires are often required and they are easily entangled and/or damaged on an active jobsite. As a general rule, hardwired systems are too clumsy to deploy and operate on most of today's jobsites, and the wires themselves may represent a separate hazard.

Regardless of the type of equipment used, it is important that signal persons remember and use the proper format for all verbal signals transmitted to a crane operator: function and direction, distance and/or speed, and function followed by a Stop command. These skills must be demonstrated during a practical examination. Signal persons should be familiar with the equipment in use, rather than try to figure it out for the first time as a lift progresses.

Remember that all electronic communication devices must be tested before a lift begins. Workers charged with the responsibility of maintaining the communications equipment should ensure that the batteries are kept freshly charged, and that spare batteries are always available on site. Shutting down a complex lift operation temporarily due to a lack of batteries is a very expensive, time-consuming, and often embarrassing way to learn this lesson.

2.3.0 Nonverbal Communications

Although OSHA's Standard Method of hand signaling is the most common type of nonverbal communication used in crane operations, several other nonverbal modes have also been used. One possible mode is the use of a distinct audible signal, such as a siren, buzzer, and/or whistle, in which the number of repetitions and duration of the sounds convey the message. An example of the use of audible signals in crane operations that can often be justified is related to work inside large tanks. Workers must often be inside the tank, receiving components lowered through an opening by a crane. Hand signals are not possible, and the tank itself might interfere with wireless communications. In this case, workers inside the tank might be forced to use a hammer to make a series of rapping sounds on the tank walls to communicate with the outside. However, a hard-wired means of verbal communication is certainly the more practical and safer alternative.

NCCER – *Mobile Crane Operations*

Another nonverbal mode, although rare, is the use of signal flags. This mode might require the use of different colored flags, or a specific positioning of the flags, to communicate the desired message. When there is considerable distance between the signal person and crane operator, this may seem to be a practical solution. However, verbal communication using reliable radio equipment likely provides a better solution, especially since a line of sight evidently does exist if flags are an option. When a line of sight exists, verbal electronic communications function at their greatest range (up to 1,600 feet or 500 meters with some DECT-compatible devices). The disadvantage of a flag-communication mode is that 29 *CFR* 1926.1419(d) requires that any new signals must be shown to be equally effective as the prescribed voice, audible, or Standard Method hand signals, or comply with an equally-effective national consensus standard. Since there is no apparent national consensus standard for flags, their use can be difficult to justify in most cases.

The most common mode of nonverbal communication is the use of the Standard Method of hand signals provided by OSHA. Although *ASME Standard B30.5* also provides a standard set of hand signals, it is the OSHA standard that is enforceable by law. The signals of the two sets are nearly identical, although the drawings used to depict the signals are drawn with slight differences. The required OSHA hand signals are shown in a sequence of drawings shown in *Figure 5A* through *Figure 5S*.

Figure 5A Hoist.

HOIST

With arm vertical, forefinger pointing up, move hand in small horizontal circle.

OPERATOR ACTION: Slowly pull the hoist control lever back, controlling the ascent speed of the load (block or ball) with the control lever position. Keep the engine rpm constant until the desired lift is complete, then slowly return the control to the center position.

If the load block or ball is near the boom point, exercise caution to avoid two-blocking.

EXPECTED MACHINE MOVEMENT: The load attached to the block or ball rises vertically, accelerating and decelerating smoothly.

Figure 5B Lower.

LOWER

With arm extended downward, forefinger pointing down, move hand in small horizontal circle.

OPERATOR ACTION: Slowly push the hoist control lever for the desired hoist forward, controlling the descent speed of the load (block or ball) with the control lever position. Keep the engine rpm constant until load lowering is complete, then slowly return the control to the center position.

Do not allow the block or ball to contact the ground or any surface that can cause slack in the load line(s).

EXPECTED MACHINE MOVEMENT: The load block or ball smoothly lowers vertically.

RAISE BOOM

Arm extended, fingers closed, thumb pointing upward.

OPERATOR ACTION: Slowly pull the boom control lever back, controlling the speed of the boom raising movement with the control lever position. Keep the engine rpm constant until the desired position is reached, then return the control to the center position.

Exercise caution to avoid any obstructions to boom movement in the vertical plane, such as trees or power lines.

If the load or load block is close to the ground in front of the machine, the operator may be required to hoist the load to avoid contacting the crane.

EXPECTED MACHINE MOVEMENT: The boom rises, increasing the hook height and reducing the overall machine height clearance. The operating radius is slowly decreased, thus possibly increasing machine capacity and stability.

Figure 5C Raise boom.

RAISE BOOM AND LOWER THE LOAD

Arm extended, fingers closed, thumb pointing up, flex fingers in and out as long as load movement is desired.

OPERATOR ACTION: This requires a two-hand operation. Use your left hand to push the hoist control lever forward, while using your right hand to pull the boom control lever back. Keep the engine rpm constant until the desired position is reached, then slowly return both controls to their center position.

Move both controls independently to maintain the load (block or ball) at an even distance from the ground, with movement horizontal to the level ground plane.

Keep the engine rpm at a high level to ensure sufficient oil flow to sustain smooth load movement. Exercise caution to avoid any obstructions to boom movement in the vertical plane, such as trees or power lines.

EXPECTED MACHINE MOVEMENT: The boom rises, reducing overall machine height clearance, as the load moves horizontally toward the crane. The operating radius is slowly decreased, thus possibly increasing machine capacity and stability.

Figure 5D Raise boom and lower the load.

NCCER – *Mobile Crane Operations*

LOWER BOOM

Arm extended, fingers closed, thumb pointing downward.

OPERATOR ACTION: Slowly push the boom control lever forward, controlling the speed of the boom lowering movement with the control lever position. Keep the engine rpm constant until the desired position is reached, then return the control to the center position.

Exercise caution to avoid any obstructions to boom movement in the vertical plane, such as trees or power lines.

EXPECTED MACHINE MOVEMENT: The boom will lower, decreasing the hook height and reducing the overall machine horizontal clearance. The operating radius is slowly increased, thus possibly decreasing machine capacity and stability.

Figure 5E Lower boom.

LOWER BOOM AND RAISE THE LOAD

Arm extended, fingers closed, thumb pointing down, flex fingers in and out as long as load movement is desired.

OPERATOR ACTION: This requires a two-hand operation. Use your left hand to pull the hoist control lever back, while using your right hand to push the boom control lever forward. Keep the engine rpm constant until the desired position is reached, then slowly return both controls to their center positions.

Move both controls independently to maintain the load (block or ball) at an even distance from the ground, with movement horizontal to the level ground plane.

The engine rpm must be kept at a high level to ensure sufficient oil flow to sustain smooth load movement.

EXPECTED MACHINE MOVEMENT: The boom lowers, increasing overall machine height clearance, as the load moves horizontally away from the crane. The operating radius is slowly increased, thus possibly reducing both machine capacity and stability.

Figure 5F Lower boom and raise the load.

Figure 5G Extend telescoping boom.

Figure 5H Retract telescoping boom.

EXTEND TELESCOPING BOOM

Both fists in front of body at waist level, with thumbs pointing outward.

OPERATOR ACTION: Using your left hand, push the telescope control lever forward, controlling the boom extension speed with the control position and keeping the engine rpm constant until the desired boom length is reached. Slowly return the controls to the center position. Boom extension may also be accomplished by using the left foot to slowly rock the left foot pedal forward until the desired boom length is reached. Then, slowly return the foot pedal to the center position.

Take care not to bring the hook block or ball too close to the boom head when extending the boom to avoid two-blocking.

EXPECTED MACHINE MOVEMENT: Boom sections telescope out. The load radius is increased, possibly decreasing machine capacity and stability. The load (block or ball) rises vertically.

RETRACT TELESCOPING BOOM

Both fists in front of body at waist level, with thumbs pointing toward each other.

OPERATOR ACTION: Using your left hand, pull the telescope control lever back, controlling the boom retraction speed with the control position and keeping the engine rpm constant until the desired boom length is reached. Slowly return the controls to the center position. Boom retraction may also be accomplished by using the left foot to slowly rock the left foot pedal rearward until the desired boom length is reached. Then, slowly return the foot pedal to the center position.

EXPECTED MACHINE MOVEMENT: Boom sections retract. The load radius is decreased, possibly increasing machine capacity and stability. The load (block or ball) lowers vertically.

SWING

Arm extended, point with index finger in direction of boom swing. (Swing left is shown as viewed by the operator.) Use appropriate arm for desired direction.

OPERATOR ACTION: Push the far left swing control lever forward to swing toward the boom, swinging left for right side operator position and right for left side operator position. For a centrally located operator position, control lever movement is the same as for the left side operator position. Pull rearward to reverse the action.

Keep the engine rpm constant. For inexperienced operators, keep the rpm at a lower level than is used for other craning operations.

Acceleration and deceleration should be slow and steady, with swing speed being adjusted near the end of the swing so that the final desired swinging position is not overrun, thus causing the load to swing like a pendulum. Note that when the signal person uses his or her left arm to swing, the swing is to the operator's right.

Exercise caution to avoid obstructions in the path of the swing. The signal person should warn the operator of obstructions, especially when the swing is to the boom side and the operator's vision is limited.

EXPECTED MACHINE MOVEMENT: The boom moves about the center of rotation with the load (block or ball) swinging in an arc, either toward the right or left, while remaining approximately equidistant to a level plane.

Figure 5I Swing.

MOVE SLOWLY
(ANY SIGNALED MOTION)

Use one hand to give any motion signal and place the other hand motionless over the hand giving the motion signal.

OPERATOR ACTION: Perform the action indicated by the hand signal in a slow manner. Pictured is hoist slowly, directing the operator to pull the hoist control lever for the desired hoist toward the operator until the load (block or ball) ascends slowly. Keep the engine rpm constant until the desired lift is complete, then slowly return the hoist control lever to the center position.

Other move slowly signals (not pictured) include hoist down slowly, raise boom slowly, etc.

EXPECTED MACHINE MOVEMENT: Machine movement will vary depending on the signal being given.

Figure 5J Move slowly.

Figure 5K Use main hoist.

USE MAIN HOIST

Tap open hand, palm down, on head and then use regular hand signals to show the desired action.

OPERATOR ACTION: Grasp the hoist up/down control lever for the main hoist and await further signaling from the signal person.

EXPECTED MACHINE MOVEMENT: None. This signal is used only to inform the operator that the signal person has chosen the main hoist for the action to be performed as opposed to the auxiliary hoist.

Figure 5L Use auxiliary hoist (whipline).

USE AUXILIARY HOIST (WHIPLINE)

Tap elbow with open palm of one hand, then use regular hand signal to show desired action.

OPERATOR ACTION: Grasp the hoist up/down control lever for the auxiliary hoist and await further signaling from the signal person.

EXPECTED MACHINE MOVEMENT: None. This signal is used only to inform the operator that the signal person has chosen the auxiliary hoist for the action to be performed, as opposed to the main hoist.

STOP

Arm extended, palm down, move arm back and forth horizontally.

OPERATOR ACTION: Return all activated controls to the center or neutral position in a smooth motion at a slow to moderate speed to prevent pendulum action of the load (block or ball).

EXPECTED MACHINE MOVEMENT: None. All movement of the machine ceases.

NOTE: The Stop and Emergency Stop signals are the only signals that may be given by anyone, and the crane operator is required to obey.

Figure 5M Stop.

NCCER – *Mobile Crane Operations*

Figure 5N Emergency stop.

EMERGENCY STOP

Both arms extended, palms down, move arms back and forth horizontally.

OPERATOR ACTION: Immediately return all activated controls to the center or neutral position as quickly as is safe.

EXPECTED MACHINE MOVEMENT: None. All movement of the machine ceases.

NOTE: The Stop and Emergency Stop signals are the only signals that may be given by anyone, and the crane operator is required to obey.

Figure 5O Dog everything.

DOG EVERYTHING

Clasp hands in front of body at waist height.

OPERATOR ACTION: Ensure that no controls are activated, then engage all positive locking devices including hoist pawls, swing brakes, and house locks.

EXPECTED MACHINE MOVEMENT: None.

Figure 5P Travel / tower travel.

TRAVEL / TOWER TRAVEL

Arms are straight and extended horizontally, moved back and forth in a pushing motion away from the body. All fingers point straight up.

OPERATOR ACTION: Move the crane in the direction indicated.

TRAVEL – BOTH TRACKS
(CRAWLER CRANE)

Position both fists in front of body and rotate them around each other, indicating the direction of travel (forward or backward).

OPERATOR ACTION: Decrease the engine speed to idle. Move the slide pinion to the travel position. Hold down the deadman control button. Move the main drive control lever forward and back slightly to fully engage the slide pinion, depending on the grade of the travel surface (uphill, downhill, or level). Position the travel locks per the manufacturer's directions. Engage the swing lock. Position the boom angle as shown in the manufacturer's travel tables. Move the steering clutch control to the "straight" position, and move the main drive control lever in the desired direction to travel forward or backward.

EXPECTED MACHINE MOVEMENT: Machine travels in the direction chosen.

Figure 5Q Travel — both tracks (crawler crane).

TRAVEL – ONE TRACK
(CRAWLER CRANE)

Lock track on side of raised fist. Travel opposite track in direction indicated by circular motion of other fist, rotated vertically in front of body.

OPERATOR ACTION: Decrease the engine speed to idle. Move the slide pinion to the travel position. Hold down the deadman control button. Move the main drive control lever forward and back slightly to fully engage the slide pinion. Depending on the grade of the travel surface (uphill, downhill, or level), position the travel locks per the manufacturer's directions. Engage the swing lock. Position the boom angle as shown in the manufacturer's travel tables. Move the steering half lock control to either engage for a gradual turn or disengage for a sharp turn. Move the steering clutch control to the desired position to turn right or left. Move the main drive control lever in the desired direction to turn right or left.

EXPECTED MACHINE MOVEMENT: Machine turns in the direction chosen.

Figure 5R Travel — one track (crawler crane).

Figure 5S Trolley travel (tower or overhead crane).

Remember that the hand-signal chart must also be posted conspicuously at the jobsite. These hand signals are recognized by the industry as the standard hand signals to be used on all jobsites. This helps ensure that there is a common core of knowledge and a universal meaning to the signals when lifting operations are being conducted. As discussed previously, this helps to eliminate a significant barrier to effective communication. These same signals apply to tower cranes, including luffing-boom tower cranes (*Figure 6*), as well as mobile cranes.

Additions or modifications may be made for crane functions not covered by the OSHA-illustrated hand signals, such as the deployment of outriggers. The operator and signal person must agree upon any signals and their meaning that are not illustrated in the Standard Method before the lift begins. These signals cannot conflict or be easily confused with any Standard Method signal.

There are two hand signals pictured in *ASME Standard B30.5* that are not included in the OSHA Standard Method. Since they are not pictured in the OSHA Standard Method set of signals, they are technically considered nonstandard. However, they are also a part of a national consensus standard and are commonly used. These two signals (*Figure 7*) are one-handed alternates for the standard telescopic boom retraction and extension signals, typically used when the signal person is also manning a tag line or has one hand otherwise occupied.

ASME Standard B30.2, Overhead and Gantry Cranes, also offers some hand signals (*Figure 8*) that are devoted to the operation of overhead cranes and similar equipment. Again, they are not part of the OSHA Standard Method set of signals, but are endorsed by a national consensus standard. Therefore, they are the best choice when such signals are needed. The signal for Bridge Travel is the same as that for Travel/Trolley Travel used with other cranes. The **bridge** of an overhead crane is defined as the part of

Figure 6 Luffing-boom tower crane.

Figure 7 Alternate one-hand signals for telescopic boom retraction and extension.

a crane consisting of one or more girders and the supporting trucks that carry the trolley(s). This basically describes the large, weight-bearing structure above the hook. The trolley assembly moves along the bridge, from side-to-side.

Remember that whenever nonstandard signals are used, OSHA requires the signal person and crane operator to agree on their use and meaning before the operation begins. If it is necessary to switch from hand signals to verbal signals during an operation, all crane motions must be stopped before doing so.

A signal person's position at a lift site is very important, especially when hand signals are used. He or she must be in full view of the lift operation, as well as in the clear view of the crane operator. Signal persons should always wear high-visibility clothing or a vest that contrasts with the surroundings. High-visibility gloves may also be in order, so the crane operator can clearly see the position and/or motion of the hands.

As a general rule, the best possible position for the signal person to be positioned is with the line of sight perpendicular to the boom. In this position the signal person is visible to the operator, as well as able to observe the load and any boom deflection that may occur. Choose the side of the boom that presents the fewest visual distractions or obstructions.

Another common nonverbal signal set is actually one used by the crane operator when moving a mobile crane. These signals are outlined in *ASME Standard B30.5*, Section 5-3.3.7. It is common for the operator to use the following audible travel signals, using the crane's horn:

- *Stop* – One audible signal
- *Forward* – Two audible signals
- *Reverse* – Three audible signals

NCCER – *Mobile Crane Operations*

(A) BRIDGE TRAVEL

(B) MULTIPLE TROLLEYS

(C) MAGNET DISCONNECTED

Figure 8 Additional signals for overhead cranes.

Additional Resources

ASME Standard B30.2, Overhead and Gantry Cranes. Current edition. New York, NY: American Society of Mechanical Engineers.

ASME Standard B30.5, Mobile and Locomotive Cranes. Current edition. New York, NY: American Society of Mechanical Engineers.

29 *CFR* 1926, Subpart CC, **www.ecfr.gov**

2.0.0 Section Review

1. In regards to a signal person signaling more than one crane at a time, ______.

   a. OSHA standards prohibit it under any circumstances
   b. crane operators must exit their crane cab when they are not the active crane
   c. the Standard Method of hand signaling cannot be used
   d. signals to identify which crane is being signaled must be determined and agreed upon

2. Which of the following is a true statement about verbal communications with a crane?

   a. Cell phones legally qualify as a verbal communication device, but reliable signal strength at every jobsite is difficult to predict.
   b. Hardwired communication equipment is no longer allowed by OSHA standards.
   c. Low-powered, inexpensive walkie-talkie equipment is the best choice for every environment.
   d. DECT-compatible devices are designed to seek out and use any frequency that shows the least amount of signal traffic.

3. OSHA's chart of Standard Method hand signals must be posted ______.

   a. in the lift plan
   b. in a conspicuous location on the jobsite
   c. on each signal person's clothing
   d. on the load being lifted

SUMMARY

There are many different aspects and variables associated with communication. To be an effective communicator, one must minimize the barriers to communication and ensure that messages are transmitted and received clearly.

OSHA standards found in the *Code of Federal Regulations* (*CFR*) apply to many industries and are enforceable by law. Although there are other national consensus standards that apply to crane operations and communication, the OSHA standards must be considered the higher priority.

The OSHA standards outline the requirements for signal persons that communicate with crane operators of all kinds, and provide the Standard Method for hand signals used. Both signal persons and crane operators must have a clear understanding of each signal, verbal or nonverbal, and its meaning in order to function safely and effectively as a team. Like crane operators, OSHA requires that signal persons possess specific qualifications and knowledge that is tested through both written and practical means.

1. Internal and external barriers to communication affect ______.
   a. only the sender
   b. only the receiver
   c. both the sender and the receiver
   d. only the party to which the message is related

2. Expressing the perceived meaning of something one has read or heard from a sender back to the sender in your own words is referred to as ______.
   a. paraphrasing
   b. passive listening
   c. message filtration
   d. negotiation

3. Verbal communication barriers include language differences, overuse of abstractions, and ______.
   a. body language
   b. feedback
   c. the use of clear language
   d. a lack of confidence

4. The Standard Method of hand signals is found in which standard?
   a. 29CFR 1926, Subpart CC, Appendix A
   b. 29CFR1926.1428
   c. *ASME Standard B30.5*
   d. *ASME Standard B30.2*

5. The main advantage of a hardwired crane-communication system is ______.
   a. low cost
   b. portability
   c. less radio interference
   d. ease of use

6. A signal person's position at the lift site should be ____.
   a. behind the crane where the operator can see the signal person in the cab's rearview mirrors
   b. in full view of the lifting operation and crane operator
   c. following the load from directly underneath
   d. in full view of the crane operator and within the swing radius of the crane

7. What action is signaled by tapping the palm of your hand on the top of your head, and then following up with signals for other functions?
   a. Lower
   b. Use Main Hoist
   c. Use Auxiliary Hoist (whipline)
   d. Raise Boom and Lower Load

8. When the signal person gives the Dog Everything signal, the crane operator ______.
   a. pulls the telescope control lever toward himself/herself
   b. decreases the engine speed to a fast idle
   c. moves the slide pinion to the travel position and holds down the deadman control button
   d. engages all positive locking devices including hoist pawls, swing brakes, and house locks after ensuring that no controls are activated

9. When preparing to drive a mobile crane in reverse, how many audible travel signal(s) must be given using the crane's horn?
   a. One
   b. Two
   c. Three
   d. Four

10. Alternate one-hand signals to signal telescopic boom retraction and extension are ______.
    a. part of OSHA's Standard Method
    b. found in *ASME Standard B30.5*
    c. found in both the OSHA and ASME crane-related standards
    d. used only if directly approved by an OSHA representative

Trade Terms Quiz

Fill in the blank with the correct term that you learned from your study of this module.

1. The trolley of a typical overhead crane moves back and forth along the __________.

2. When a signal person continuously holds a radio's Transmit button or switch down, even during pauses in speech, it is referred to as an __________.

3. Repeating the perceived meaning of a received message using different words describes __________.

4. A __________ describes the lifting of a load that is out of sight of the crane operator.

5. Giving someone a thumbs-up signal is a good example of __________.

6. Giving the verbal order, "Swing right a hair" is a type of communication __________.

7. A __________ is responsible for managing the umbilical of a submerged diver.

8. Effective nonverbal communications between the signal person and the crane operator depend on a clear __________.

9. The moving, upright portions of an overhead crane that support the bridge are referred to as __________.

10. When working near power lines with a crane, a __________ is often needed to ensure that the minimum approach distance is not compromised by the crane, the load line, or the load.

11. *ASME Standard B30.5, Mobile and Locomotive Cranes is an example of a(n)* __________.

Trade Terms

Abstraction	Dedicated spotter	Open mike
Blind lift	Diver tender	Paraphrasing
Bridge	Line of sight	Trucks
Consensus standard	Nonverbal communication	

Robert Capelli

Senior Health, Safety, and Environmental Manager, Orion Marine Construction

With a wealth of experience, insight, and extensive training from the U.S. Marine Corps, Robert continues to inspire excellence in his role as Senior Health, Safety and Environmental Manager with a large marine construction company.

Please give a brief synopsis of your construction career and your current position.

My first job after the Marine Corps was as a safety specialist in a roadway maintenance facility operated by the Florida Department of Transportation, which was a great segue into construction. The knowledge and experience I gained there was invaluable. From that foundation, I then worked for a company that had road construction, utility, demolition, and mining divisions. The diversity of work and operations in that one company alone gave me experience that would have taken four times as long to obtain elsewhere and has served me well in dealing with industry business owners and regulators over the years. I now work for a marine construction company as their safety manager.

How did you get started in the construction industry?

After I left active duty in 1991, I planned on going to school to become a firefighter/paramedic, but I ran into problems during school enrollment and was forced to look for work. As it turned out, the safety position I was offered with the Florida Department of Transportation used portions of the risk assessment training program I experienced in the Marines and the fire service.

Who or what inspired you to enter the construction industry?

The need to provide for my family was the motivation and inspiration to accept a position in the construction industry. I had a completely different career path in mind and was not even considering construction as an option, but because of the income and benefits I would lose if I didn't take the job I was offered, the choice was easy to make. That was my inspiration, and it was a good choice.

How has training in construction impacted your life and career? What types of training have you completed?

My training certificate folder is several inches thick, and I have been very fortunate that my careers and volunteer work have overlapped and provided training not available to most safety professionals. The knowledge gained also revealed opportunities to me that were not available to others. After active duty, I continued serving in the Marine Corps Reserve and was the officer in charge of an environmental response team that performed hazardous material storage and remediation activities. Additionally, I was the fire chief of a local volunteer fire department, providing fire and rescue services.

In 2015, I earned a Master of Public Health degree, with a concentration in health, safety and environment. Many courses I have taken applied to more than one field. Hazardous material and environmental classes transcended all three fields. Urban search and rescue technician, confined space, fall protection, rigging, rope rescue, emergency medical technician, pump operator, and fire inspector classes all have applications in both the fire service and construction industries.

One may have twenty years of experience in construction, but is that one year of experience twenty times over, or twenty years of progressive experience with training and growth in the industry? The answer to this question is fairly evident when I perform the rigging practical examinations, where we always screen the candidates to ensure they are not overstating their qualifications. Many of them have struggled even on the basic rigger practical exam because they have not taken any additional training classes and have stayed within their industry segment. This leads to workers that cannot solve problems any more difficult than being confronted with a two-legged bridle sling.

Why do you think credentials are important in the construction industry?

Credentials are important in any profession as a way to validate experience and knowledge in a given field. In many cases, employers seek out credentialed people because they have verifiable knowledge and skills the employer needs.

The construction industry is very transient, with construction workers chasing work from one project to another. A person's reputation and who they know is often the key to getting hired for that next project. That works well if hiring is performed at the local level, but when hiring is centralized or the project is very large, recruiting managers have to sort through hundreds of applications to fill slots. Having an industry recognized credential, such as the NCCER Certified Plus credential, makes an individual stand out to a hiring manager.

The value of the right credentials presented to acknowledge your skills is huge. Obtaining industry-recognized credentials, such as those provided by NCCER, will help you get ahead.

What do you enjoy most about your career?

I thrive on the variety of work, with every project having its own challenges, and the need to solve problems and develop unique solutions is what keeps the job interesting. I have worked with some really great craft professionals who taught me the business of construction as well as the mechanics of the field, and every one of them had great problem-solving skills.

Would you recommend construction as a career to others? Why?

Absolutely, yes! This industry offers everyone who applies themselves to the effort an opportunity to provide for their family and continue growing as a professional. Academia has perhaps done a generation of Americans a disservice by saying you can only get ahead in life with a college degree. The theory that a four-year college degree is the one true path to success has been debunked over the years, and there are many unskilled workers with college degrees that are not gaining significant value from that education.

Starting out in construction with a broad orientation to the trades and applying yourself over the years will get you ahead. At some point, vocational school training or a bachelor's degree may be necessary to help you progress to the next level, but it is unnecessary at the start of your career. The opportunities to travel, learn new crafts, and move into management positions are unlimited.

What advice would you give to someone who is new to the construction industry?

Apply yourself, learn, seek opportunities, and don't become stagnant in your job. Approach every task by asking yourself, "What could go wrong here?". If you see something wrong or unsafe, fix it or tell a supervisor who can fix it. There is nothing so important that it has to be done right now at the expense of injuring someone, or worse. We always seem to find time to correct things, so why not do it right the first time? In addition, your reputation will follow you forever, so protect it.

How do you define craftsmanship?

Craftsmanship encompasses the physical result of applying your focus, training, and skills to the work effort in the best way possible.

Trade Terms Introduced in This Module

Abstraction: Any form of verbal, graphical, or written communication representing a generalized and nonspecific idea or quality of a thing, action, or event.

Blind lift: Any lift involving a load that is out of the direct view of the operator. Blind lifts are generally always categorized as critical lifts.

Bridge: In relation to overhead cranes, the part of an overhead crane consisting of one or more girders or beams and the supporting trucks. The bridge is the overhead, weight-bearing structure along which the trolley(s) and load block assembly travels.

Consensus standard: A set of proprietary guidelines published and agreed to by a consensus (representative majority) of members of a given industry. While not legally binding, they are often cited in governmental regulations, such as OSHA standards.

Dedicated spotter: An individual qualified as a signal person who is charged with monitoring the separation between power lines and the equipment, load line, and load, so that the minimum approach distance is not compromised per OSHA standards.

Diver tender: One or more individuals assigned to attend to a diver's needs, including providing assistance in equipment preparation and managing the diver's cables and hoses.

Line of sight: The straight-line path between an observer's eyes and the thing being observed.

Nonverbal communication: All communication that does not use words. This includes appearance, personal environment, use of time, and body language.

Open mike: In electronic communications, the condition where a radio's Transmit button or switch is held continuously without releasing it, even during pauses in speaking.

Paraphrasing: Expressing the perceived meaning of something read or heard in one's own words, generally to ensure clarity. Paraphrasing is an important component of active listening.

Trucks: In relation to overhead cranes, a mechanical assembly consisting of a frame, wheels, bearings, and axles that support the bridge of an overhead crane and provide the ability for it to move along a set of parallel tracks.

Additional Resources

This module presents thorough resources for task training. The following reference material is recommended for further study.

ASME Standard B30.2, Overhead and Gantry Cranes. Current edition. New York, NY: American Society of Mechanical Engineers.

ASME Standard B30.5, Mobile and Locomotive Cranes. Current edition. New York, NY: American Society of Mechanical Engineers.

Interplay: The Process of Interpersonal Communication, Ronald Adler, Lawrence Rosenfeld, and Russell Proctor. 13th Edition. New York, NY: Oxford University Press.

NCCER Module 00107, *Basic Communication Skills*.

29 *CFR* 1926, Subpart CC, **www.ecfr.gov**

Figure Credits

© iStock.com/VasilySmirnov, Module opener

Carolina Bridge Co., Figure 2

Motorola Solutions, Inc., Figure 3

Sonetics Corporation, Figure 4

Roy Laney, SME, Figure 6

Answer	Section Reference	Objective
Section One		
1. b	1.1.2	1a
2. c	1.2.2	1b
Section Two		
1. d	2.1.1	2a
2. a	2.2.0	2b
3. b	2.3.0	2c

NCCER CURRICULA — USER UPDATE

NCCER makes every effort to keep its textbooks up-to-date and free of technical errors. We appreciate your help in this process. If you find an error, a typographical mistake, or an inaccuracy in NCCER's curricula, please fill out this form (or a photocopy), or complete the online form at **www.nccer.org/olf**. Be sure to include the exact module ID number, page number, a detailed description, and your recommended correction. Your input will be brought to the attention of the Authoring Team. Thank you for your assistance.

Instructors – If you have an idea for improving this textbook, or have found that additional materials were necessary to teach this module effectively, please let us know so that we may present your suggestions to the Authoring Team.

NCCER Product Development and Revision

13614 Progress Blvd., Alachua, FL 32615

Email: curriculum@nccer.org
Online: www.nccer.org/olf

❏ Trainee Guide   ❏ Lesson Plans   ❏ Exam   ❏ PowerPoints   Other ___________________

Craft / Level: ____   Copyright Date: ______________

Module ID Number / Title: ____

Section Number(s): ____

Description: ____

____

____

____

Recommended Correction: ____

____

____

____

____

Your Name: ____

Address: ____

____

Email: ____   Phone: __________________________

Crane Safety and Emergency Procedures

OVERVIEW

Cranes are used to accomplish very important tasks in various construction and industrial settings. When working with or near cranes, safety is always the highest priority. Crane operators and other members of the lift team must embrace their responsibility as the manager of a powerful machine that can both accomplish great things and destroy property and lives. Thousands of successful crane operations occur each day without incident; all of the lifts in the future can end the same way. The goal of this module is to present a wide variety of safety information related to crane operation and prepare lift team members for their role in a safe workplace.

Module 21106

Trainees with successful module completions may be eligible for credentialing through the NCCER Registry. To learn more, go to **www.nccer.org** or contact us at 1.888.622.3720. Our website has information on the latest product releases and training, as well as online versions of our *Cornerstone* magazine and Pearson's product catalog.

Your feedback is welcome. You may email your comments to **curriculum@nccer.org**, send general comments and inquiries to **info@nccer.org**, or fill in the User Update form at the back of this module.

This information is general in nature and intended for training purposes only. Actual performance of activities described in this manual requires compliance with all applicable operating, service, maintenance, and safety procedures under the direction of qualified personnel. References in this manual to patented or proprietary devices do not constitute a recommendation of their use.

21106 V3

From *Mobile Crane Operations, Trainee Guide*. NCCER.
Copyright © 2018 by NCCER. Published by Pearson. All rights reserved.

CRANE SAFETY AND EMERGENCY PROCEDURES

Objectives

When you have completed this module, you will be able to do the following:

1. Identify relevant OSHA and ASME standards and general crane safety considerations.
   a. Identify safety standards relevant to mobile cranes and their operation.
   b. Identify general safety considerations for mobile crane operation.
2. Identify mobile-crane operation considerations related to specific applications and explain how to respond to various incidents.
   a. Describe the purpose of pre-lift meetings and identify the topics of discussion.
   b. Identify safety considerations related to power lines.
   c. Identify safety considerations related to weather conditions.
   d. Describe safety considerations related to specific crane functions and how to respond to various incidents.

Performance Tasks

This is a knowledge-based module; there are no Performance Tasks.

Trade Terms

Avoidance zone	Prohibited zone
Competent person	Recloser
Critical lift	Shock loading
High-voltage proximity warning device	Standards
Insulating link	Standard lift
Minimum clearance distance	

Industry Recognized Credentials

If you are training through an NCCER-accredited sponsor, you may be eligible for credentials from NCCER's Registry. The ID number for this module is 21106. Note that this module may have been used in other NCCER curricula and may apply to other level completions. Contact NCCER's Registry at 888.622.3720 or go to **www.nccer.org** for more information.

Contents

1.0.0 CRANE SAFETY

Objective

Identify relevant OSHA and ASME standards and general crane safety considerations.

a. Identify safety standards relevant to mobile cranes and their operation.
b. Identify general safety considerations for mobile crane operation.

Trade Terms

Competent person: As defined by OSHA, an individual who is capable of identifying existing and predictable hazards in the surroundings or working conditions which are unsanitary, hazardous, or dangerous to employees, and who has the authorization to take prompt corrective measures to eliminate such hazards.

Shock loading: A sudden, dramatically increased load imposed on a crane and rigging, usually as the result of momentum from the load that occurs due to swinging side-to-side, dropping the load and then stopping it suddenly, and similar actions that create momentum.

Standards: As defined by OSHA, statements that require conditions, or the adoption or use of one or more practices, means, methods, operations, or processes, that are reasonably necessary or appropriate to provide safe or healthful employment and places of employment. Standards developed by some organizations are voluntary in nature, while OSHA standards and those they incorporate by reference are enforceable by law.

Equipment can be damaged and people can be severely injured or killed in crane accidents. Lives, careers, and companies can all be lost as the result of a crane accident. This module provides common safety guidelines for the operation of mobile cranes. In addition, it provides an overview of situations that can occur on the jobsite and the steps to take in response to various situations.

Injuries and fatalities related to crane operations happen for a variety of reasons. The vast majority of incidents are preventable. Many incidents fall into the category of struck by / caught between incidents. One of the leading causes of fatalities in crane operations over the years has been electrocution. These are not cases of electrical problems developing in the cranes; they are electrocutions resulting from power line contact. For this reason, there are a number of OSHA directives related to operations around power lines that must be followed without compromise.

Because working on or around mobile cranes can be dangerous (*Figure 1*), members of a lift team must understand that the first responsibility on the job is safety. This responsibility includes personal safety, the safety of others, and the safety of the equipment and materials on the jobsite. One must know the safety requirements for each jobsite and be aware of any unique hazards that may be associated with the work. Accidents can happen anywhere at any time, but they can be prevented when safety is always the first priority.

1.1.0 Safety Standards

The mobile crane industry is very large and complex. The possibility of major damage and loss of life demands that the industry be monitored. There are several groups that monitor and regulate crane operations. The three primary organizations to be introduced here are the following:

- The Occupational Safety and Health Administration (OSHA)
- The American Society of Mechanical Engineers (ASME)
- The American National Standards Institute (ANSI)

ASME and ANSI are nonprofit professional organizations. The principal difference in the two is that ASME does the bulk of the technical research and offers guidance (standards) based on this research and engineering studies. ANSI is more focused on embracing and supporting standards of their choosing through a specific set of procedures to gain consensus; ANSI does not develop standards independently. Both have an international component as well as a domestic function and provide guidance that serves to reduce or eliminate hazards and the resulting accidents. OSHA, on the other hand, is an office of the US Government, established under the Occupational Safety and Health Act of 1970.

There are several terms often used in discussions of publications from these organizations that should be defined, beginning with the term *standard*. A standard, as it applies to the crane industry, outlines specific conditions, or the adoption or use of one or more practices and methods, necessary or appropriate to provide safe, functional products and/or places of employment. Some standards may be referred to as *national consensus standards*. This means several things:

Figure 1 Crane accidents can result in property damage, serious injuries, and fatalities.

- The standard has been adopted by one or more nationally recognized organizations under a specific set of procedures whereby it can be determined that persons affected by it have reached substantial agreement on its adoption.
- The standard was created in a manner that afforded an opportunity for diverse views to be considered.
- The standard has been designated as a national consensus standard by the US Secretary or the Assistant Secretary of Labor, after consultation with other appropriate federal agencies.

OSHA and ASME standards frequently refer to other standards, often informing the reader that those standards are incorporated by reference. This means that the standards referred to must also be considered and recognized as if they were a part of the text. This helps to reduce the duplication of effort to develop a similar piece of work.

ASME develops many standards, some of which apply to cranes and the crane industry. However, ASME and ANSI do not have the power to enforce standards. ANSI selects standards from organizations like ASME and works to gain consensus and acceptance on a national or international scale. However, ANSI standards are considered voluntary and are not written as laws or regulations, since ANSI is also a nonprofit organization.

OSHA, however, is different. OSHA was created under federal law, and employers are required to follow the standards they develop, adopt from others, or incorporate by reference.

OSHA often creates their own enforceable standards, some of which are based on the work of organizations like ASME and ANSI.

In some cases, OSHA may simply adopt or incorporate the standards of other respected organizations by reference. A good example can be found in this statement from 29 *CFR* 1926.1433(b): "Mobile (including crawler and truck) and locomotive cranes manufactured on or after November 8, 2010 must meet the following portions of *ASME B30.5-2004* (incorporated by reference, *see* §1926.6)…" This section of the OSHA standards goes on to list specific passages from the ASME standard that apply. Incorporation by reference into an OSHA standard effectively makes those standards enforceable as well.

It is very important to understand that OSHA standards are not just suggestions; they are enforceable laws. Employers can be punished for failing to comply with OSHA standards. As an employee, you have a responsibility to your employer, as well as to yourself, to follow the OSHA standards. Remember that the standards of other organizations that are incorporated by reference into OSHA standards and are made mandatory by their language also become legally enforceable.

1.1.1 Crane-Industry Safety Standards

Mobile crane operations are governed primarily by several standards. It is important to note that all standards related to crane operations typically

NCCER – *Mobile Crane Operations*

contain safety information. That is their priority and reason for existence. Even standards such as *ASME Standard B30.10, Hooks* are based on safety. Consider that the standards related to hooks and their design and fabrication are created to ensure that they are safe and reliable when used properly. Of course, this is also true of OSHA standards, all of which exist to support OSHA's mission of safety in the workplace.

29 *CFR* 1926, Subpart CC, *Cranes and Derricks in Construction* is arguably the most important set of standards in the crane industry. This is especially true since OSHA standards are enforceable by law. Subpart CC includes 29 *CFR* 1926.1400 through 1926.1442, plus a listing of other standards that are incorporated by reference. Topics covered in this standard include, but are not limited to, the following:

- Ground conditions for crane support
- Assembly and disassembly of equipment and attachments
- Powerline safety
- Equipment inspections
- Wire rope inspection, selection, and installation
- Required safety devices and operational aids
- Crane operation
- Signaling
- Work area control
- Operator, signal person, and crane maintenance personnel qualifications
- Hoisting personnel
- Equipment modifications
- Crane operation and safety when used on floating barges

ASME Standard B30.5, Mobile and Locomotive Cranes is the most important ASME standard relevant to crane operators. Note that some portions of this standard have been incorporated into the OSHA standards by reference. Again, this means they are not simply suggestions, but are legally enforceable. Topics covered in this standard include, but are not limited to, the following:

- *Personnel competence* – "Persons performing the functions identified in this Volume shall meet the applicable qualifying criteria stated in this Volume and shall, through education, training, experience, skill, and physical fitness, as necessary, be competent and capable to perform the functions as determined by the employer or employer's representative."
- *Crane construction and characteristics* – Items covered here include: crane load ratings; boom hoists and telescoping boom mechanisms; crane travel; controls; cabs; and structural performance.
- *Inspection, testing, and maintenance* – This section covers the crane as well as the inspection and replacement of the wire ropes.
- *Operation* – The specific qualifications of crane operators are provided here, including the physical requirements as well as those related to testing. The specific requirements are covered in NCCER Module 21101, "Orientation to the Trade," from *Mobile Crane Operations Level One*. The role and responsibilities of each individual that is part of a typical lift crew, including the operator, are also outlined. Following this information, the standard addresses a wide variety of common crane movements, such as attaching, lifting, and swinging the load, and provides safety guidelines specific to each action. Crane hand signals are also found in the Operation section of the standard.

Another important ASME standard is *ASME Standard P30.1, Planning for Load Handling Activities*. The standard documents lift-planning considerations that extend beyond cranes to other load-handling equipment as well. Guidance is divided into two categories—Standard Lift Plans and Critical Lift Plans— based on the degree of exposure to hazards. Lift planning will be covered in detail in *Mobile Crane Operations Level Three*.

Throughout this module, standards are referenced where appropriate. Note that this text attempts to present the standards as accurately

Glossaries

Some OSHA and ASME standards contain glossaries to clearly define important terms used in the text. Although every trade has its own verbiage that changes over the years, it is helpful to be familiar with the definition of terms as the standard-setting organizations see them. Many of the OSHA and ASME standards that apply to the mobile crane industry have a glossary at the beginning to ensure that the meaning of a given term is not misunderstood or misapplied.

as possible, but does not present all relevant standards or the requirements they contain. It is the responsibility of crane operators, riggers, signal persons, and all other members of a lift team to directly review and follow the appropriate standards. Requests for interpretations and clarifications of the various standards can be addressed directly to OSHA, ASME, or other issuing authority.

1.2.0 Mobile Crane Safety Considerations

Mobile-crane operators must be aware of the unavoidable hazards associated with the trade. Lifted loads will be moved above and around other workers, and such loads represent an extreme hazard to workers in the area (*Figure 2*). You may also work during inclement weather conditions where wind, slippery surfaces, and other hazards exist. When working near mobile cranes, look up and be mindful of the hazards above and around you, but do not forget the potential hazards that exist on the ground.

As a result of the industry's efforts and losses experienced by employers, construction-trade contractors have made the development of a safety culture in the organization a priority. However, it isn't just about complying with the laws—most employers truly care about the lives of their employees and their families, and developing a safety culture on the job supports their concerns for employee safety and welfare.

Safety consciousness and helping to build a culture that promotes safety from within is extremely important. The earning ability of injured employees may be reduced or eliminated for the rest of their lives. The number of injured employees can be significantly reduced if each employee is committed to safety awareness and exhibits that attitude in their daily work. Full participation in the employer's safety program is a matter of personal responsibility. Making safety the first priority is the key to reducing accidents, injuries, and fatalities on the job. Most accidents can be avoided, because most result from human error. Show that you are a team player by helping to establish and support a safety culture within your organization and on the jobsite.

1.2.1 Personal Protection

Hard hats, safety shoes, safety glasses, and barricaded cranes to discourage personnel entry into the area are among the personnel-protection requirements for almost every jobsite (*Figure 3*). Gloves are also required in many cases, especially when working with rigging equipment such as wire rope. Other personal protective equipment (PPE) may be required at specific jobsites, such as those that produce hazardous chemicals that could be released to the environment. It is essential that every worker be familiar with the requirements of each individual jobsite, and embrace those requirements consistently.

1.2.2 Basic Rigging Safety

Riggers and other members of lift teams must be capable of selecting suitable rigging and lifting equipment, as well as directing the movement of the crane to assure the safety of all personnel and the load itself. All rigging operations must be planned, supervised, and accomplished by qualified and competent personnel. [OSHA defines competent person in 29 *CFR* 1926.32(f).]

One very important rigging requirement is to determine the weight of all loads before attempting to rig and lift them. Crane operators must

Figure 2 Lifting and positioning large loads is hazardous work.

Figure 3 Wearing the correct PPE is a crucial first step toward a safe working environment.

NCCER – *Mobile Crane Operations*

know the weight of the load to ensure it is within the rated load capacity of the crane under the circumstances of the lift. Riggers, however, must also know the weight of the load to ensure that the rigging equipment and techniques used are suitable for the task.

The following rigging-related safety practices should be followed at all times:

- Determine the weight of the load before rigging. If this is not possible, the load can be lifted slightly while the crane operator monitors the instrumentation and determines the weight. If it does not exceed 75 percent of the crane's rated load capacity at the operating radius required, the lift can be made. If it does exceed that value, the load must be set down and the weight reevaluated per 29 *CFR* 1926.1417(o)(i). Note that rigging components or techniques may need to be changed based on any new weight information that is discovered.
- Ensure that the appropriate rigging equipment and components are available. Using an inappropriate piece of equipment due to an equipment shortage can lead to rigging failures. Know the rated load capacity of the rigging equipment and never exceed the limit.
- Ensure that the rigging equipment has been properly inspected and is in good working condition. Remove any damaged or defective equipment from service.
- Always maintain the manufacturer's information for the rigging equipment in an easily accessible location. The literature provides information on hitch configurations, lift angles, and similar information that may be needed as the rigging process proceeds.
- Recognize factors in the lift that can reduce rigging equipment capacity. Remember that the rated load capacity of all hoisting and rigging equipment is based on ideal conditions; lifts often involve conditions that are less than ideal.
- Use proper padding and protection to protect slings as well as the surface of the load.
- Never place loads on the tip of the hook, where it is weakest and most likely to fail.

Gloves for Everyone

There is a time and place in every construction trade for gloves. Injure your hands, and you have damaged the most versatile and important construction tool you will ever own. Craft professionals are observed every day doing tasks without gloves where it is clear that the protection is needed. There has been a long history of workers rejecting gloves for any number of reasons. Many workers have rejected the use of gloves in years past because they felt too restrictive and awkward.

There are more work gloves on the market today than ever before. Today's gloves are miles ahead of the work gloves of old that fit poorly, were unnecessarily bulky and clumsy, and were constructed only of relatively simple materials such as leather and canvas. Although wearing gloves on the job may seem awkward at first, there is a glove out there that fits you well and offers essential protection for your hands without restricting movement. If you work with the right gloves for a while, you will soon feel naked without them. Look for reasons to wear them instead of reasons to reject them, and find your pair of gloves.

- Observe the area where loads will be placed and ensure that it is clear of obstructions and properly prepared for load placement. Preparing the target area while the load is suspended represents poor planning. The practice of lowering the load just above the landing zone and then placing needed blocking is hazardous since the riggers are forced to work beneath the load. Riggers need to think ahead of the crane.
- When serving as part of the rigging team, do not assume that the crane operator and other members of the lift team see the same potential hazards that you do. They may also see something that you do not see. Discuss the lift prior to beginning and share any concerns about obstructions, power lines, and other hazards with the rest of the team. Also listen to and consider any concerns that other members of the team may share. Every lift offers its own unique hazards and obstructions (*Figure 4*).
- Rig the load and connect it to the crane with the center of gravity directly below the hook.
- Always consider where your fingers, hands, and feet are in relation to pinch points. Wear appropriate gloves when handling rigging equipment.
- Never ride a load or the hook.
- Remain outside the load's fall zone at all times unless it is required to guide or receive a load.
- If there is ever any doubt about the reliability or arrangement of the rigging or you observe something unexpected, stop the lift, lower the load, and report it to lift supervision.

Figure 4 A complex jobsite with multiple cranes.

1.2.3 Pre-Lift Considerations

Careful planning, detailed inspections, and timely maintenance help prevent accidents. Crane operators must demonstrate their attention to detail as they prepare for a lift as well as during the lift. Prior to operating a crane, the operator should accomplish the following:

- Determine if there are any locally established restrictions placed on crane operations, such as traffic considerations or time restrictions for noise abatement.
- Ensure that a complete operating manual is in the crane cab. The manual should remain with the crane at all times.
- Accurately determine the weight of the load. Regardless of the perceived accuracy, begin every lift slowly to ensure there are no surprises in the load weight.
- Confirm that load charts in one form or another are readily available for use and review them. Refer to the load charts for every lift and always remain within the capabilities of the crane.
- Determine the deductions to be made from the rated load capacity due to attachments or other crane-related factors.
- Look for documentation of recent crane inspections, as well as any deficiency-correction statements, and review the results.
- Examine the site and ensure that it is suitable to support the crane. Ask questions and seek information about the presence of underground utilities such as gas, oil, electrical, and telephone lines; sewage and drainage piping; and underground tanks. Also ask if the area has been recently excavated. Determine if the lifting operation is limited in some way by stability or structural concerns.
- Evaluate the weather conditions and be familiar with the wind speed limitations of the crane.
- Determine how close the crane or load path may be to power lines throughout the lift and whether the clearance is sufficient.
- Confirm that the crane boom is assembled correctly or extends as designed.
- Determine the hoist line pull and the maximum permissible line pull.
- Ensure there is a safe path to move the crane around, if point-to-point movement on site is necessary.
- Make sure that the crane can rotate unobstructed in the required quadrants for the planned lift.
- Complete the required daily pre-start inspection.
- Ensure that the crane is level before lifting.

1.2.4 Load-Handling Safety

The safe and effective control of the load involves the strict observance of load-handling safety requirements by the entire lift team, including the crane operator. This includes making sure that the swing path of the crane upperworks remains clear of personnel and obstructions any time the crane is in operation (*Figure 5*). Barricades or other visual barriers are required. Also keep the path of any planned load movement clear. Many people tend to watch the load when it is in motion, which prevents them from watching for hazards on the ground.

Here are a few additional precautions related to crane operation and load handling to consider in every lift:

- Consult and follow all applicable safety standards (29 *CFR* 1926.1402 and *ASME Standard B30.5*) and manufacturer guidelines to properly stabilize mobile cranes. Most modern cranes are equipped with outriggers. To be properly stabilized, the outriggers must be used to relieve the weight from the tires of most truck cranes (*Figure 6*). Crawler cranes can usually operate directly from their tracks, but they may also use outriggers in certain conditions.
- When computing equipment loads, the blocks, hooks, slings, equalizer beams, lifting components, and other equipment below the hook must also be taken into consideration. Crane load ratings only extend to the hook.
- Avoid allowing a suspended load to swing more than necessary. This subjects the equipment to additional side loading that can cause a failure of a component or tip the crane. Keep the load directly below the boom.
- Crane operators must avoid snatching or stopping the descent of a suspended load suddenly. Rapid acceleration and deceleration results in

Figure 6 Extended and lowered outriggers.

shock loading, greatly increasing the stress on equipment and rigging.
- Physical control of the load beyond the ability of the crane operator may be required. Tag lines are used to limit the unwanted movement of the load as it reacts to the motion of the crane, wind, or other external influences. They are also used to allow the controlled rotation of the load for final positioning in the landing zone (*Figure 7*). Tag lines are attached after the rigger verifies that the load is balanced.

1.2.5 Signaling

Topics of signaling and communication with a crane operator are covered in 29 *CFR* 1926.1419 through 1926.1422. According to these standards, a signal person is required in the following situations:

- When the points of operation, including the load travel path or the area in the vicinity of the load and its landing place, are not in full view of the crane operator
- When the crane will travel and the view in the direction of travel is obstructed
- Any time the crane operator or workers handling the load determine that it is necessary

Voice signals can be used as well as hand signals. When voice signals are used, the crane operator, signal person, and lift director must all agree on the voice signals to be used. 29 *CFR* 1926.1421 requires that voice signals contain three elements, provided in the following order: a function with direction, such as hoist up or boom left; the distance and/or speed of the function; and a command to stop the function. If electronic devices such as radios are to be used, they must be tested at the site before beginning the operation.

Figure 5 Crane barricading is required.

Crane Safety and Emergency Procedures

Figure 7 Using taglines.

The crane operator's version of any radio or telephone used must allow for hands-free reception.

Using hand signals to communicate with a crane operator is also very common. There are established hand signals used for communicating load navigation directions. The required hand signals are referred to as the Standard Method in 29 *CFR* 1926.1419(c)(1) and they are pictured in the appendix of the publication. In addition, 29 *CFR* 1926.1422 requires that hand signal charts be posted in a conspicuous location near the lift operation.

Standard hand signals, when used correctly and known by both parties, provide the needed information to the crane operator. Nonstandard hand signals may be developed and used by a lift team when standard hand signals are not feasible for some reason or the use and operation of a crane attachment is not provided for in the standard hand-signal set.

Serving in the role of a signal person requires qualification, as outlined in 29 *CFR* 1926.1428. However, it is important to note that any member of a lift team that becomes aware of an issue that affects safety is authorized to display or speak the Stop or the Emergency Stop signals (*Figure 8*). Crane operators are required to obey these two signals regardless of their source. To build flexibility into lift teams, it is not unusual for a crane operator or rigger to also seek certification as a signal person.

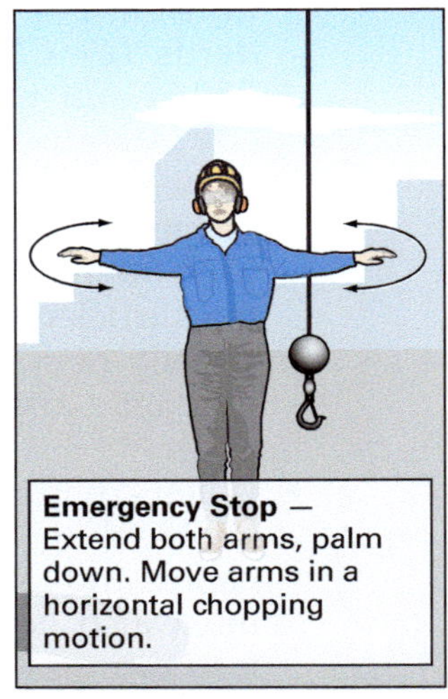

Figure 8 The Stop and Emergency Stop hand signals.

Additional Resources

ASME Standard B30.5, Mobile and Locomotive Cranes. Current edition. New York, NY: American Society of Mechanical Engineers.

ASME Standard B30.20, Below-The-Hook Lifting Devices. Current edition. New York, NY: American Society of Mechanical Engineers.

ASME Standard P30.1, Planning for Load Handling Activities. Current edition. New York, NY: American Society of Mechanical Engineers.

29 *CFR* 1926, Subpart CC, **www.ecfr.gov**

29 *CFR* 1926.251, **www.ecfr.gov**

29 *CFR* 1926.753, **www.ecfr.gov**

Mobile Crane Safety Manual (AEM MC-1407). 2014. Milwaukee, WI: Association of Equipment Manufacturers.

The following websites offer resources for products and training:

American National Standards Institute (ANSI), **www.ansi.org**

The American Society of Mechanical Engineers (ASME), **www.asme.org**

Occupational Safety and Health Administration (OSHA), **www.osha.gov**

North American Crane Bureau, Inc., **www.cranesafe.com**

Electronic Code of Federal Regulations, **www.ecfr.gov**

1.0.0   Section Review

1. Which of the following statements is *not* a required characteristic of a national consensus standard?

    a. The standard has been adopted by one or more nationally recognized organizations under a specific set of procedures.
    b. The standard was created in a manner that afforded an opportunity for diverse views to be considered.
    c. The standard has been adopted or incorporated through reference by the American National Standards Institute (ANSI).
    d. The standard has been designated as a national consensus standard by the US Secretary or the Assistant Secretary of Labor.

2. Positioning a load just above a landing zone and then placing any needed blocking or support for the load is ______.

    a. considered the safest way to do it
    b. considered hazardous
    c. required by law
    d. a function of the load owner

2.0.0 SITE SAFETY AND EMERGENCIES

Objective

Identify mobile-crane safety considerations related to specific applications and explain how to respond to various incidents.

a. Describe the purpose of pre-lift meetings and identify the topics of discussion.
b. Identify safety considerations related to power lines.
c. Identify safety considerations related to weather conditions.
d. Describe safety considerations related to specific crane functions and how to respond to various incidents.

Trade Terms

Avoidance zone: An area both above and below one or more power lines that is defined by the outer perimeter of the prohibited zone. As the name implies, any part of the crane should avoid this area whenever possible, and may not enter the area except under special circumstances.

Critical lift: As defined in ASME Standard B30.5, a hoisting or lifting operation that has been determined to present an increased level of risk beyond normal lifting activities. For example, increased risk may relate to personnel injury, damage to property, interruption of plant production, delays in schedule, release of hazards to the environment, or other significant factors.

High-voltage proximity warning device: An early-warning device that senses the electric fields created by high-voltage power lines and alerts the crane operator and/or the lift team to the hazard.

Insulating link: An electrical insulating device used on the crane hook to protect workers in contact with the load from the danger of electrocution in the event the crane contacts a power line. The link can also provide some level of protection for the crane if the load alone contacts a power line.

Minimum clearance distance: The OSHA-required distance that cranes, load lines, and loads must maintain from energized power lines. This OSHA term is synonymous with the ASME term prohibited zone.

Prohibited zone: An area of specific dimensions, based on the voltage of a power line(s) that no part of the crane is allowed to enter during normal operations. Special considerations and preparations are required if the crane's task must place any part of it within the prohibited zone. The prohibited zone is a term used by ASME that is synonymous with the term minimum clearance distance used by OSHA.

Recloser: A device that functions much like a circuit breaker, or in conjunction with a circuit breaker, in power distribution and transmission systems that automatically recloses the circuit after a fault has been detected and the circuit has been opened. Reclosers allow the power system to be re-energized quickly after a transient (temporary) condition, such as a tree limb falling across power lines and then falling to the ground, has occurred. If the fault reoccurs upon closure, the circuit will typically remain open until the situation has been addressed by power line workers or operators.

Standard lift: A lift that can be accomplished through standard procedures, allowing load-handling and lift team personnel to execute it using common methods, materials, and equipment.

One of the ways to avoid accidents and incidents in crane operations is to discuss the plan in detail in a pre-lift meeting. However, in spite of such meetings and a consistent focus on safety, accidents and equipment failures can and do happen. This section focuses on some of the common environmental hazards that crane operators encounter and how to respond to a variety of emergency situations.

One of the keys to a successful response to an emergency is the crane operator's intimate knowledge of the equipment and its controls. There is rarely time for research and a great deal of thought when an operator is confronted with an emergency situation. It is far better to consider how you should react to a given problem throughout the process in order to be well prepared for an unexpected event.

2.1.0 Pre-Lift Meetings

One of the best ways to avoid incidents and hazards is to plan fir the lift carefully. *ASME Standard P30.1, Planning for Load Handling Activities* provides guidance in lift planning. The first requirement is to evaluate a load-handling activity and place it in a category based on the following characteristics:

> **NOTE**
>
> Pre-lift planning is discussed in detail in NCCER Module 21304, "Lift Planning," from *Mobile Crane Operations Level Three.*

- The potential hazard to people that the operation represents
- Hazards that exist in close proximity to the operation
- The complexity of the activity
- The potential for problems that may be caused by the weather or other environmental conditions
- The capacity and ability of the load-handling equipment to cope with the stresses involved
- The potential for an adverse commercial impact, such as the loss of a unique or irreplaceable load, or the costly delay of a major project
- Site requirements that are unique, such as the effect on roadways or other infrastructure

The standard does not limit the evaluation process to these areas; other areas of concern can also be factors. Documentation of this evaluation process is not required, but it is certainly a good idea to do so. Once the evaluation is complete, the activity is then placed in one of two major categories—a **standard lift** or a **critical lift**.

A standard lift, per *ASME Standard B30.5*, is one that "can be accomplished through standard procedures, and that the load-handling activity personnel can execute using common methods, materials, and equipment." A critical lift, again per the ASME standard, is one that has been evaluated and it has been determined that the activity "exceeds standard lift plan criteria and requires additional planning, procedures, or methods to mitigate the greater risk." 29 *CFR* 1926.751, *Steel Erection*, defines a critical lift as "a lift that (1) exceeds 75 percent of the rated capacity of the crane or derrick, or (2) requires the use of more than one crane or derrick." Although the term is not defined in 29 *CFR* 1926, Subpart *CC*, both of the above definitions are widely applied in the industry. In fact, both the OSHA and ASME definitions are applied by most organizations. A load that is over 75 percent of the rated capacity becomes a critical lift automatically, based on the OSHA standard, but other conditions can also lead to a lift being labeled as critical. These other conditions represent the influence of the ASME definition.

2.1.1 Standard Lift Planning

A standard lift plan can be written or verbal. There is no OSHA requirement for a standard lift plan to be documented, although its documentation is a good idea. However, many employers require both the evaluation process and the standard lift plan to be documented. A standard lift plan should address the following:

- The load, its center of gravity, and available points of attachment
- Confirmation that the load is within the crane's rated load capacity
- Rigging
- Movement of the crane and/or the load
- The personnel required
- Site conditions such as weather, crane support and ground conditions, and utilities
- Communication method
- Site control of non-essential personnel and pedestrians
- Contingency plans
- Emergency action plans
- Equipment inspection during repetitive processes

Any time the operation is not going according to plan, the operation should stop and the situation evaluated. Any changes should be clearly communicated to all members of the lift team.

2.1.2 Critical Lift Planning

Unlike standard lift plans, critical lift plans are required to be in written form. An example of a critical lift planning worksheet is available in the *Appendix* for review. *ASME Standard P30.1* provides examples and templates for the evaluation process as well as for lift planning.

Essentially, a critical lift plan addresses the same topics as a standard lift plan. However, each topic is considered at a deeper level. Since the lift has been classified as critical, there is at least one element of the lift that requires additional planning and possibly a deviation from normal procedures. This information must be carefully considered and documented in the plan.

The lift director typically schedules a pre-lift meeting to construct, discuss, and review the details of the plan and ensure all personnel involved understand their role. The lift director has overall responsibility for the lift from start to finish. He or she ensures that all the appropriate preparations have been made, the plan is executed as scheduled, and that the lift is stopped if it is not going as planned. Once the activity is stopped for any reason, only the lift director can restart it. Upon completion, a post-lift review is common to assess the process and determine what can be done better in the future. Any recommendations are shared with the lift team and others involved in the process.

2.2.0 Working Around Power Lines

Operating mobile cranes where they can become electrified by power lines is an extremely hazardous practice, although it is sometimes necessary. Work must be performed so that there is no possibility of the crane, load line, or load contacting an energized component and becoming a conductive path. Contact with high-voltage power sources is a major cause of fatalities associated with crane operations. However, these accidents can be prevented.

It is important to note that contact between the crane and power line is not always necessary to initiate an incident. Due to the high voltage carried by some power lines, electricity can jump across an open gap and create a sustainable arc between the energized power line and any path to ground—in this case, a metal crane. Moist air masses allow a larger gap to be crossed. Establishing an arc across a gap is the principle on which automotive spark plugs are based.

Surrounding every energized power line is an area referred to as the prohibited zone (*Figure 9*). The prohibited zone is an area around an energized power line that no part of a crane, boom, load, or load line is allowed to enter. The extent of the prohibited zone is shown in *Table 1* for various line voltages. The table reflects the clearance requirements established by OSHA in 29 *CFR* 1926.1408, Table A, and 1926.1411, Table T. They are identical to the values in tables provided in *ASME Standard B30.5* at the time of this writing (2017). OSHA uses different terminology however, using the term minimum clearance distance in place of prohibited zone. The line voltage determines how much clearance is required. Note that there are different values for cranes in operation versus those that are in transit with no load and the boom lowered.

Figure 9 also shows an avoidance zone. The avoidance zone is the area above and below the prohibited zone, defined by imaginary vertical lines. The distance of the vertical lines from the power lines is determined by the outer edge of the prohibited zone. The avoidance zone exists due to the increased probability of accidental power line contact when working in the area. The prohibited zone, or minimum clearance distance, extends away from the power lines in all directions, so there is a prohibited area above and below the lines as well as to their left and right. The avoidance zone is above and below the prohibited zone.

Note that the term *avoidance zone* is not a term used by ASME or OSHA, but it is commonly used to identify these hazardous areas. OSHA does not address the area above power lines that is beyond the minimum clearance distance. However, 29 *CFR* 1926.1408(d) does address crane operation in the avoidance zone below the power lines. Cranes cannot be used to operate under energized power lines (which is within the avoidance zone) unless any part of the machine or load cannot physically reach the prohibited zone. Although the crane operator may have no intention of fully extending a boom to accomplish the desired task, it could be extended and therefore poses too great a risk.

However, there are times when a task must be done inside the avoidance zone, and sometimes

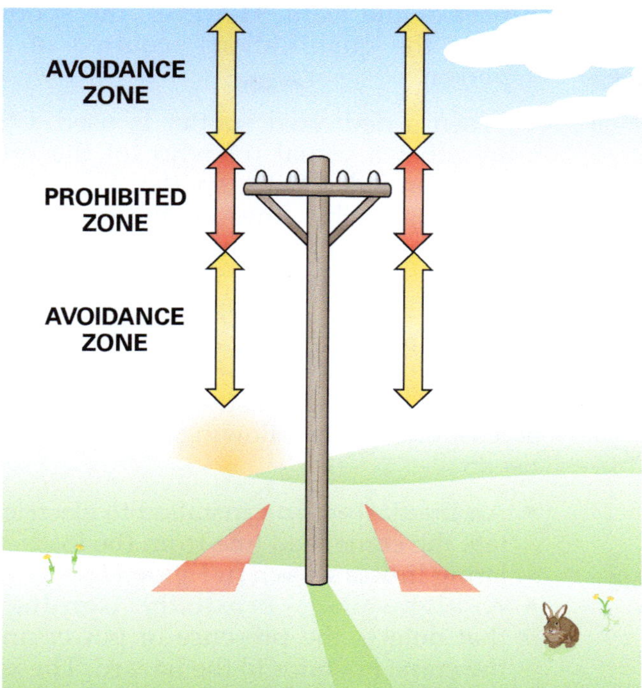

Figure 9 The prohibited zone and the avoidance zone.

Table 1 Minimum Clearance Distance for High-Voltage Power Lines

CRANE IN OPERATION	
POWER LINE VOLTAGE IN	MINIMUM CLEARANCE IN FEET (METERS)
0 to 50	10 (3.05)
50 to 200	15 (4.60)
200 to 350	20 (6.10)
350 to 500	25 (7.62)
500 to 750	35 (10.67)
750 to 1,000	45 (13.72)
Over 1,000	Distance established by the utility owner / operator or registered professional engineer who is a qualified person in power transmission and distribution

CRANE IN TRANSIT (with no load and the boom or mast lowered) [1]	
POWER LINE (kV)	MINIMUM CLEARANCE IN FEET (METERS)
0 to 0.75	4 (1.22)
0.75 to 50	6 (1.83)
50 to 345	10 (3.05)
345 to 750	16 (4.87)
750 to 1,000	20 (6.10)
Over 1,000	Distance established by the utility owner / operator or registered professional engineer who is a qualified person in power transmission and distribution

(1) Environmental conditions such as fog, smoke, or precipitation may require increased clearances.

within the prohibited zone as well. 29 *CFR* 1926.1410 addresses these situations and outlines the requirements that must be met. Some of these situations are addressed in the sections that follow. However, it is important to read and study the OSHA standards whenever work in the vicinity of power lines is planned, especially when they are energized.

Before operations begin near power lines, the owner of the power lines or an authorized representative must be notified and provided with all relevant information. In addition, the cooperation of the owner must be requested. Any overhead line must be considered to be electrically energized unless and until the owner or utility confirms that it is not energized, per 29 *CFR* 1926.1408(e).

There are four scenarios to consider when operating a mobile crane near power lines:

- Power lines de-energized and grounded
- Power lines energized and the crane operating near the prohibited zone
- Power lines energized and the crane operating within the prohibited zone
- Crane in transit with no load and the boom lowered

Each of these scenarios will be discussed further in the sections that follow.

2.2.1 Near De-Energized, Grounded Power Lines

Working in the vicinity of power lines that have been de-energized and grounded is always the preferred situation, since the vast majority of the hazard has been removed. To ensure power lines have in fact been de-energized, the following steps must be taken:

- The utility or owner of the power lines must be contacted first. When possible, they will de-energize the lines.
- The lines must be visibly grounded to avoid electrical feedback and be appropriately marked at the jobsite location (*Figure 10*).
- A qualified representative of the power line owner or the electrical utility must be on site to verify that the first two steps have been completed and that the lines are not energized.

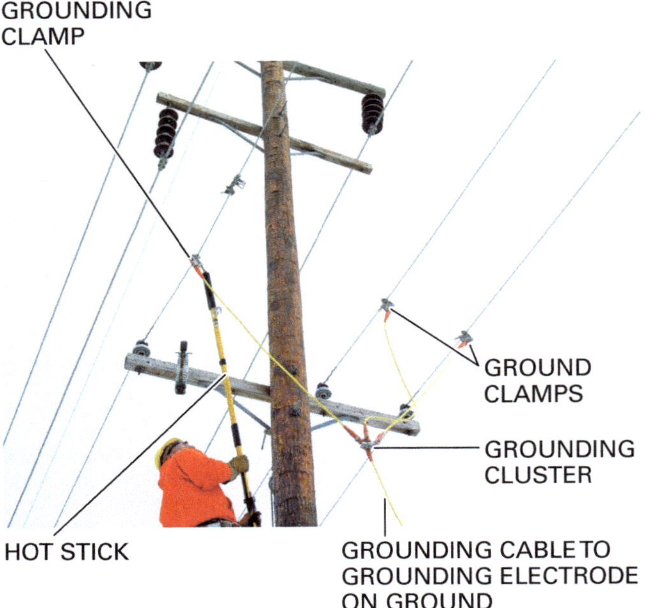

Figure 10 A worker installs temporary grounds on power distribution lines.

Of course, power lines and other lines such as those providing cable and telephone service still represent a significant hazard and an obstruction that can easily become entangled in the boom or hoist lines. Even when they are de-energized, all overhead lines must be treated with great respect.

2.2.2 Near Energized Lines and Near Prohibited Zone

When work must be conducted in the vicinity of the prohibited zone, specific precautions must be taken to ensure that the crane and load do not enter the hazardous areas. Per 29 *CFR* 1926.1408:

- A planning meeting must be held with the lift team and other involved workers to review the plan to avoid encroachment.
- Nonconductive tag lines must be used.
- A visual aid must be erected to aid the crane operator in determining where the prohibited zone boundary begins. For voltages up to 350 kilovolts (kV), or 350,000 volts, the lift team can choose whether to use the minimum clearance distances from the OSHA tables, or to simply use the listed minimum clearance dimension for 350kV lines of 20 feet (6.1 meters). The visual aid must be elevated and can be fabricated several different ways. A suspended warning line with flags attached is one example. If the crane operator cannot clearly see the visual aid, a dedicated spotter must be used to monitor the crane, load line, and load to ensure no part enters the hazardous areas. This individual is in addition to any signal person that may be part of the team. However, the spotter must have the qualifications required of a signal person.

When a dedicated spotter is used, OSHA requires that a visual reference of the prohibited zone be provided to the spotter. This can be done by placing visible lines on the ground, aligned with the edge of the zone; placing a line of small posts or stanchions in a row; or using existing points of reference as a sight line, such as a sign post behind the spotter and a distant object that aligns with the edges of the zones. If the use of a spotter is chosen instead of an operator's aid, at least one of the following devices must also be used:

- An insulating link installed to electrically isolate the crane load line from the load. An insulating link is shown in *Figure 11*.
- A high-voltage proximity warning device that detects the presence of power and alerts the crane operator to the hazard. These devices sense the electric fields around power lines to determine their presence. The alert feature is usually a series of lights—green, yellow, and red—that indicates how close the hazard is to the instrument. The components of the system are shown in *Figure 12*.
- A device that automatically warns an operator to stop movement of the boom or load line.
- A device that automatically limits the movement of the crane.

> **CAUTION**
>
> Many high-voltage proximity warning devices used are unable to sense the presence of direct current (DC) power. Although DC power lines are rare in the United States, they are used in a few power transmission systems, and are more likely to be encountered in Europe, South America, and Asia. DC power is also often associated with power generated through wind and solar systems. When working near energized power lines and related systems, be sure that the equipped proximity alarm system is capable of detecting the power in use.

Keep in mind that power lines tend to move and sway with the wind. Essentially, wind causes the prohibited zone to be in motion. The anticipated horizontal and/or vertical movement of power lines due to the wind must be added to the clearance distances in OSHA's tables. The utility or owner's representative must be consulted to determine the specific distance to be added for a given situation.

Figure 11 An insulating link.

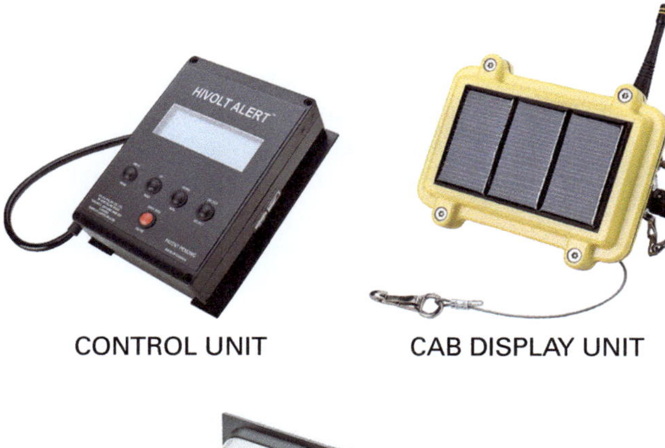

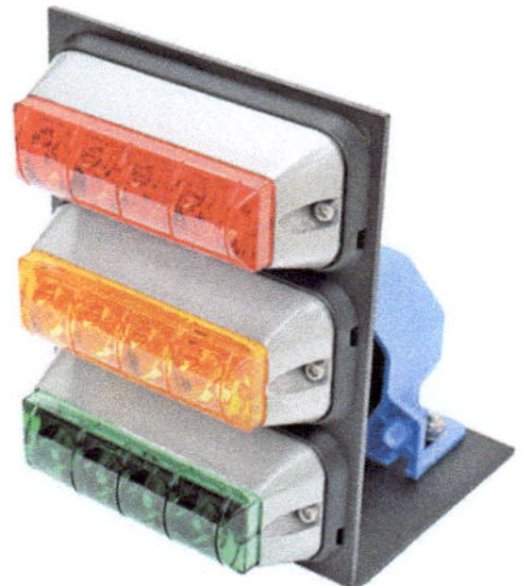

Figure 12 High-voltage proximity warning system.

2.2.3 Within Prohibited Zone and Power Lines Energized

Crane operations can be performed within the prohibited zone if the task is absolutely necessary and the crane employer can show that the task cannot be done in any other practical and safe way. Such work places the crane and lift team in close proximity to a major hazard. 29 *CFR* 1926.1410 provides the majority of the guidance for this situation.

All of the steps associated with operating a crane within reach of the prohibited zone are also required in this case. In this case, the required planning meeting must include the utility operator or owner's representative who is a qualified person in the field of electrical power transmission and distribution. The procedures and techniques that help avoid an incident are determined and documented. The completed documentation must be readily available at the jobsite. Per the OSHA standard, the resulting plans and procedures must include the following:

- Any device such as a **recloser** that can automatically re-energize a power line after a fault occurs must be disabled if possible.
- A dedicated spotter that can communicate with the crane operator directly and is provided with the aforementioned visual references must be in place.
- An elevated warning line to act as a visual aid for the crane operator must be erected.
- An insulating link must be installed to isolate the load from the load line. No worker other than the crane operator can be allowed to contact any part of the crane or load line above the insulating link.
- If any of the rigging devices, such as slings, will be within the prohibited zone, they must be nonconductive. Tag lines must also be nonconductive.
- A perimeter at least 10 feet (3 meters) away from the crane must be established with barricades to prevent workers and others from getting too close to the crane. If there are structures around the crane that prevent placing barricades that far away, they must be placed as far away as possible. All persons must be kept away from the crane and the work area except those that are essential to the task.
- The crane and any other involved equipment must be properly grounded.
- Power line insulating hose and/or blankets (*Figure 13*) must be applied by the utility if such products are available for the voltage of the lines. Note that such products cannot provide effective protection when the voltage is very high.

Figure 13 Installing insulating line hose and blankets on power lines.

In addition to the initial planning meeting, the utility or owner's representative must also meet with the work team(s) at the site to review the procedures. 29 *CFR* 1926.1410(h) directs that involved employers, as well as the utility/owner's representative, together identify an individual that will be responsible for implementing the procedures that have been developed. The individual has the authority to stop work at any time for safety reasons. If the procedures are not working out as planned, the process stops until new procedures can be developed and implemented.

OSHA also requires that all crane operators and crew members involved with lifts in the prohibited zone of power lines must be specifically trained. 29 *CFR* 1926.1408(g) provides a list of the training topics to be covered.

Note that the guidance regarding work around power lines in *ASME Standard B30.5* is slightly different from the OSHA directives. In this case, the related ASME standards have not been incorporated by reference into the OSHA standard, but the ASME standard contains valuable guidance that should be followed regardless.

2.2.4 Crane Transit with No Load and Boom Lowered

While in transit with no load and the boom lowered, the minimum clearance as specified in *Table 1* must be maintained. You will recall that the lower half of the table provides a separate set of clearance requirements for cranes in transit. Consider however, that a crane bouncing along rough terrain or crossing a rise under a power line can become taller than its specified height. The condition can be momentary, but a moment is all it takes to make contact with a power line. When moving around the site, the effect of speed and terrain on the height of the crane must be considered when evaluating the minimum clearance distance. Additional clearance may be in order to accommodate these factors.

2.2.5 Power Line Contact Emergency Procedures

In spite of everyone's best efforts, something has gone wrong and the boom or load has made contact with an energized power line. Now what? First and foremost, the crane operator must not panic.

Did You Know?

Power Lines That Re-Energize Automatically

At some point in your life, you have likely experienced a power outage in your home or place of business. Sometimes the power goes off, but returns a moment later. Sometimes it quickly goes off again and stays off for a significant period of time, while on other occasions it may return a second time and stay on. Did you ever wonder what's going on at those moments?

Lots of things can disturb power lines and cause what is known as a fault to occur. For example, a wet tree limb can fall across the lines. This causes a short circuit and the current flow becomes extraordinary. The power line must be de-energized when this happens to prevent serious damage to the system or its many components. In many cases though, the fault is temporary – the limb simply falls to the ground after making initial contact.

Many power distribution systems are protected by a special type of circuit breaker called a recloser (shown here). When a fault is sensed, the recloser opens like a circuit breaker, but is also programmed to close the circuit again after a very short interval. If the fault has cleared itself, power is restored and the event is over. If not, a fault is again detected and the recloser opens the circuit. Reclosers may be programmed for one or more attempts to clear the fault, depending upon the utility's policies and the type of customer being served by the circuit. Leaving a recloser active on a circuit near a crane lift could cause a serious incident such as the boom or other object touching the power line to go from bad to much worse very quickly, as it may try repeatedly to re-energize the line.

If power line contact occurs, first try to gently reverse the action that caused the contact. Snatching or grabbing at the controls in response can cause the load to swing out of control, making matters worse. Side loading or another condition that upsets the balance of the crane can then occur, causing the crane to tip. If the contact between the crane and the power line can be broken, the immediate danger to the operator and lift team has been resolved. However, a power outage may have resulted, and the crane may have sustained damage. The crane must be carefully inspected for damage caused by the electrical contact. Wire rope should be replaced if it touches an energized line since the arc is easily sufficient to melt and/or badly scorch the rope. Arcing can occur in a number of other areas in the crane as well. Assuming all is well because the crane still functions is a dangerous practice.

If there is no immediate sign of fire or explosion, remain inside the cab. The operator is usually safest inside the cab at this moment (often safer than any other team member). The crane operator is at the same electrical potential as the equipment and is not in the path of the power as it seeks a path to ground. Note, however, that this is not true when operating a boom truck with standing controls. In this case, the operator might be standing on the ground while in contact with the controls or the frame of the vehicle. This creates a very dangerous situation for the crane operator. For cranes with cabs, however, unless an extreme emergency such as an explosion or fire involving the crane presents itself, operators should remain in the cab and avoid touching the ground.

If you must exit the crane due to fire or an explosion, try to jump off from the lowest point of the crane. As you jump, make no further contact with the crane in an attempt to stabilize yourself. If contact with the crane is made after your feet contact the ground, you become a path to ground and create a complete circuit. Land on the ground with your feet close together and make no further contact with the crane. Do not exit the crane one foot at a time while holding onto the crane.

While moving away from the crane, do not run or take long strides. Instead, shuffle your feet along in very small steps (about six inches, or 15 cm) or hop away with your feet together until you are a safe distance from the crane. High-voltage current transmitted from the power lines through the crane to the ground energizes the ground around the crane (*Figure 14*). As the distance from the crane increases, the voltage and difference in electrical potential decreases. The rate of the decrease varies depending on the resistance of

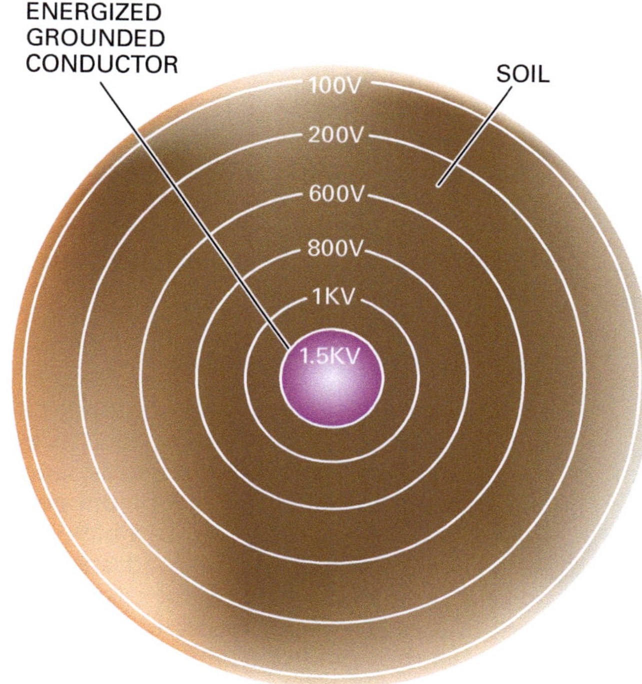

Figure 14 Voltage applied to the ground diminishes with distance.

the surrounding soil. If you take large steps, it is possible for one foot to be in a high-voltage area and the other to be in a lower-voltage area. This increases the difference in electric potential between the two feet, possibly initiating the flow of electricity through the body.

> **WARNING!**
>
> Do not casually step down from an electrically energized crane. Both feet need to touch the ground at the same time to minimize the potential for serious injury or death. Also, if an electrical circuit is created by making contact with the crane and the ground at the same time, the possibility of electrocution is high.

Instruct all other personnel on the site to stay away from the crane and anything connected to it. If they are close to the crane, instruct them to move away in short steps as well.

If you can remain in the crane, wait until the electrical authorities de-energize the circuit and confirm that the crane is no longer energized. It is likely that a substation circuit breaker or a recloser detected the fault and has opened to a point where it must be manually reset. However, this cannot be assumed until qualified personnel provide that information through testing.

Report every incident involving contact with power lines to the electrical authority and safety officer. If there was ever a chance of such an incident occurring due to work within the prohibited zone, utility personnel should have already been on site. As soon as possible, ensure that someone on the site has called for medical assistance or any other help needed on the scene. Lifts are made with cranes in the vicinity of power lines on a daily basis without incident. Take every precaution to ensure your encounters with work in this area are safe and successful as well.

2.3.0 Hazardous Weather

Mobile crane operators work outdoors. Under certain environmental conditions, such as extremely hot or cold weather or in high winds, work can become uncomfortable and maybe dangerously so. For example, snow and rain can have a dramatic effect on the weight of the load and on ground compaction. During the winter, the tires, outriggers, and crawlers can freeze to the ground. This may lead the operator to the false conclusion that the crane is on stable ground. As weight is then added during the lifting operation, an outrigger

float, tire, or crawler track may sink into mushy ground below the frozen surface. Heavy rain can also cause the ground under the crane to become unstable. The crane set-up site must be carefully evaluated to ensure that there is sufficient stability, including the condition of the soil below the surface.

High winds and lightning represent significant hazards on the jobsite (*Figure 15*). Both must be taken seriously. Crane operators must be prepared to respond appropriately to weather changes in order to avoid accidents and injuries. Fortunately, it is relatively rare for high winds or lightning to arrive without at least being reported as a possibility in weather reports. As a general rule, the lift team and crane operators have time to react appropriately. The site supervisor or lift director (possibly the same individual, depending on the lift characteristics) is responsible for ensuring that factors such as wind, heavy rain, fog, and the soil conditions that change as a result of weather are properly considered and addressed.

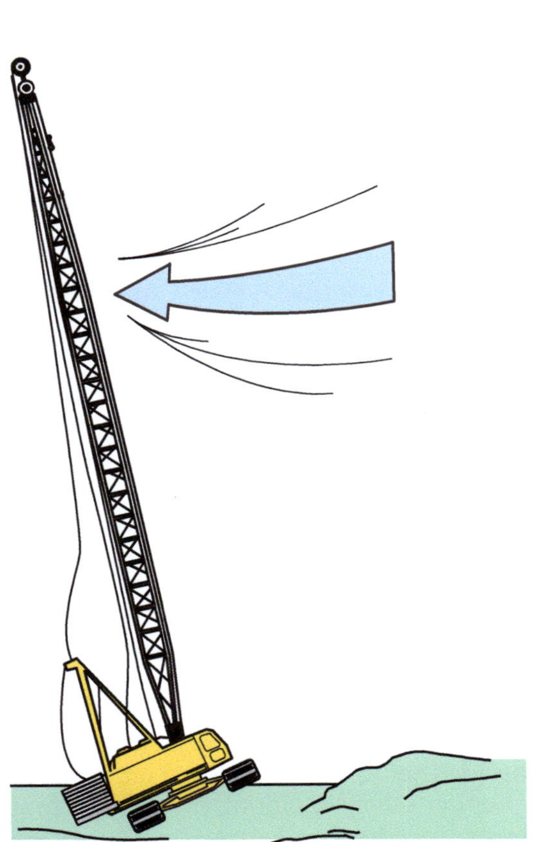
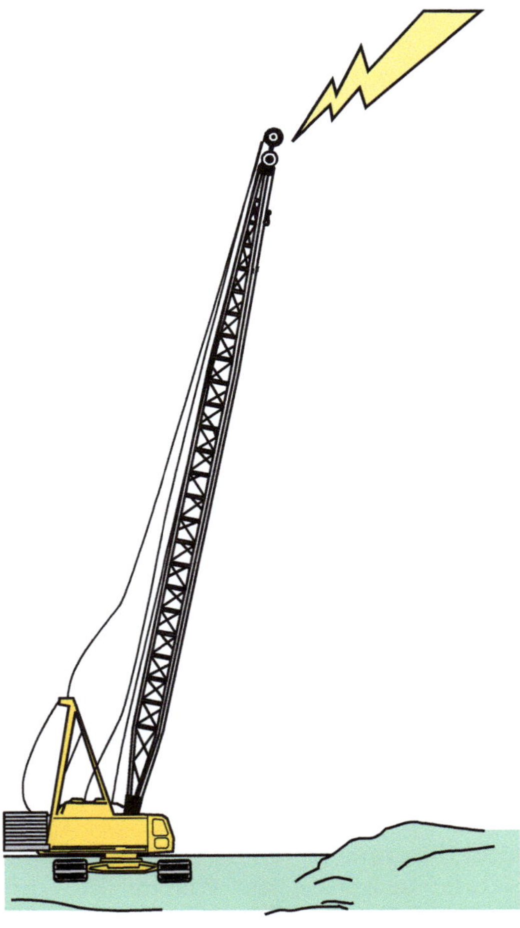

Figure 15 Wind and lightning hazards.

2.3.1 Wind

High winds typically start out as less dramatic gusts. Operators must be keenly aware of changing wind speeds. The crane boom should be down and stowed before the wind becomes too strong, not after the threshold has been passed. Keep in mind that the wind speed can be dramatically higher at the tip of the boom than it is at ground level. The operator must end and secure crane operations as soon as possible when the wind speed is increasing. This involves placing the boom in the lowest possible position and securing the crane. However, since wind speed affects capacity, a lift may have to be postponed due to a loss of crane capacity at wind speeds that are still acceptable for operation in general.

<table>
<tr><td>CAUTION</td><td>The wind chart shown in Table 2 is for a specific crane model. The information provided cannot be applied to all cranes. Always check and follow the manufacturer's wind chart speed for the specific crane in use.</td></tr>
</table>

Crane operators should be familiar with and follow the crane manufacturer's guidance related to wind. Their guidelines will differ from model to model. It is important to have the correct information for the specific crane in use. *Table 2* provides an example of a crane manufacturer's wind chart, showing how various wind speeds affect the rated load capacity of the crane. Note that the rated load capacity shown on the load charts is valid through wind speeds of 20 mph (32.2 kph). The chart also shows different capacity deductions for boom lengths less than or greater than 250 feet (76 meters). These reductions are applied to the load chart being used. This particular crane cannot be operated at all when the wind speed is above 45 mph (72.4 kph).

It is also important to point out that wind affects the load as much as the crane. It is not unusual for a crane to have the capacity to lift a given load at a wind speed of 30 mph (78.3 kph). But many loads have a very low weight-to-surface area ratio. A sheet pile is a good example. With a great deal of surface area and limited weight, a sheet pile is easily blown around by winds of 30 mph. In some cases, even though the crane is capable of making the lift, the wind's effect on the load has to be a significant factor in the decision. Perhaps additional tagline personnel can be put on the job, but the safer thing to do is wait until the wind speed has dropped to a more acceptable level. In this example, the wind hazard is not about the crane—it is about the load, the workers that must handle it, and other equipment or property in the area that could be damaged by it.

It is important to make this point regarding changing wind speeds. Assume that you are lifting a load with a wind speed of 25 mph. If the wind speed chart in *Table 2* applied, the rated load capacity of the crane must therefore be reduced by 20 percent. Now assume that the load represents 50 percent of the normal rated-load capacity, so the load weight now represents 70 percent of the capacity; still within the parameters of a standard lift plan. However, if the wind speed increases just 6 mph to 31 mph, the operation must be stopped and the crane secured. Although the crane itself can operate at this wind speed, the reduction of capacity is now 40 percent, and the load weight now represents 90 percent of the crane's capacity. Under these conditions, the lift becomes a critical lift that requires a documented lift plan. The alternative is to simply wait until

Table 2. Example of a Wind Speed Chart

Boom and Boom + Jib Lengths up to 250'	
Description	**Allowable Windspeeds in Miles Per Hour (mph)**
Boom and Boom + Jib Lengths Greater than to 250'	
1. Normal Lifting Operation. (See Capacity Charts.)	0–20 mph
2. Reduced Operation. Capacities must be reduced by 20%.	21–30 mph
3. Reduced Operation. Capacities must be reduced by 40%.	31–40 mph
4. Reduced Operation. Capacities must be reduced by 70%.	41–45 mph
5. No Operation. Store attachment on ground.	Over 45 mph
1. Normal Lifting Operation. (See Capacity Charts.)	0–20 mph
2. Reduced Operation. Capacities must be reduced by 35%.	21–30 mph
3. Reduced Operation. Capacities must be reduced by 60%.	31–40 mph
4. Reduced Operation. Capacities must be reduced by 70%.	41–45 mph
5. No Operation. Store attachment on ground.	Over 45 mph

the wind speed is lower. This type of scenario and the related decision-making process is repeated daily in the crane industry.

2.3.2 Lightning

Because crane booms extend so high and are made of metal, they are easy targets for lightning. Operators must be constantly aware of this threat. Lightning can usually be detected when it is several miles away. As a general rule of thumb, sound travels near the ground about 1 mile in 5 seconds, or about 1 kilometer in 3 seconds. Therefore, if there is a five-second delay between the flash of lightning and the sound of thunder, the lightning strike was roughly one mile away. Be aware, however, that successive lightning strikes can touch down up to 8 miles apart. That means once you hear thunder or see lightning, it is close enough for the next strike to present a hazard.

In some high-risk areas, local proximity sensors provide warnings when lightning strikes occur within a 20-mile radius. Once a warning is given, lightning is spotted, or thunder is heard, the crane operator must secure crane operations as soon as practical.

There have been many cases when the warning signs have not been taken seriously enough. Even if the above common rule of thumb were completely accurate, the process provides no information allowing one to determine when and where the next lightning strike will occur. If lightning is seen or thunder is heard, it is time to secure the crane and ignore the math.

Crane operators and the rest of the lift team must pay attention to the signs of thunderstorms and other weather events developing. It is best not to attempt operations that require a significant amount of time, such as a lengthy concrete pour, when there is a possibility that the operation may have to stop before it is complete. Doing so encourages the team to rush or remain in operation longer than it should once the warning signs are evident, in an attempt to complete the task. Both responses raise the potential for an accident.

Once crane operations have been shut down, all personnel should seek indoor shelter away from the crane. Even with the boom in the lowest position, it may be taller than surrounding structures and could still be a target for lightning strikes. Always wait a minimum of 30 minutes from the last instance of lightning or thunder before resuming work.

If lightning strikes a crane, a thorough inspection of the crane will be required. If lightning strikes the wire rope, for example, it may be damaged beyond safe use. All electrical systems need to be tested before the crane is returned to service, in addition to a thorough visual inspection.

2.4.0 Other Operational Safety Topics and Incidents

There are several other issues that affect the safety of a crane operation, as well as specific incidents to which a crane operator may need to respond. These issues are presented in the sections that follow.

2.4.1 Manufacturer's Requirements and Guidance

To operate a mobile crane safely, the operator must use the manufacturer's data and documentation provided for the specific crane in use. These manuals provide information on required startup checks and periodic inspections, as well as inspection guidelines. These manuals also provide many safety precautions and restrictions of use. Ignorance of any of these requirements or precautions is hazardous to the safe operation of the crane and could make the operator liable if an accident should occur. Operators should always read and follow the manufacturer's instructions. Crane manuals also provide information related to certain types of equipment failures and error messages that may present themselves. The manufacturer has reasons for any specific responses they provide. Follow the manufacturer's recommendations in all such cases.

2.4.2 Moving Cranes Safely

During the course of a job, cranes and other heavy equipment are moved to, around, and away from the site. Many accidents and injuries happen during the movement of heavy equipment. It is important to be especially safety conscious whenever equipment is moving.

Always follow these guidelines when driving equipment on public roads:

- Know and obey all state and local laws.
- Secure all attachments and loose gear.
- Use proper warning signs and flags per state and federal Department of Transportation (DOT) requirements.
- Drive slowly and never speed.
- Allow extra time to enter traffic.
- Stay in the extreme right lane on multi-lane highways.
- Travel with your lights on, day or night.
- Be aware of the crane's turning radius.
- Turn cautiously; allow for extensions or attachments and for structural clearances. Some

equipment is top-heavy and will tip over if a turn is made too fast.

- Be aware of the crane's stopping distance. Due to their size and weight, cranes can develop a great deal of momentum. Be especially careful when driving downhill.

When driving on the job site, follow these guidelines:

- Never drive a machine on a job site, in a congested area, or around people without a spotter or flagger to guide you. The spotter or flagger is responsible for determining and controlling the driver's speed.
- Be sure everyone is in the clear while backing up, hooking up, or moving attachments. When backing, allow a few moments for the back-up alarm to announce your intention before putting the crane in motion.
- If you cannot see your area clearly from the operator's seat and have no spotter, dismount and examine the site for possible hazards before proceeding.
- Wait for an all-clear signal from spotters before moving.
- Signal a forward move with two blasts of the horn; signal a reverse move with three blasts of the horn.

- Yield the right-of-way to moving equipment on haul roads and in pits.
- Maintain a safe distance from all other vehicles.
- When moving, keep the crane in gear at all times; never coast.
- Maintain a speed consistent with ground conditions.
- Pass only when necessary; use caution.
- Watch for overhead electrical power lines and ensure you have sufficient clearance. Refer to the lower portion of *Table 1*.
- Watch for flags indicating buried utilities (*Figure 16*).

2.4.3 Using Cranes to Lift Personnel

Although using a crane to hoist personnel is generally discouraged, it can be done safely with the correct equipment (*Figure 17*) and procedures. There are many personnel-platform styles to choose from, and they can be custom-made by several vendors to suit unique needs. Using a crane to lift personnel is prohibited by 29 *CFR* 1926.1431, unless the employer can demonstrate that the erection and use of a more conventional means to access an area is more hazardous than using the crane. When it is allowed, a personnel platform that meets the requirements of the

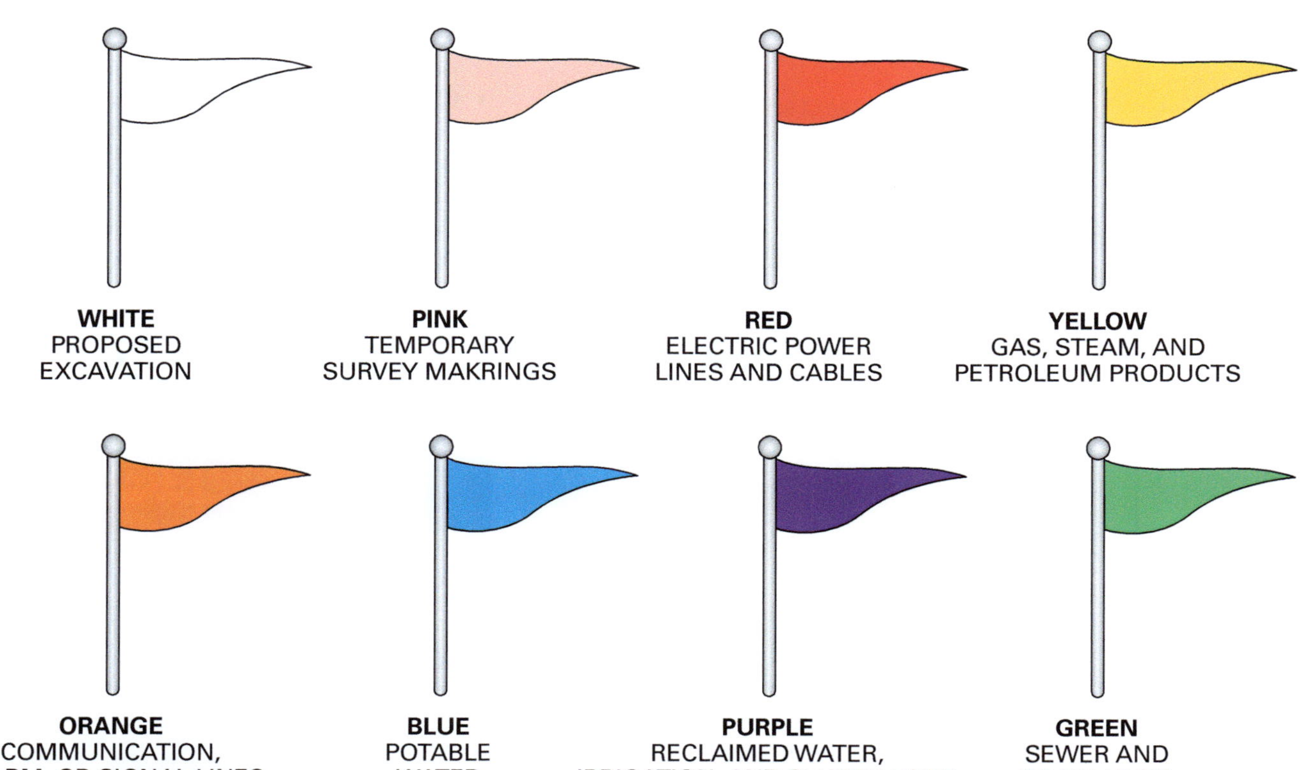

Figure 16 Flag colors used to identify underground utilities and related areas.

OSHA standard must be used (with some special exceptions).

ASME Standard B30.23, Personnel Lifting Systems, is devoted exclusively to the topic of hoisting personnel and the equipment requirements, as the name implies. Note, however, that this ASME standard has not been incorporated by reference into the OSHA standards, but the two standards do have a lot of similarities. Some requirements found in the standards include the following:

- The crane must be level within 1 percent, and outriggers must be used if the crane has them.
- The load cannot be more than 50 percent of the rated load capacity of the crane for the configuration and operating radius.
- If the personnel platform is in a stationary working position, the primary and secondary boom and vehicle braking systems and locking features must be engaged.
- The load line hoist drum must be equipped with a system that regulates lowering speed. If the crane has the capability to allow free-fall of the load line, it cannot be used for hoisting personnel.
- A boom angle indicator, a boom hoist limiting device, and anti-two-blocking devices are all required. If a luffing jib is in use, a jib angle indicator and a jib hoist limiting device are also required.
- Trial lifts, using an equal or greater weight than the expected load of the personnel and

their equipment, must be made. The crane must lift the load and move it to each location that the personnel will need to access. Trial lifts must continue to be conducted at the beginning of each shift, any time the crane is moved to a new location, and whenever the lift route changes in a way that adds new hazardous factors to the task.

- Proof testing is also required. The platform and rigging must be tested at 125 percent of the platform's rated load capacity and then inspected by a competent person. This testing must be done at each new jobsite and after any repairs or alterations have been performed on the personnel platform. Proof testing can be done in conjunction with the trial lifts. Some manufacturers of personnel platforms have developed a simple system of attaching a weight that equals 125 percent of the capacity to the bottom of the platform for this purpose (*Figure 18*).
- Unless the personnel platform is equipped with crane controls, the operator must remain at the controls in the cab at all times.
- Personnel being lifted must remain in contact with either the crane operator or the signal person (if used) at all times.
- If the wind speed exceeds 20 mph (32.2 kph), a qualified person must determine if it is safe to lift personnel for the required task, or whether the lift should be ended or postponed. Other weather issues may also prompt a qualified person to stop or postpone the lift.
- Occupants of the personnel platform must be equipped with personal fall-arrest equipment, with the lanyard attached to a structural member of the platform.
- Any other lift lines on the crane may not be used for lifting other items while lifting

Figure 17 Enclosed round personnel platform.

Figure 18 Personnel platform with detachable weight for proof testing.

NCCER – *Mobile Crane Operations*

personnel. Pile-driving operations are an exception to this rule.

- Unless the task directly involves work on a power line, hoisted personnel cannot be placed within 20 feet (6.1 meters) of lines up to a voltage of 350 kV, or within 50 feet (15.2 meters) of power lines over 350 kV.

Note that there are a number of other requirements, especially for special situations such as pile driving, lifting personnel in and out of drilled shafts, and transferring personnel to the site of a task in a marine environment. The above list does not represent all of the requirements and conditions found in the OSHA or ASME standards. If hoisting personnel is part of your work schedule, it is important to review the requirements that apply to your particular situation.

2.4.4 Incidents During Lifting Operations

Mechanical malfunctions or lapses in judgement during a lift can be very serious. If an equipment failure or operator error causes the operating radius to increase unexpectedly, the crane can tip or the structure could collapse. Loads can also be dropped during a mechanical malfunction. A sudden loss of load on the crane can cause a whiplash effect that causes the crane to tip or the boom to fail. The chance of these types of incidents occurring in modern cranes is greatly reduced because of system redundancies and safety backups. However, failures do happen, so the operator must stay alert at all times.

If a mechanical problem occurs, the operator should attempt to lower the load immediately. Next, the operator should secure the crane, tag the controls indicating the crane is out of service, and report the problem. The crane should not be operated until it is checked and repaired if necessary, by a qualified technician.

Carelessness by the operator can lead to accidents other than those associated with overloading the crane. These incidents include the following:

- *Striking the boom* – The operator must never allow the boom to strike any structure or load. Even what seems to be mild contact can dent, bow, or bend the lower boom chords and compromise the integrity of the boom. A serious incident may result in total boom collapse. If the boom touches or rests on another structure, the boom-loading changes from a compression force to a bending force. The boom is very strong in compression but weak in bending.

If the boom, mast, or jib is struck or damaged, stop the lift and leave the boom where it is, unless it is creating a new significant hazard by remaining in position. The load on the boom increases as the boom is lowered. As a result, a damaged boom or boom suspension could collapse during the lowering process. A second crane may be required to help lower a significantly damaged boom. The site supervisor or lift director will generally make this decision.

- *Backward collapse of a boom* – When operating near the minimum radius, with the boom at its highest angle, boom down as you set the load down. This will compensate for the tendency of the boom to move or jump back against the boom stops when the load is released, especially if the load on the boom is relieved too quickly. This action occurs because of the elasticity in the boom and boom hoist systems (*Figure 19*), and it can result in a backward collapse of the boom.

Another factor that may cause the backward collapse of a boom or upset the balance of the crane is high winds. Consult and follow the wind speed charts for the specific crane in use. Other factors that may contribute to boom collapse include the following:

- Continuing to pull on the hoist line after two-blocking has occurred if the crane is not equipped with a functional anti-two-blocking device
- Starting or stopping a swing suddenly if the boom is at a high angle
- Sudden forward movement of a crane that can send the boom over backward if it is being carried at a high boom angle
- Snubbing the hook block to the boom foot, then pulling it up tight
- Instability in certain positions when the crane is traveling on an incline

- *Two-blocking* – Two-blocking refers to a situation that results in the load block or hook assembly contacting an upper load block (if equipped) or the boom-point sheave assembly. Damage can occur to the sheaves, block, and/or wire ropes. However, if the load block makes contact and the operator continues to wind the drum, the crane is essentially pulling against itself. This can result in serious damage and boom failure. The devices shown in *Figure 20* offer protection against two-blocking. The dangling weight that encircles the hoist rope is attached to a switch. If the load of the weight is removed from the switch due to the load block contacting and raising it up, the switch opens and stops the hoist drum.

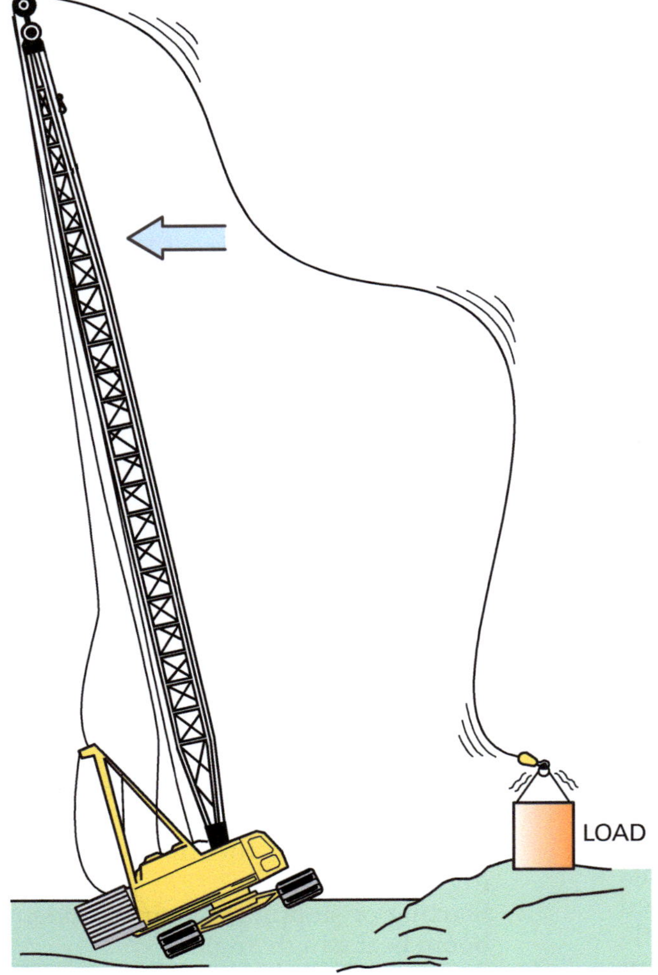

Figure 19 Crane response to the sudden release of a load.

In *Figure 20*, note the red devices that are inserted into the anti-two-block switches near the sheaves. These devices can be inserted into the switch to disable the anti-two-blocking safety feature. This may be necessary when stowing a mobile crane and preparing it for travel, so that the load block can be drawn up closer to the sheaves. They must not remain in place when the crane begins normal operations.

Crane operators must avoid two-blocking by being attentive to the position of the load block and boom. Do not rely on anti-two-blocking devices as control devices. They are there for safety purposes only.

2.4.5 Fire

The operator's judgment is crucial in determining the correct response to fire. The preferred first response is to cease crane operation, lower the load if practical, and secure the crane. In all cases of fire, evacuate the area even if the load cannot be lowered or the crane secured. After emergency

Figure 20 Anti-two-blocking devices.

services have been notified, a qualified individual may judge if the fire can be combated with a fire extinguisher. A fire extinguisher can be successful at fighting a small fire in its beginning stage, but a fire can get out of control very quickly. The operator must keep in mind that the highest priority is preventing loss of life or injury. Do not be overconfident in your ability to control a fire. Even trained firefighters using the best equipment can be overwhelmed and injured by fires.

According to 29 *CFR* 1926.1433(d)(6), all mobile cranes are required to have an accessible fire extinguisher on the equipment. *ASME Standard B30.5* indicates that it should be a minimum of a 10BC-rated portable fire extinguisher (*Figure 21*). Note that this has not been incorporated by reference into the OSHA standards. The operator should be trained in its use. This type of extinguisher is designed to combat Class B (flammable liquids) and Class C (electrical) fires. A BC extinguisher is typically charged with dry chemicals. The number 10 is an indicator that the extinguisher should be sufficient to fight a fire that covers 10 square feet (0.9 square meters). The effective range of this type of extinguisher is generally 5 to 20 feet (1.5 to 6 meters).

The extinguisher is constructed so that the extinguishing agent becomes pressurized when the pin is pulled and the handle is compressed. Remember to use the PASS method to fight a fire with a fire extinguisher:

- **P**ull the pin from the handle, breaking the tamper seal.
- **A**im the nozzle at the base of the fire while 8–10 feet (2.5–3 meters) away.
- **S**queeze the discharge handle.
- **S**weep the nozzle back and forth at the base of the flames.

ASME standards require that refueling of the crane be done when the engine is not running and in the absence of smoking, open flames, or any other sources of ignition. If refueling is being done with a portable container, the container must be a safety-type fuel can equipped with an automatic-closing cap and a flame arrester.

The best way to prevent a fire is to make sure the three elements needed for fire—fuel, oxygen, and a source of ignition—are never present in the same place at the same time. Oxygen in the atmosphere is impossible to eliminate, leaving only the fuel source and a source of ignition within human control. Here are some basic safety guidelines for fire prevention:

- Always operate in a well-ventilated area, especially when flammable materials such as shellac, lacquer, paint stripper, or construction adhesives are in use.
- Never smoke, strike a lighter, or light a match when you are working with flammable materials.
- Dispose or store oily rags only in approved, self-closing metal containers. Store other combustible materials only in approved containers.
- Know where to find additional fire extinguishers, what kind of extinguisher to use for different kinds of fires, and how to use the extinguishers when called upon.
- Make sure all extinguishers are fully charged. Never remove the tag from an extinguisher; it shows the date the extinguisher was last serviced and inspected.
- Keep open fuel containers away from any sources of sparks, fire, or extreme heat.
- Do not fill a gasoline or diesel fuel container while it is resting on a truck bed liner or other

ungrounded surface. The flow of fuel creates static electricity that can ignite the fuel if the container is not grounded.

- Be prepared at all times to call on professional firefighters, either by using the 911 system or through a direct call to the nearest fire department, even for a small fire. If you are in unfamiliar territory, ask for the location and contact information of the nearest fire department. Time permitting, this call is best made before a personal attempt to fight the fire begins.

2.4.6 Accident Investigations

Lifting operations must always be conducted in the safest manner possible to prevent accidents. However, should an accident occur, the operator should follow the required emergency procedures. Your employer is also likely to have specific procedures to be followed in the event of an accident. The crane operator must have an understanding of the type of information that may be requested on a typical accident investigation checklist. A great deal of information is typically required, including the following:

- Company name and mailing address
- Person receiving the report (name, title, and phone number)
- Investigator's name, title, and company
- Date and time of investigation
- Location and exact time of accident
- Description of the equipment involved, including the manufacturer, model, serial number, age of machine, unit number, and configuration
- Summary of the accident as recalled by witness(es)
- List of people who will make formal statements, including witnesses to the events from one hour before the accident to the end of the accident sequence, all personnel who perform maintenance on the involved equipment, and all personnel involved in determining and planning the lift
- Sketches of the accident scene to scale with as much detail as possible
- Photographs from as many angles as possible
- Weather conditions at the time of the accident
- Ground conditions
- Boom length
- Operating radius
- Actual load weight including load block, sling, and boom attachments
- Background information on the crane operator, riggers, and other members of the lift team – name, age, training, and experience

Figure 21 10BC-rated fire extinguisher.

- Summary of the accident and sequence of events
- Any conclusions based on findings

Normally, a safety officer or inspector from the company responsible for the jobsite and/or crane will conduct the accident investigation and complete the required forms. After the investigation is completed, the report is filed by the responsible company with the proper state and federal authorities.

If the accident results in one or more fatalities that are immediate or occur within 30 days of the incident, the accident must be reported to OSHA within eight hours. If the accident causes inpatient hospitalization, amputation, or the loss of an eye, it must be reported to OSHA within 24 hours. One minor exception relates to the timing of the hospitalization, amputation, or eye loss: If it occurs more than 24 hours after the accident, it is not required to be reported. After a report is submitted, investigators from OSHA and/or the National Institute for Occupational Safety and Health (NIOSH) will also conduct investigations into the accident.

Employers may require the crane operator and others involved in the incident to complete controlled substance and/or alcohol testing immediately following the event.

ASME Standard B30.5, Mobile and Locomotive Cranes. Current edition. New York, NY: American Society of Mechanical Engineers.

ASME Standard B30.23, Personnel Lifting Systems. Current edition. New York, NY: American Society of Mechanical Engineers.

ASME Standard P30.1, Planning for Load Handling Activities. Current edition. New York, NY: American Society of Mechanical Engineers.

OSHA Standard 1926, Subpart CC, Cranes and Derricks in Construction. **www.ecfr.gov**

Mobile Crane Safety Manual. 2014. Milwaukee, WI: Association of Equipment Manufacturers.

The following websites offer resources for products and training:

American National Standards Institute (ANSI), **www.ansi.org**

The American Society of Mechanical Engineers (ASME), **www.asme.org**

Occupational Safety and Health Administration (OSHA), **www.osha.gov**

North American Crane Bureau, Inc., **www.cranesafe.com**

Electronic Code of Federal Regulations, **www.ecfr.gov**

2.0.0 Section Review

1. Which of the following characteristics would most likely place a lift in the category of a critical lift?

   a. The load weight exceeds 60 percent of the crane's rated load capacity at the required operating radius.
   b. Winds at the time of the lift are forecast to be as high as 15 mph.
   c. A load must be lifted between two occupied buildings that are very close together.
   d. The boom of a hydraulic crane will need to be fully extended.

2. The minimum clearance to be maintained from power lines carrying up to 50 kilovolts (kV) when the crane is operating nearby is ______.

   a. 10 feet (3.05 meters)
   b. 15 feet (4.60 meters)
   c. 20 feet (6.10 meters)
   d. 25 feet (7.62 meters)

3. Which of the following lift factors is most likely to prompt a decision to stop a lift due to wind, even if the load weight is well within the limitations for the situation and the crane in use?

   a. The load has to be lifted over 25 feet up (7.6 meters).
   b. The load is primarily constructed of wood.
   c. The load has a high weight-to-surface area ratio.
   d. The load has a low weight-to-surface area ratio.

4. When driving a mobile crane, the crane operator signals a forward move with ______.

   a. one horn blast
   b. two horn blasts
   c. one forward hand signal
   d. two forward hand signals

5. A crane accident must be reported to OSHA within eight hours if the accident results in ______.

   a. injuries to two or more workers
   b. one or more fatalities
   c. a fire inside the crane cab
   d. property damage exceeding $5,000

SUMMARY

The size of a crane and the huge loads it handles expose the crane operator and anyone in the vicinity of the crane to the potential for serious injury and even death. However, crane operations are carried out thousands of times a day without incident. Mobile crane operators must be skilled, knowledgeable, and mentally and physically fit to reduce the potential for accidents. Other members of a lift team must be equally fit for their role. There are many rules and guidelines that apply to crane operations. The vast majority of accidents occur when those rules aren't followed.

Mobile crane operators carry a great deal of responsibility on their shoulders. It is imperative that they embrace safety as the number priority on every job, every day. This module is just the beginning of what you will learn about safety as you progress in your training. Decide now to be a part of establishing and supporting a safety culture within your organization, and commit yourself to the task of learning how to keep yourself and your co-workers safe in the most effective way possible.

1. Which of the following is the most important ASME standard for crane operators?

   a. *ASME Standard B30.20*
   b. *ASME Standard P30.1*
   c. *ASME Standard B30.5*
   d. *ASME Standard B30.23*

2. Information that must be kept in the crane cab at all times includes ______.

   a. the complete operating manual
   b. a copy of *ASME Standard B30.5*
   c. a copy of 29CFR1910.333
   d. the title to the crane, confirming the owner

3. Pre-lift planning guidance is provided by ______.

   a. *ASME Standard B30.5*
   b. *ASME Standard P30.1*
   a. 29CFR1926.1431
   b. 29CFR1910.180

4. The prohibited zone is defined as ______.

   a. any quadrant the crane is not designed to work in according to the manufacturer's guidance
   b. an area around an energized power line that no part of the crane or load should enter
   c. the length and operating radius at which the crane cannot operate if it is making a lift on uneven soil
   d. a circle defined by the outermost point of the crane's outriggers when extended

5. If a visual aid to identify the prohibited zone is not provided for the crane operator when operating near energized power lines, the crane operator must be in constant contact with a(n) ______.

   a. professional engineer qualified in power distribution and transmission
   b. designated representative of the electrical utility or power line owner
   c. dedicated spotter whose sole responsibility is to monitor the required clearance
   d. site supervisor and the local fire department

6. If it becomes necessary to jump off a crane that has contacted a power line, you should keep your feet ______.

   a. as far apart as possible
   b. about one foot apart, side-by-side
   c. about one foot apart, one in front of the other
   d. close together as you land

7. The rate of voltage decrease in the ground surrounding a crane after it has contacted a power line varies depending on the ______.

   a. type of boom
   b. length of the boom
   c. voltage of the power line
   d. resistance of the soil

8. The wind speed at which the operation of a given crane must be limited or halted is determined by ______.

   a. the crane manufacturer
   b. the crane operator
   c. OSHA
   d. crane manufacturer

9. When personnel are being hoisted on a personnel platform, the total weight of the load, basket, and rigging may not exceed ______.

   a. 100 percent of the crane's rated load capacity
   b. 75 percent of the crane's rated load capacity
   c. 50 percent of the crane's rated load capacity
   d. 25 percent of the crane's rated load capacity

10. If the boom, mast, or gantry is struck or damaged, the operator should first ______.

    a. lower the boom
    b. raise the boom
    c. bring the boom to the ground
    d. stop the lift

Trade Terms Quiz

Fill in the blank with the correct term that you learned from your study of this module.

1. A(n) __________ presents an increased level of risk beyond normal lifting activities.

2. Although the ones developed by many organizations are voluntary in nature, the __________ established by OSHA are enforceable by law.

3. An individual that can identify existing and predictable hazards and has the authority to take prompt corrective measures is referred to by OSHA as a(n) __________.

4. As a general rule, cranes are not allowed to handle materials on the ground beneath power lines because this area is part of the __________.

5. The area of specific dimensions that surrounds energized power lines that no part of a crane is allowed to enter is called the __________.

6. A device that senses the electric fields created by power lines and alerts the crane operator to the danger is a(n) __________.

7. A rigger or crane operator can use a(n) __________ to help protect workers in contact with a load from the danger of electrocution in the event the crane contacts a powerline.

8. When a lift can be accomplished using common procedures, methods, and equipment, it is referred to as a(n) __________.

9. If a crane operator lowers a load quickly and then brings it to a sudden stop in the air, the crane and rigging will experience __________.

10. OSHA refers to the space around energized power lines that no part of a crane or load can enter as the __________.

11. A brief power outage that is followed by power being quickly restored is probably the result of a __________ doing its job.

Trade Terms

Avoidance zone	device	Recloser
Competent person	Insulating link	Shock loading
Critical lift	Minimum clearance distance	Standards
High-voltage proximity warning	Prohibited zone	Standard lift

NCCER – *Mobile Crane Operations*

Appendix

EXAMPLE — CRITICAL LIFT PLANNING WORKSHEET

POWER CONSTRUCTION

CRITICAL LIFT PLANNING WORKSHEET

PROJECT: _______________________ SUBCONTRACTOR: _______________________

COMPETENT PERSON: _______________________ QUALIFIED RIGGER: _______________________

IS OPERATING ENGINEER CITY OF CHICAGO CERTIFIED? ☐ YES ☐ NO ☐ N/A

IF MULTIPLE CRANES ARE REQUIRED FOR THE LIFT, A SEPARATE WORKSHEET IS REQUIRED FOR EACH CRANE.

CRANE INFORMATION

CRANE OWNER / SUPPLIER: _______________________

BOOM TYPE: ☐ TELESCOPING BOOM ☐ LATTICE BOOM

CRANE BASE: ☐ ON RUBBER TIRE ☐ OUTRIGGERS ☐ CRAWLER ☐ AIRCRAFT

BOOM LENGTH: _______________________

JIB LENGTH: _______________________

COUNTERWEIGHT: _______________________

CAPACITY OF CONFIGURATION: _______________________

ANNUAL CERTIFICATION DATE: _______________________

LOAD DATA & RIGGING

WHAT IS BEING HOISTED? (TYPE OF MATERIAL / PRODUCT) _______________________

HOW WILL THE LOAD BE HOISTED? (RIGGING CONFIGURATION) _______________________

WILL ENGINEERED PICK POINTS BE UTILIZED? ☐ YES ☐ NO

WHAT TYPE(S) OF RIGGING IS NEEDED? _______________________

WHO IS PROVIDING THE RIGGING? _______________________

HAS THE RIGGING BEEN INSPECTED? ☐ YES ☐ NO

TAG LINES UTILIZED (IF NOT, WHY) ☐ YES ☐ NO

WEIGHT OF LOAD: _______________________

RIGGING WEIGHT: _______________________

BLOCK & LINE WEIGHT: _______________________

TOTAL LOAD WEIGHT: (RIGGING + BLOCK & LINE + LOAD WEIGHT) _______________________

IS LOAD GREATER THAN 75% OF CHART? ☐ YES ☐ NO

Critical Lift Planning Worksheet Revised October 2012

Figure A01A Critical Lift Planning Worksheet (1 of 2)

Figure A01B Critical Lift Planning Worksheet (2 of 2)

Avoidance zone: An area both above and below one or more power lines that is defined by the outer perimeter of the prohibited zone. As the name implies, any part of the crane should avoid this area whenever possible, and may not enter the area except under special circumstances.

Competent person: As defined by OSHA, an individual who is capable of identifying existing and predictable hazards in the surroundings or working conditions which are unsanitary, hazardous, or dangerous to employees, and who has the authorization to take prompt corrective measures to eliminate such hazards.

Critical lift: As defined in ASME Standard B30.5, a hoisting or lifting operation that has been determined to present an increased level of risk beyond normal lifting activities. For example, increased risk may relate to personnel injury, damage to property, interruption of plant production, delays in schedule, release of hazards to the environment, or other significant factors.

High-voltage proximity warning device: An early-warning device that senses the electric fields created by high-voltage power lines and alerts the crane operator and/or the lift team to the hazard.

Insulating link: An electrical insulating device used on the crane hook to protect workers in contact with the load from the danger of electrocution in the event the crane contacts a powerline. The link can also provide some level of protection for the crane if the load alone contacts a power line.

Minimum clearance distance: The OSHA-required distance that cranes, load lines, and loads must maintain from energized power lines. This OSHA term is synonymous with the ASME term prohibited zone.

Prohibited zone: An area of specific dimensions, based on the voltage of a power line(s) that no part of the crane is allowed to enter during normal operations. Special considerations and preparations are required if the crane's task must place any part of it within the prohibited zone. The prohibited zone is a term used by ASME that is synonymous with the term minimum clearance distance used by OSHA.

Recloser: A device that functions much like a circuit breaker, or in conjunction with a circuit breaker, in power distribution and transmission systems that automatically recloses the circuit after a fault has been detected and the circuit has been opened. Reclosers allow the power system to be re-energized quickly after a transient (temporary) condition, such as a tree limb falling across power lines and then falling to the ground, has occurred. If the fault reoccurs upon closure, the circuit will typically remain open until the situation has been addressed by power line workers or operators.

Shock loading: A sudden, dramatically increased load imposed on a crane and rigging, usually as the result of momentum from the load that occurs due to swinging side-to-side, dropping the load and then stopping it suddenly, and similar actions that create momentum.

Standards: As defined by OSHA, statements that require conditions, or the adoption or use of one or more practices, means, methods, operations, or processes, that are reasonably necessary or appropriate to provide safe or healthful employment and places of employment. Standards developed by some organizations are voluntary in nature, while OSHA standards and those they incorporate by reference are enforceable by law.

Standard lift: A lift that can be accomplished through standard procedures, allowing load-handling and lift team personnel to execute it using common methods, materials, and equipment.

Additional Resources

This module is intended as a thorough resource for task training. The following reference works are suggested for further study.

ASME Standard B30.5, Mobile and Locomotive Cranes. Current edition. New York, NY: American Society of Mechanical Engineers.

ASME Standard B30.20, Below-The-Hook Lifting Devices. Current edition. New York, NY: American Society of Mechanical Engineers.

ASME Standard B30.23, Personnel Lifting Systems. Current edition. New York, NY: American Society of Mechanical Engineers.

ASME Standard P30.1, Planning for Load Handling Activities. Current edition. New York, NY: American Society of Mechanical Engineers.

29 CFR 1926, Subpart C, **www.ecfr.gov**

29 CFR 1926.251, **www.ecfr.gov**

29 CFR 1926.753, **www.ecfr.gov**

Mobile Crane Safety Manual (AEM MC-1407). 2014. Milwaukee, WI: Association of Equipment Manufacturers.

The following websites offer resources for products and training:

American National Standards Institute (ANSI), **www.ansi.org**

The American Society of Mechanical Engineers (ASME), **www.asme.org**

Occupational Safety and Health Administration (OSHA), **www.osha.gov**

North American Crane Bureau, Inc., **www.cranesafe.com**

Electronic Code of Federal Regulations, **www.ecfr.gov**

Figure Credits

Link-Belt Construction Equipment Company, Module opener, Figures 4, 6, Table 2

© Gary Whitton/Shutterstock.com, Figure 1

Carolina Bridge Co., Figures 2, 3

Mechanix Wear, SA01

© Donvictorio/Dreamstime.com, Figure 7

Salisbury Electrical Safety, Figure 10

Insulatus Company, Inc., Figure 11

Atlas Polar Company Ltd., Figure 12

Topaz Publications, Inc., Figures 13, 20, SA02

Lifting Technologies, Figures 17, 18

Courtesy of Amerex Corp., Figure 21

Power Construction Company, LLC, Appendix

Answer	Section Reference	Objective
Section One		
1. c	1.1.0	1a
2. b	1.2.2	1b
Section Two		
1. c	2.1.0	2a
2. a	2.2.0; Table 1	2b
3. d	2.3.1	2c
4. b	2.4.2	2d
5. b	2.4.6	2d

NCCER CURRICULA — USER UPDATE

NCCER makes every effort to keep its textbooks up-to-date and free of technical errors. We appreciate your help in this process. If you find an error, a typographical mistake, or an inaccuracy in NCCER's curricula, please fill out this form (or a photocopy), or complete the online form at **www.nccer.org/olf**. Be sure to include the exact module ID number, page number, a detailed description, and your recommended correction. Your input will be brought to the attention of the Authoring Team. Thank you for your assistance.

Instructors – If you have an idea for improving this textbook, or have found that additional materials were necessary to teach this module effectively, please let us know so that we may present your suggestions to the Authoring Team.

NCCER Product Development and Revision

13614 Progress Blvd., Alachua, FL 32615

Email: curriculum@nccer.org
Online: www.nccer.org/olf

❏ Trainee Guide ❏ Lesson Plans ❏ Exam ❏ PowerPoints Other ___________________

Craft / Level: ____ Copyright Date: ___________

Module ID Number / Title: ____

Section Number(s): ____

Description: ____

____

____

____

____

Recommended Correction: ____

____

____

____

____

Your Name: ____

Address: ____

____

Email: ____ Phone: ______________

Operating a Crane

OVERVIEW

This module reinforces and elaborates on information learned in previous modules and prepares mobile crane operator trainees for their first opportunity at the controls of a running crane. The details of site evaluation and its suitability for the lift are presented, as well as the inspections and preparations necessary to ready the crane itself. Trainees will operate a crane to lift, move, and land a load in response to the appropriate signals.

Module 21105

Trainees with successful module completions may be eligible for credentialing through the NCCER Registry. To learn more, go to **www.nccer.org** or contact us at 1.888.622.3720. Our website has information on the latest product releases and training, as well as online versions of our *Cornerstone* magazine and Pearson's product catalog.

Your feedback is welcome. You may email your comments to **curriculum@nccer.org**, send general comments and inquiries to **info@nccer.org**, or fill in the User Update form at the back of this module.

This information is general in nature and intended for training purposes only. Actual performance of activities described in this manual requires compliance with all applicable operating, service, maintenance, and safety procedures under the direction of qualified personnel. References in this manual to patented or proprietary devices do not constitute a recommendation of their use.

21105 V3

OPERATING A CRANE

Objectives

When you have completed this module, you will be able to do the following:

1. Describe lift preparations related to the site and the environment.
   a. Describe lift considerations and preparations related to the site.
   b. Identify considerations for operation around the public and other equipment.
2. Describe crane operational aids, inspection requirements, and general operating practices.
   a. Describe basic mobile crane operational aids and controls.
   b. Describe the different types of required crane inspections.
   c. Describe general crane operating practices.

Performance Tasks

Under the supervision of your instructor, you should be able to do the following:

1. Survey and select a lift position based on the adequacy of ground support.
2. Analyze a lift site for room to assemble/disassemble a crane and boom.
3. Perform pre-operation and post-operation crane inspections.
4. Safely start a mobile crane.
5. Safely set up a mobile crane, including the outriggers, blocking, and barricades.
6. Manipulate crane controls to smoothly lift, position, and land a load.
7. Determine and report the boom angle to the instructor during operations upon request.
8. Safely shut down a crane.

Trade Terms

Blind lift	Initial inspection
Cofferdam	Periodic inspection
Fall zone	Pressure transducer
Frequent inspection	Regular inspection
Hoist pawl	

Industry Recognized Credentials

If you are training through an NCCER-accredited sponsor, you may be eligible for credentials from NCCER's Registry. The ID number for this module is 21105. Note that this module may have been used in other NCCER curricula and may apply to other level completions. Contact NCCER's Registry at 888.622.3720 or go to *www.nccer.org* for more information.

Contents

1.0.0 LIFT PREPARATIONS

Objective

Describe lift preparations related to the site and the environment.

  a. Describe lift considerations and preparations related to the site.
  b. Identify considerations for operation around the public and other equipment.

Performance Tasks

1. Survey and select a lift position based on the adequacy of ground support.
2. Analyze a lift site for room to assemble/disassemble a crane and boom.

Trade Terms

Blind lift: A lift in which the crane operator cannot see the load during any portion of the lift.

This module is designed to prepare crane operators for their first opportunity to operate a mobile crane. Note that the Performance Tasks associated with this module should not be attempted without first completing Module 21106, *Crane Safety and Emergency Procedures*, and Module 53101, *Crane Communications*.

Although one can learn a great deal about operating a crane from a book, there is far more to learn inside the crane cab. Operating a crane requires coordination, a complete understanding of the crane and its specific controls, and often, a gentle touch. Proficiency comes from practice and experience.

1.1.0 Site Conditions and Setup

According to *ASME Standard B30.5, Section 5-3.1.3.2.1*, the site supervisor is ultimately responsible for the site conditions and their suitability for a crane lift. Depending on the project, the site supervisor and lift director may be the same person. The crane operator must have a thorough understanding of the site-related issues that can negatively affect the crane and the lift, and communicate with the site supervisor as necessary to ensure the site is adequate. It is important that there be an agreement on the suitability of the site between the two parties.

On many job sites, the operator will be told to perform a lift at a specific place on a designated load with limited guidance. Everyone pays close attention to the big, complex lifts, but little attention is given to smaller lifts that should be straightforward. If the operator chooses to perform such jobs without an experienced lift director in place, then he or she needs to personally analyze the site conditions for safety in relation to crane operation. 29 *CFR* 1926.1402(d) states that "If there is no controlling entity for the project, the requirement in paragraph (c)(1) of this section must be met by the employer that has authority at the site to make or arrange for ground preparations needed to meet paragraph (b) of this section." In other words, in the absence of a site supervisor and lift director, the crane operator is likely the only individual with the authority and knowledge to determine if the ground conditions are safe for crane assembly and/or lifting. The crane operator cannot ignore the responsibility of ensuring that the ground conditions are appropriate for the work. Operators must understand all aspects of crane setup, site preparation, and site evaluation to ensure the crane and other workers are not placed in a hazardous situation.

The sections that follow present considerations for the lift site, assembly and disassembly of the crane, and getting the crane to the site. Crane transit is presented after the other two topics for a reason; unless it is a simple routine lift in predictable conditions, the lift should be planned and the site analyzed before the crane ever leaves the lot.

1.1.1 Site Assessment and Ground Conditions

Safe crane operation starts with a thorough site assessment. The site assessment is a methodical process used to identify hazards, evaluate ground conditions, and devise plans to overcome or eliminate all hazards which may interfere with safe crane operation. This process could require an investigation by personnel from many different trades and technical backgrounds, and may be quite time consuming. As the size of the crane and the complexity of the lift increase, the attention given to the site assessment also increases.

As operators gain experience and progress through their careers, they are likely to become involved in the site assessment process as experts in their own right. Other professionals who may be involved in the assessment process might be riggers, structural engineers, geotechnical engineers, and crane manufacturer representatives, to name a few. This review is critical in ensuring that a crane can be safely transported into and used in various areas of the jobsite. On the other hand, the site assessment process is quite simple for routine lifts of light loads on sturdy ground.

The site assessment is a compilation of questions that must be answered with priority given to safety. To perform an adequate assessment, the right questions must be asked and correctly answered to ensure that the appropriate safety precautions will be provided for planned crane operations. Questions that should be resolved include, but are not limited to, the following:

- Is the ground firm? Has there been any recent excavations?
- Is the area well-drained? Any standing water?
- Are there any potential voids underneath a concrete slab or other type of pavement from which the crane may be lifting?
- Does the ground slope such that the crane cannot be easily leveled?
- Are there any underground utilities, such as gas or water mains, that could be damaged by the weight of the crane?
- Are there any nearby power lines? If so, what is the voltage?
- Does the site allow proper clearance for the crane to rotate freely?

Remember that, on routine lifting jobs, the crane operator may be responsible for the site assessment.

If the location of any underground utilities is a concern, the best thing to do is to have any buried utilities located and marked. Each state has its own agency to coordinate this service. However, there is a nationwide connection between them, reached by dialing 8-1-1 in any state. The 8-1-1 service puts callers in touch with their state agency to arrange for site surveys. Expect some lead time before the survey is completed. Most states require 3–10 days of notice to ensure the task is done before the work begins. If a local utility is to be contacted directly about the location of buried utilities, several days of lead time is required for them as well. Expect longer delays from utilities and the 8-1-1 service following a natural disaster or similar event.

When preparing for any type of lifting operation, consider the effects that different ground conditions can have on a crane and its stability. One basic requirement is that a crane be within one percent of level. Another way to think of this is that the drop or rise (slope) of the crane cannot exceed one foot for every 100 feet. Failure to meet

this requirement can move the tipping point of the crane closer to its center of gravity, reducing stability. Trenches, cellars, embankments, flowing water, and standing water have a direct effect on the compaction of the ground. These situations require special attention.

When assessing the ground conditions, carefully note the location of any potential hazards. *Figure 1* shows examples of hazardous ground conditions. Where there is loosely compacted fill or recent excavations, including trenches dug for the installation of piping and other utilities, the crane must be positioned a distance away from the edge of the excavation. As a general rule, crane mats, crawler-crane track edges and tips, and outriggers must be placed 1.5 times the depth of the trench away from the edge. For example, if the trench is 6 feet (2 meters) deep, these parts of the crane must be positioned at least 9 feet (3 meters) away. However, each situation should be evaluated independently with consideration given to the crane size, load, blocking material, and soil type.

Outrigger floats require blocking or pads to increase the surface area to which the weight of the crane and load is applied (*Figure 2*). Outrigger float blocking must be done properly to gain the desired effect, as shown in *Figure 3*. If the operator has any doubts about the compaction of the soil, crane mats (*Figure 4*) are the safest alternative, as they maximize the surface area across which the load is applied. The extra time and money required to set up and use crane mats is certainly less than the cost of recovering an overturned crane. It is important, however, to note that most lifts that require crane mats are not considered routine jobs where a crane operator represents the organization alone. Transporting and assembling a crane mat is a significant task of its own.

In addition to the ground conditions, a site assessment must also include the available clearance, including the space required for the crane to safely rotate. This clearance serves two functions:

- It ensures that there is sufficient space around the crane as it rotates, to prevent workers from being inadvertently crushed between the crane and another object.
- It ensures that there is sufficient clearance for the crane to rotate without striking another object as the lift progresses.

During lifting operations, the natural movement and flexing of the crane structure makes it important for the moving parts of the crane to have at least a minimum clearance of 2 feet (0.6 meters) to avoid creating a hazard. This includes any counterweights mounted; they typically extend farther

NCCER – *Mobile Crane Operations*

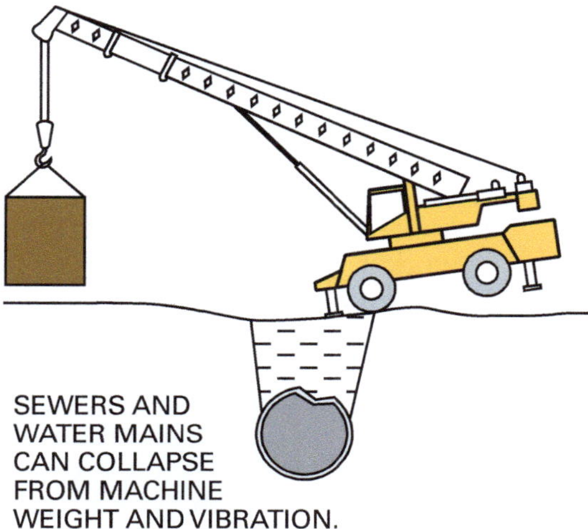

Figure 1 Potentially hazardous ground conditions.

Figure 2 Manufactured outrigger float pad.

from the center of the superstructure than any other component. As a result, it is the portion of the crane that is most likely to contact a nearby obstruction. Contacting a wall or other obstruction during the lift could have devastating results. Neither the OSHA or ASME standards identify a minimum clearance to nearby obstructions. OSHA does, however, identify some operator practices to maintain personnel safety in the immediate area of the crane. Those requirements are reviewed in the sections that follow.

1.1.2 Assembly/Disassembly Clearance

When a crane arrives at a job site, it may need to be assembled in order to prepare it for lifting operations. This is especially true of lattice-boom cranes, but any crane may need a jib installed, for example. When crane assembly or disassembly is required, there must be sufficient clearance available. This can only be determined through a site assessment. Keep in mind that a jobsite can change significantly from day-to-day, and obstructions may not be in the same place each day. Assembly and disassembly processes require an assembly/disassembly (A/D) director, according to 29 *CFR* 1926.1404. Per the standard, the A/D director must be both a qualified person and a competent person. On a large lattice-boom crane, a significant team may be required. A second crane to assist in manipulating the boom sections is often required. On routine lifts, however, especially those involving telescopic-boom cranes, the crane operator may be the only company representative on site. In this case, the crane operator also serves in the role of A/D director. This is acceptable as long as the operator meets the OSHA description of both a qualified and competent person. Limited assembly and disassembly of many cranes can be done by one person, especially telescopic-

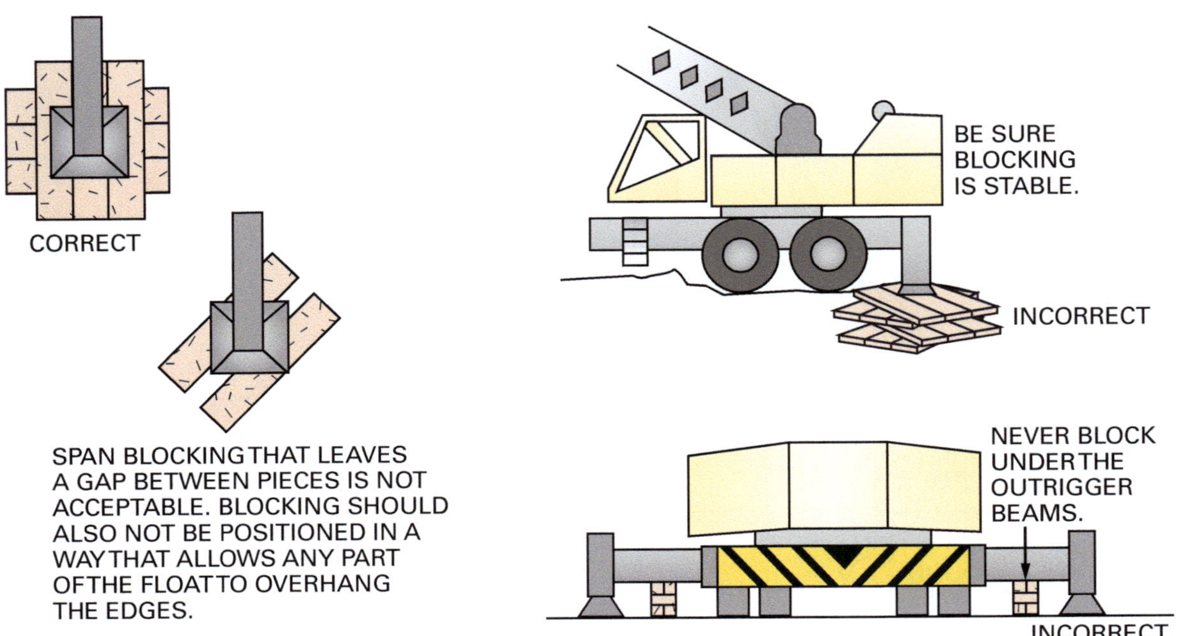

Figure 3 Outrigger blocking installation.

boom models. In crane rental operations, this is a common requirement.

Lattice-boom cranes require room to fully extend the boom on the ground as well as room for the support vehicles that carry the boom sections and counterweights. The assembly and disassembly process requires the orderly staging of materials to support the boom and individual sections. The sections must be organized in the manufacturer-directed sequential order.

Boom assembly typically requires the installation of large pins to attach sections together. A typical double-taper boom pin is shown installed in *Figure 5*. Note that there are two safety pins shown; one on each end of the pin. If the boom pins are being secured after assembly with cotter pins that must be bent to lock them in place, the cotter pins should be replaced after each use. The safety pins shown here can be reused and are far more convenient. Although double-taper boom pins are popular today, there are many single-taper pins in service as well. *Figure 6* shows an example of both a double-taper and a single-taper pin, stowed in place for transportation of the boom section. The single-taper pin has only one safety pin, as the head is enlarged on the opposite end to keep it from pulling through.

Double-taper boom pins can be driven in or out from either side of the connection. This eliminates the need for a worker to crawl inside the lattice to install or remove the pin. This is a dangerous position, as boom sections that are not yet connected can move at any time. The hazard is especially high during disassembly, as the

Figure 4 Crane mat in place.

boom section often moves unexpectedly when the boom pin is driven out. Due to the size and weight of boom sections, the potential for serious injury is significant. A single-taper pin, however, must come out in the same direction from which it was installed. If it is driven in from the outside, a worker will have to drive it out from inside the lattice. Alternatively, it can be driven in from inside the lattice, which allows it to be driven out from outside of the boom. Although the latter choice may be somewhat safer, neither choice represents an ideal situation. For this reason, the double-tapered pins are generally favored. Regardless of which type of pin is used, follow the manufacturer's guidance for assembly and heed all warnings offered in the instructions.

Figure 5 A double-taper boom pin installed.

Figure 6 Double-taper and single-taper boom pins, stowed for travel.

1.1.3 Transit

One of the concerns in crane operation is getting the crane to the site. Those involved with the transport of the crane must know the safest and most acceptable travel route. Consideration must be given to the type of roadway, its width and grade, and any overhead obstructions. For example, if the only travel route to the jobsite has a bridge with a limited weight capacity and the crane exceeds that capacity, the crane cannot travel this route. Situations like this may result in the use of another crane that is already on the other side of the bridge or obstruction, or selection of an entirely different type or configuration of crane capable of crossing the bridge. Overhead clearance at overpasses and under utility lines on the way must also be considered. Poorly planned moves can cause extensive damage to cranes, roadways, and utilities. This, in turn, causes disruption of traffic flow and great financial liability for those involved and found responsible.

Crane operators must consider whether the jobsite on which the crane will work has adequate access to enter and move to the needed positions. An alternate access route may be needed in some cases. Large wheeled cranes and trucks carrying crawler cranes cannot maneuver in exceptionally tight spaces. The planned quadrant of operation may even be a factor in how the crane is driven into the site if it is unable to turn around or be repositioned easily.

Severe road conditions can cause crane damage, especially when speed is added. A combination of rough roads and the weight of the crane can severely damage the suspension. Jobs at remote sites, such as wind turbine farms, can challenge even the toughest crane. This can result in expensive repair bills and project delays created by repair work.

When there is any doubt about the route, always consult maps, GPS systems, and local utility companies about the location of hazards. This service is provided free of charge by most utilities if they have advance notice. Ignoring this procedure can cost your company a lot of money for the restoration of service by the utility company and equipment repairs.

1.2.0 Lifting Operation Considerations

There are three crucial issues that must be considered throughout lifting operations: the operation of a crane in close proximity to another crane; safety precautions that must be observed when lifting in areas of public access; and the safety of on-site personnel.

1.2.1 Proximity to Other Equipment

Multiple cranes are often needed in the same area (*Figure 7*). Cranes operating near each other can pose a hazard on the jobsite. When several cranes are operating close together, either independently or in a cooperative effort, there are some basic rules

Figure 7 Cranes working in close proximity.

to follow in order to minimize the hazards. One example of a hazard is the entanglement of hoist lines, with or without a suspended load to complicate matters. Whenever possible, operators should be in eye contact with each other and with a qualified signal person. Voice communication for cooperative work instead of hand signals is generally preferred.

Cranes are not the only type of equipment that may be nearby. Trucks carrying materials may come and go beside the crane so the loads can be picked and placed in a staging area or positioned for assembly. Personnel lifts may also be nearby; striking the boom of an extended personnel lift can have devastating consequences.

1.2.2 Protecting the Public

Not all crane operations take place in isolated areas. The probability of lifts taking place in public areas is high (*Figure 8*). As a result, there are many concerns that must be addressed when the public has access to an area where crane operations are planned. To protect the public, a few simple rules must be followed.

A qualified signal person and warning signs must be used during the operation of the crane. Also, the area under the crane's required operating radius must be isolated with barricades and access must be controlled. This area is particularly hazardous to the public. If working from or adjacent to a public street, highway, or thoroughfare, it may be necessary to obtain a permit from

Figure 8 Crane on city streets.

the proper authorities. This is common when working in congested city areas. Advance notice of any highway closure must be provided to the proper authorities. Law enforcement officers may be required to control traffic.

Note that many cities, such as Chicago and New York City, require crane operators working in the city to have a license issued by a governing authority, such as city government. Ensure that you are aware of and familiar with the licensure requirements of the local area. Cities such as New York may also have regulations regarding crane operation that go well beyond the OSHA requirements.

1.2.3 The Lift Team and Team Safety

When assessing a crane operation, personnel requirements must be addressed (*Figure 9*). It is critical that enough personnel be available to handle the lift. If a **blind lift** is being conducted, or if the crane is operating near power lines, an additional qualified signal person may be required to assist in the lifting operation. Some lifts also require tagline handlers to control the load during movement and placement.

Qualified riggers, lift directors, and/or other craft professionals are generally involved in a lift process. Always keep an eye out for personnel that could be injured if rigging or crane functions fail. Crane operators have a unique perspective from the cab that other workers do not share. Use that perspective to monitor the work area and ensure workers are not wandering into hazardous areas without realizing it.

Remember that the swing radius of the crane superstructure must be barricaded to keep personnel out of a hazardous area. If a member of the lift team must enter that area for any reason and will be out of the clear view of the operator, the operator must be informed. Once informed, the operator cannot rotate the superstructure of the crane until being notified that the individual has returned to a safe area. The crane operator's responsibilities for personnel safety within the swing radius of the crane are found in 29 *CFR* 1926.1424.

Figure 9 Personnel needs for the lift must be considered.

Changes in New York City

In response to a number of accidents and casualties related to cranes, New York City made some changes to the regulations governing cranes. As of March, 2016:

- Cranes that must be shut down and secured when winds exceed 20 mph (32.2 kph), according to their manufacturer, can no longer operate in the city. The large crawler crane that collapsed in Tribeca in February, 2016 had no ratings above 20 mph.
- A licensed engineer must be present for any crawler crane operation if the crane is rated for less than 30 mph (48.3 kph) winds. The engineer must monitor wind speeds and bring the operation to a halt when necessary. He or she must also ensure that the crane is properly secured after work has stopped.
- Crane operations are prohibited entirely whenever winds exceed 30 mph (48.3 kph) or are forecast to do so.

Rules for controlling pedestrians and vehicular traffic also became more stringent, and the base fine for improperly safeguarding a crane operation more than doubled from $4,800.00 to $10,000.00. Additional research and discussion may lead to more changes to improve crane safety.

Additional Resources

ASME Standard B30.5, Mobile and Locomotive Cranes. Current edition. New York, NY: American Society of Mechanical Engineers.

29 CFR 1926, Subpart CC, *Cranes and Derricks in Construction.* **www.ecfr.gov**

Mobile Crane Safety Manual. 2014. Milwaukee, WI: Association of Equipment Manufacturers.

Cranes: Design, Practice, and Maintenance. Second Edition, 2002. Ing J. Verschoof. Hoboken, NJ: John Wiley and Sons, Inc.

North American Crane Bureau, Inc. **www.cranesafe.com**

1.0.0 Section Review

1. A crane's suspension system can be severely damaged by the combination of ______.

    a. slick roads and axle weight
    b. steep grades and crane weight
    c. rough roads and crane weight
    d. crane weight and boom type

2. Which of the following is *not* required when lifting in a public area?

    a. A signal person
    b. Warning signs posted
    c. Isolation of the area under the crane's operating radius
    d. Covered walkways for pedestrians to pass under the boom area

2.0.0 CRANE INSPECTION AND OPERATING PRACTICES

Objective

Describe crane operational aids, inspection requirements, and general operating practices.

a. Describe basic mobile crane operational aids and controls.
b. Describe the different types of required crane inspections.
c. Describe general crane operating practices.

Performance Tasks

3. Perform pre-operation and post-operation crane inspections.
4. Safely start a mobile crane.
5. Safely set up a mobile crane, including the outriggers, blocking, and barricades.
6. Manipulate crane controls to smoothly lift, position, and land a load.
7. Determine and report the boom angle to the instructor during operations upon request.
8. Safely shut down a crane.

Trade Terms

Cofferdam: A watertight and generally temporary enclosure built in a body of water or wet area, designed so that water can be pumped out to create a dry area for work below the water line.

Fall zone: Defined in 29 *CFR* 1926.1401 as the area beneath a load in which it is reasonably foreseeable that partially or completely suspended materials could fall in the event of an accident.

Frequent inspection: Required recurring crane inspections conducted at daily-to-monthly intervals, directed by *ASME Standard B30.5*. Frequent inspections are limited in scope and focused primarily on safety.

Hoist pawl: A hinged or pivoting device designed to interlock in the hoist mechanism and physically prevent its movement.

Initial inspection: An inspection required by *ASME Standard B30.5* on new or significantly altered cranes, conducted by a qualified person, to ensure the crane's compliance with ASME standards.

Periodic inspection: Required recurring inspections conducted at monthly to annual intervals, directed by ASME Standard B30.5. The scope of these inspections is more complex than the scope of frequent inspections.

Pressure transducer: A device that measures pressure and then converts the pressure value to a proportional analog electrical signal; also referred to as a pressure transmitter.

Regular inspection: Inspections required by *ASME Standard B30.5* on all mobile cranes in service. Regular inspections include both frequent and periodic inspections.

There are a number of operational aids with which a crane operator must be familiar. Although operational aids may function in much the same way and provide the same information from crane to crane, how that information is displayed or provided to the operator can vary significantly.

All mobile cranes must be inspected consistently to ensure they are fully functional and safe to operate. Both 29 *CFR* 1926, Subpart CC and *ASME Standard B30.5* outline the inspections required and provide guidance as to what items and systems are to be inspected. These two standards are both very important, but remember that the OSHA standard is enforceable by law. While they share some guidelines related to crane inspections, they are different and use different terminology.

Understanding crane inspection requirements and the operational aids available are two topics that are important to review as trainees prepare to operate a crane for the first time in this training program.

2.1.0 Operational Aids and Controls

As technology has advanced and cranes have become more sophisticated, operational aids have become more deeply integrated into the crane operating system. Cranes are designed to accomplish tasks within certain parameters. If the crane is operated outside of its design parameters, the possibility of an accident and/or crane failure increases dramatically. For this reason, systems have been developed to automatically monitor and control critical crane operating parameters. More cranes now have controls that can intervene in the process for safety's sake, beyond simply illuminating a warning light or sending an audible signal to the operator.

29 *CFR* 1926.1415 and 1926.1416 list the safety devices and operational aids, respectively, that are required on mobile cranes. Of course, many more aids and devices are often applied to cranes; the OSHA standards outline only the minimum requirements. The OSHA standards place safety devices and operational aids into two separate categories. As a brief review of information presented in a previous module, the list of safety devices required by 29 *CFR* 1926.1415 include the following items:

- A crane level indicator, either built into the equipment or readily available on it
- Boom stops, except for derricks and telescopic booms
- Jib stops, except for derricks
- A brake-locking feature on cranes equipped with foot-pedal brakes
- An integral holding device or hydraulic check valve for hydraulic outrigger and stabilizer jacks
- Rail clamps and stops on railcar-mounted cranes
- A working horn

Like safety devices, the specified operational aids must be in proper working order; their failure during a lift requires the operator to bring the operation to a safe halt to assess the situation. However, the standard does allow limited periods of time for repairs to be made as long as specific alternative measures are taken to replace the function of the aid. For example, if a boom hoist limiting device has failed, the standard lists three acceptable alternative practices to be used while repairs are planned and executed.

OSHA places the required operational aids into two categories, designated Category I and Category II. Category I devices must be repaired within seven calendar days; Category II devices must be repaired within 30 calendar days. For both categories, if the owner has documentation showing that parts were ordered within seven days of the failure, then repairs must be completed within seven days of receiving the parts. The list of Category I operational aids includes the following:

- Boom-hoist limiting device
- Luffing jib limiting device
- Anti-two-blocking device

Category II operational aids include the following:

- Boom angle or radius indicator
- Jib angle indicator (if a luffing jib is installed)
- Boom length indicator (for cranes with a telescopic boom)

- Load-weighing device
- Outrigger/stabilizer position monitor (only required for equipment manufactured after November 7, 2011)
- Hoist drum rotation indicator (for hoist drums not visible to the operator; only required for equipment manufactured after November 7, 2011)

There are a number of exceptions in the standards related to the inclusion of these devices, including those related to the age of the crane. For additional details on these exceptions, consult 29 *CFR* 1926.1416 directly.

Crane operators must consistently consult the manufacturer's operation and maintenance manual. A crane manual includes information such as the following:

- Equipment designation or type
- Name of the equipment manufacturer and/or designer
- Equipment model number and serial number
- Total weight of the unit and ground bearing pressures on tracks or tires and outriggers
- Weight of the boom sections
- Copies of the load charts in all the rated combinations and configurations
- Inspection and maintenance procedures
- Welding specifications on booms and outriggers
- Bolting and torque specifications
- Location and purpose of operating controls and indicators
- Operating instructions
- Emergency procedures

As you can see, a great deal of information is found in the manual. However, the operating manual is not intended to be a crane operator's training program in a convenient package. Without proper training, it is unlikely that much of the material found in the manual can be properly interpreted and applied. Any operational aids that the crane is equipped with will, however, be presented in the operator's manual.

2.1.1 Boom-Hoist Limiting Device

A boom-hoist limiting device is defined by OSHA as one that disengages power to the boom hoist when the boom reaches a predetermined angle. In addition, it sets a brake or closes hydraulic valves to prevent the boom from lowering once the crane has been shut down. The primary purpose of the device is to ensure that the boom is not raised too far up. All cranes manufactured after 1969 must be equipped with a boom-hoist limiting device.

If this device fails, it must be repaired, or the parts ordered for repair, within seven days of failure (a Category I device). 29 *CFR* 1926.1416 allows the operator to temporarily use a boom angle indicator to prevent raising the boom too high. Another allowed alternative is to mark the boom hoist cable clearly where it can be seen by the operator or a spotter as it approaches the minimum angle.

Cranes with luffing jibs installed must have the same type of limiting device for the jib.

2.1.2 Load Moment Indicators

The load moment indicator (LMI) is a common device found in cranes. 29 *CFR* 1926.1401 defines a load moment indicator as a "system which aids the equipment operator by sensing (directly or indirectly) the overturning moment on the equipment, *i.e.*, load multiplied by radius. It compares this lifting condition to the equipment's rated capacity, and indicates to the operator the percentage of capacity at which the equipment is working. Lights, bells, or buzzers may be incorporated as a warning of an approaching overload condition." The load moment indicator is a Category II device. Cranes manufactured after March 29, 2003 that have a rated load capacity over 6,000 pounds (2,722 kg) must be equipped with an LMI or similar device that can report the weight on the hook. Derricks, articulating cranes, and cranes manufactured prior to that date are exempt. This is just one of the exemptions that can be found in the OSHA standard related to operational aids.

<table>
<tr><td>**WARNING!**</td><td>Overriding or bypassing the LMI is only allowed through a manufacturer's direction. Otherwise, overriding these devices can result in an unstable and hazardous operating condition. Do not attempt to circumvent the safety features of the LMI.</td></tr>
</table>

The LMI combines various input signals and compares the data to a defined set of parameters based on the machine's load chart and design limitations. It then provides the operator visual feedback as shown in *Figure 10*. The type of feedback provided will vary depending on the manufacturer. The feedback may include colored lights, buzzers, horns, sirens, and/or an instantaneous digital and analog display of the load.

The device may also be integrated into the controls in a way that allows it to stop the process before a limit is surpassed. For example, lowering the boom increases the operating radius and the leverage of the load, placing additional stress on the crane. If the crane is approaching its limitations, the boom would be prevented from lowering any further with the present load. Note that OSHA defines this type of device as a load moment limiter, rather than an indicator. The unit shown in *Figure 10* offers information such as the following:

- Boom length
- Boom angle
- Boom tip height
- Operating radius
- Maximum rated load for the configuration
- Actual load on hook
- Load as a percent of maximum
- Telescopic-boom sequencing
- Outrigger monitoring
- Range-limiting option
- Data transmission option
- Wind speed option
- Self-diagnostic function

Various display screens can be selected at any time. Load moment indicators and limiters may be hard-wired into the crane or operate wirelessly.

2.1.3 Anti-Two-Blocking Devices

Anti-two-blocking devices (*Figure 11*) provide warnings and prevent two-blocking from occurring. Two-blocking occurs when the component that is uppermost on the load line, such as the load block or headache ball, makes contact with the boom tip, fixed upper block, or a similar component. OSHA places this operational aid in Category I.

There are both passive and active versions of these devices. The passive version is a warning system. Both the active and passive versions use a weighted switch that provides a visual and/or audible warning. As the hook block gets close to the boom tip, a weight attached to the hoist rope is lifted along with it, taking the weight off of the switch arm. This causes spring pressure to close a switch, which activates the alarm.

An active two-block warning system works similarly to a warning system but takes it a step further. It locks out the crane's boom and load lift controls before two-blocking occurs, preventing damage to the hoist rope or machinery parts.

<table>
<tr><td>**CAUTION**</td><td>If the two-blocking alarm is initiated during operation, immediately stop operation and lower the hook block or headache ball.</td></tr>
</table>

Figure 10 Load moment indicator.

Most cranes manufactured today are equipped with an active system, with few allowable exceptions. Telescopic-boom cranes manufactured after February 28, 1992 must have active devices that lock out the controls. Lattice-boom cranes manufactured after February 28, 1992 may be equipped with either an active or passive system. Those manufactured after November 7, 2010 must be equipped with active anti-two-blocking equipment. As a result, it is likely that you will operate cranes equipped with both passive and active systems during your career. Remember that an active anti-two-blocking system that automatically locks out the controls is required when hoisting personnel, regardless of when the crane was manufactured, per 29 *CFR* 1926.1431(d)(5)(v). Hoisting personnel during pile-driving operations is an exception to that rule.

2.1.4 Selected-Controls Lockout System

A selected-controls lockout system allows the operator to inhibit the operation of selected crane functions. The goal of this system is to prevent the operator from making an error by locking out those crane controls that could make a situation worse. For example, the ability to move the boom to the left might be manually locked out if, under the current circumstances, it could potentially lead to a significant incident. Note that a selected-controls lockout system is not an OSHA-required aid.

To accomplish this, the system makes extensive use of solenoid-operated hydraulic blocking and bypass valves, as well as electrically operated links and pins. The system is normally controlled by a signal provided from the central processing unit (CPU). The processed signal operates a solenoid that opens or closes a valve in the hydraulic

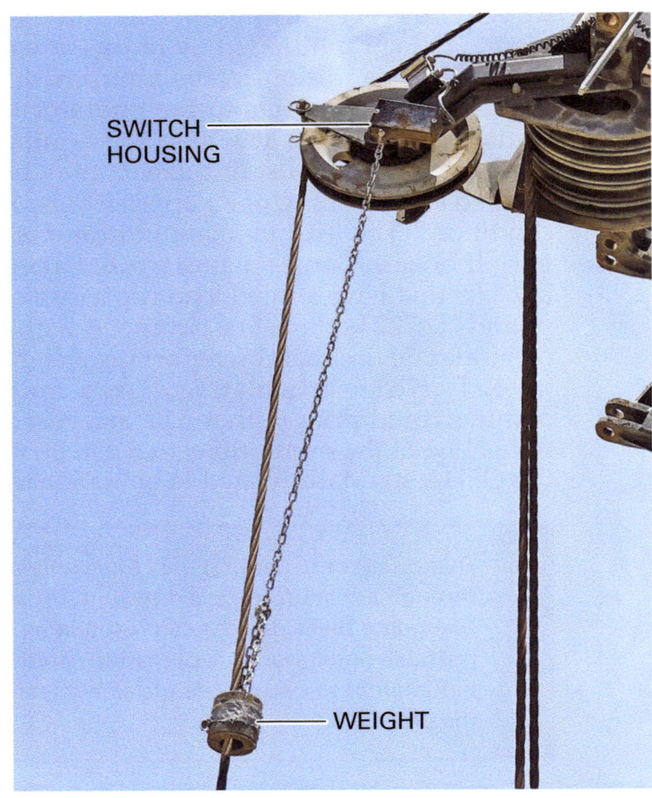

Figure 11 Anti-two-blocking device components.

system that inhibits operation of the component. When the valve is activated, it ports the hydraulic oil on the pressure side of the operating cylinder back to the sump, diverting it from the component actuator. Another method used for locking out controls is by using a solenoid that moves a pin into the mechanical linkage of the operating lever.

2.1.5 Boom Angle Indicators

A boom angle indicator detects and reports the angle of the boom. This information is often obtained from a sensor mounted on the boom housing. On many cranes, this device is located in the same housing as the boom length sensor (*Figure 12*).

A boom angle sensor generates an electrical signal proportional to the boom angle as the boom is raised and lowered. This variable signal is then transmitted to the indicator and displayed. The sensor is integrated into a central processing unit (CPU) on cranes with more sophisticated controls.

OSHA standards require mobile cranes to have a boom angle indicator that is visible from the operator's station. However, there is no requirement that it be electronic. Mechanical boom angle indicators (*Figure 13*) are installed on most cranes. A crane may be fitted with both an electronic and a mechanical boom angle indicator. This device is identified by OSHA as a Category II device.

Figure 12 Boom angle and length sensors in a shared housing.

Figure 13 Mechanical boom angle indicator.

2.1.6 Boom Length Indicators

The boom length indicator measures and reports the length of the boom or, in the case of a telescopic-boom crane, how far the boom is extended or retracted. This information may be obtained in several ways. A common type of input device is a cable reel (*Figure 14*). It consists of a reel mounted on a telescopic boom with cable wrapped around it. The free end of the cable is attached to the end of the boom. As the boom is extended and retracted, the cable pays out, rotating the drum. As the drum rotates, an electrical signal proportional to the length of the boom is sent to the CPU or to a standalone display unit. The boom length indicator is a Category II device.

Another type of sensing system uses switches located in the housing of the telescoping boom sections. As the boom sections move in and out, they activate switches which, in turn, send a signal to the CPU or display unit.

Other systems use a **pressure transducer** located on the boom hydraulic cylinders (*Figure 15*). Both wired and wireless transducers are used. Pressure sensors detect the difference in pressure in the hydraulic cylinder(s) relative to the position of the boom. The transducers then convert the measured pressure differential to a proportional electrical signal, and the transmitter sends the signal to the CPU or display device in the crane cab.

2.1.7 Operator Controls

All mobile cranes are operated through the use of joysticks, foot pedals, levers, and switches (*Figure 16*). The arrangement of the basic controls is suggested in diagrams found in *ASME Standard B30.5*. The layouts are based on the use of four independent levers used to control the boom and the hoist drum(s). However, following the layout

Figure 14 Cable-type boom length sensor.

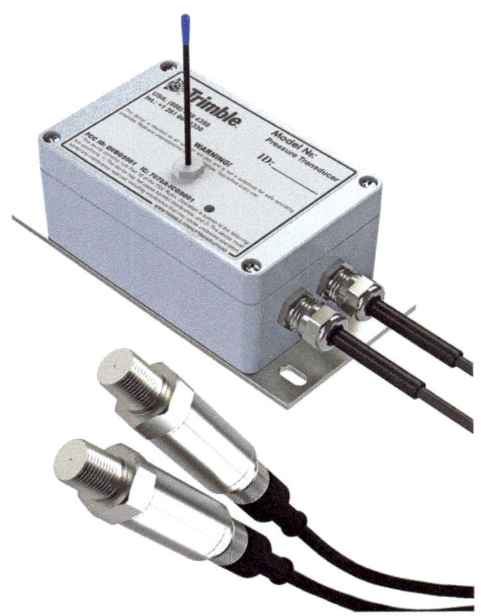

Figure 15 Pressure transducers with wireless transmitter.

is not required of crane manufacturers, and with the introduction of joystick controls that integrate many functions into a control held in one hand, many large and sophisticated cranes do not conform to the suggested layouts.

Per the ASME standard, controls must be located within the reach of the operator while at the operator's station. The controls for hoisting the load, raising the boom, swinging the boom, and telescoping the boom (when applicable) must have a means for holding them in the neutral position without the use of positive latches.

ASME Standard B30.5 does include required standards for a variety of control characteristics, although the precise layout is not one of the requirements. For example, the travel distance on hand levers may not be greater than 14 inches (35.6 cm) from the neutral position on two-way levers, and not greater than 24 inches (610 mm) on levers that move one way only. The travel distance on foot pedals may not be greater than 10 inches (25.4 cm). The force required to move hand levers cannot be greater than 35 pounds (156 Newtons) and the force required on the foot pedals cannot be greater than 50 pounds (222 Newtons). These standards are in place to ensure that mobile cranes have a consistent level of operability. As a result, operators should not see a significant difference in these characteristics from crane to crane, assuming they are in proper working condition.

Figure 16 Cab of a Link-Belt 238HSL lattice-boom crawler crane.

2.2.0 Crane Inspections

29 *CFR* 1926, Subpart CC addresses crane inspections. The information is contained in 29 *CFR* 1926.1412. OSHA requires that cranes be visually inspected prior to use on each shift. For example, if a crane is operating around the clock with three shifts of operators, then it will be inspected three times in the 24-hour period. OSHA further requires monthly and annual inspections of cranes in service. The crane must be disassembled at least annually for the inspection of certain items.

ASME Standard B30.5 defines two primary types of inspections, one of which is broken down into two subtypes. The wording is significantly different than that of the OSHA standard. It is important for crane operators to understand the difference between them.

- Initial inspection – All new and altered cranes shall be inspected by a qualified person to verify that it is functional and its features comply with ASME standards.

- **Regular inspection** – Regular inspections apply to cranes that are in service. This category is further broken down into two subtypes—**frequent inspection** and **periodic inspection**. The interval between inspections is based on the nature of the work being performed and the exposure of crucial crane components to wear and tear. Frequent inspections are conducted by a designated person at daily-to-monthly intervals. Periodic inspections are deeper and more structured in scope than frequent inspections and are conducted on one- to twelve-month intervals, or as specified by the manufacturer.

2.2.1 Inspection Documentation

Both the OSHA and ASME standards provide a list of items to be inspected, at a minimum, for shift/frequent inspections as well as the more comprehensive periodic and annual inspections. Although the two lists have many inspection items in common, there are some subtle differences. Manufacturers provide inspection checklists to ensure that certain items specific to the crane model are inspected. A well-prepared checklist should reflect both the OSHA and ASME minimum requirements. Note, however, that the ASME standards related to crane inspections have not been incorporated by reference into the OSHA standard. It is ultimately the responsibility of the crane owner to ensure and document compliance with the OSHA standards.

The shift inspections required by OSHA are not required to be documented. However, most crane owners do choose to document them at some level, often through a completed and signed checklist (*Figure 17*). 29 *CFR* 1926.1412 does require that monthly and annual inspections be documented. Monthly inspection documentation must be retained by the employer conducting the inspections for a minimum of three months; annual inspection documentation must be retained for a minimum of twelve months.

ASME Standard B30.5 recommends that a crane inspection and repair log be maintained for each crane, and that the log be kept in the crane. Although it is not a firm requirement, it is a best practice to do so. Since OSHA requires that documentation be maintained for monthly and annual inspections, a copy of the inspection results should naturally be kept in the crane along with shift/frequent inspection checklists. Some examples of inspection checklists for the Grove RT family of rough-terrain mobile cranes are provided in the *Appendix*. Any deficiencies found during a crane inspection should be reported

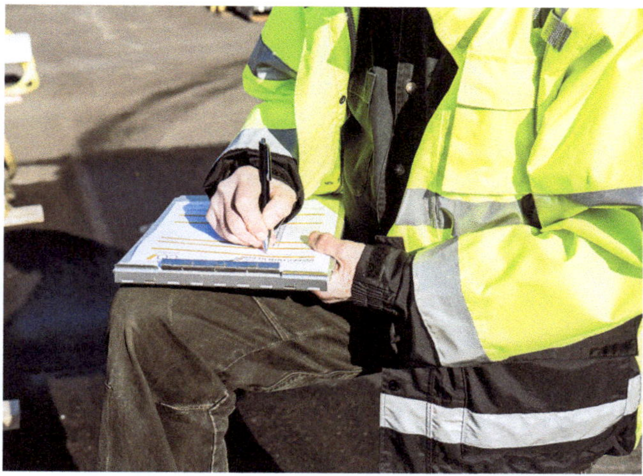

Figure 17 Using a checklist during a shift inspection.

immediately to your supervisor or instructor. On the job, report any issues to your immediate supervisor or other designated person in the organization.

To satisfactorily complete this module, trainees will be required to complete a shift or frequent inspection of a crane before operating it. Your instructor will review and complete the requirements of a shift/frequent crane inspection, likely based on the same crane that will be operated to complete other tasks required to complete the Performance Tasks for this module. That checklist may differ substantially from the examples found in the *Appendix*, as there is no standard design for inspection checklists or their entries.

Before starting a crane, the operator should review the crane inspection log or the inspection records. Operators should also pay attention to any repairs that have been made and what systems or components were involved or affected. After the crane is started, these components should be given a little extra attention to ensure they are working properly.

2.2.2 Shift Inspection Requirements

As stated previously, OSHA requires an inspection of the crane at the beginning of each shift. This is often once daily, but may be required multiple times per day if the crane is in use for more than one shift. The list of inspection items to be accomplished at a minimum is found in 29 *CFR* 1926.1412(d). To complete a shift inspection before operating a crane, your instructor will likely provide a checklist for the specific crane to be used and review the inspection requirements. Crane manufacturers may be more specific in their recommendations and direct the inspection of items not mentioned in the standards. As part of this training, a complete module is dedicated to preventive maintenance and inspections in *Mobile Crane Operations Level 2.*

Hydraulic reservoirs may contain hot oil and be pressurized. Engine coolant systems may also be extremely hot and under pressure. Wear appropriate protective gear and carefully depressurize the systems before checking the oil and coolant levels.

Any deficiencies that are identified by the operator during shift/frequent inspections should be documented on the checklist and reported to an instructor or supervisor before using the crane. Crane owners and employers have their own internal policies for operators to follow for such issues.

2.2.3 Miscellaneous Inspection Requirements

The following are some of the additional requirements from 29 *CFR* 1926.1412(d) that are worthy of mention here:

- Equipment that has been idle for three months or more must receive the equivalent of a monthly inspection, at a minimum, before use.
- Cranes that have been exposed to severe service, such as loading that might have exceeded the rated load capacity or exposure to a corrosive atmosphere, must be inspected by a qualified person before their use can continue. The inspector is directed to focus on systems and components that were likely affected by the incident.
- If a manufacturer's recommendations for inspection of any system or item that relates to safe operation is more comprehensive or exceeds the recommendations of the OSHA standard, the manufacturer's recommendations must be followed. Note that this means a manufacturer's recommendations for inspecting certain items can become enforceable by law through the OSHA standard.

OSHA also requires a crane be inspected by a qualified person any time it is modified, repaired, or adjusted in a way that could affect safety or lifting capacity. The inspection must include functional testing of the component or related system.

2.3.0 General Crane Operating Practices

Before starting a crane, it is the operator's responsibility to ensure that all the controls are in their required positions. This information will be found in the operator's manual for the specific crane in use. Before operating any crane for the first time, properly prepare yourself by studying the operator's manual for it. Studying this information will also help prepare for the required pre-start inspection.

Adjust the operator's seat to ensure that you are comfortable and can easily reach the necessary controls. Once comfortable, many cranes require that an electrical system master switch be turned on to provide electrical power to the instruments. In some cases, the keyed ignition switch serves as the electrical master switch, as it does in a standard car or truck. When the ignition switch is also the master switch, instrument power is generally provided when the key is turned to the first position from Off. With electrical power turned on, the operator can check the instruments and warning lights to ensure they operate properly before starting the engine. Again, follow the manufacturer's guidelines for starting the crane in use, and ensure there is a sufficient amount of fuel (typically diesel fuel).

Once the crane is operating, if there is reasonable cause to believe that a lift might be unsafe, it is the responsibility of the operator to refuse to lift until the concern has been reported to the lift director or immediate supervisor. As lifting proceeds, the operator should also be aware of signs that often indicate a problem is developing, such as the following:

- Unusual noises, loss of power, or controls responding poorly
- Gauges showing incorrect or zero readings
- Improper operation of the master clutch and brake units
- Sticking or binding controls
- The display or speaking of the Stop or Emergency Stop signals

2.3.1 Operators-in-Training

There are a few situations that an operator-in-training must avoid. Since an operator-in-training should always be monitored by an appropriate trainer, the trainer is primarily responsible to ensure that an operator trainee is not placed in the following situations, as outlined in 29 *CFR* 1926.1427:

- Lifts where any part of the equipment, load line, or load could reach within 20 feet (6.1 meters) of a power line up to 350 kilovolt (kV), or within 50 feet (15.2 meters) of a power line carrying a higher voltage.
- Personnel hoisting
- Multiple-equipment lifts
- Lifts that occur over a shaft (such as a mine shaft), cofferdam, or in a tank farm

Operator trainees are also prohibited from being involved in multiple-lift rigging operations unless the trainer is confident that the trainee is proficient enough to participate.

2.3.2 Handling the Load

The operator must know the weight of the load to be lifted. It is the crane operator's responsibility to verify the load weight and to ensure that it does not exceed the capacity rating for the crane at the working load radius. When the weight of a load to be lifted cannot be verified through a reliable resource before the lift begins, the crane operator must slowly pick the load and monitor the weight and instruments to determine the weight. If it is over 75 percent of the crane's capacity at the required operating radius, the lift must stop and the weight of the load verified through industry-recognized sources.

Even when the weight of the load has been determined beforehand, it is necessary to slowly transfer the weight to the crane to verify that the information is accurate. This also avoids the potential for shock-loading the crane. Sources can be incorrect and additional weight could have been added without being documented. Bring the hook over the load gently to minimize swinging and lower the hook so riggers can make the final connection. Carefully remove the slack in the wire rope and slowly lift the load until it just clears the ground or other means of support. Watch the instruments to monitor the growing load imposed on the crane as the weight is transferred to it.

Eyes on the Prize

Today's technology has made a dramatic difference in the operation of cranes, making them safer as well as more flexible and productive. One of the tools now available due to advanced technology is the crane-mounted camera.

Cameras can be mounted on various areas of the crane, but they are most often mounted on the load block/hook assembly or on the boom tip. Powerful magnets often serve as the primary means of attachment, with a lanyard securing it. The cameras can often operate wirelessly, transmitting the camera's view to a display in the operator's cab. The operator then has a bird's eye view of the load (from the hook) or the overall lift area (from the boom tip). More sophisticated models allow the crane operator to zoom the lens in and out. Cameras improve safety and have proven to be effective in increasing production as well. Consider the value of a camera on the hook or boom tip when an operator must make a blind lift, or a camera on the superstructure to view blind spots before rotating the crane.

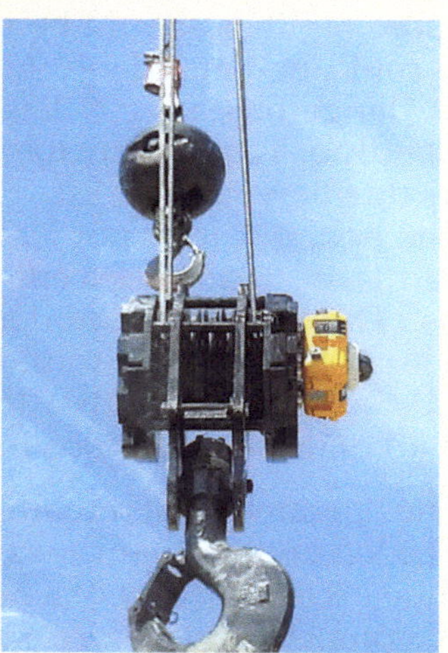

(A) CAMERA MOUNTED TO CRANE HOOK

(B) CAB DISPLAY OF THE AREA DIRECTLY BELOW THE HOOK

If a crane is being operated from its tires, do not lift the load from any quadrant or use an operating radius that is not allowed by the crane manufacturer. If using outriggers or operating from crawler tracks, be sure to follow the manufacturer's guidance as well.

During the lifting operation, move the controls smoothly and evenly so that there is no sudden acceleration or deceleration of the moving load. Follow the directions of the signal person unless you see that the requested move would result in an unsafe condition or cause an accident. For example, lowering the boom increases the operating radius, increasing the leverage of the load and decreasing the rated load capacity of the crane. If the signal person requests the boom to be lowered, but doing so would result in exceeding the rated capacity of the crane, do not comply. Stop the process and discuss the problem. Make sure the load and boom do not strike any obstructions, and do not carry the load over people.

Avoid any sudden starts and stops when rotating the crane. Rotate slowly enough to prevent the load from lagging behind or swinging out beyond the radius of the boom tip (*Figure 18*). One or more taglines should be used when load rotation could be hazardous. A rotating load can be very dangerous as it is lowered toward riggers for landing. Your instructor may use tagline handlers during your first lift.

The operator may not leave the controls while a load is suspended, except under special conditions. The conditions under which an operator can leave the controls, found in 29 *CFR* 1926.1417, are listed here. Note that the standard requires that all of these conditions must be met:

- The operator must remain adjacent to the crane.
- The load will remain suspended for a period of time exceeding that of a normal lift.
- A competent person determines it is safe to do so and makes sure that all of the appropriate precautions are taken. Precautions would include but not be limited to restraining the boom hoist, telescoping, load, swing, and outrigger functions.
- Barricades, caution lines, or similar deterrents are erected around the area beneath the load, referred to as the fall zone. No one, including qualified riggers that are prepared to receive the load, can be in the fall zone while the load is suspended and the operator is away from the controls.

As you prepare for your first opportunity to operate a crane, it is highly unlikely that there will be any need to leave the controls while the load is suspended.

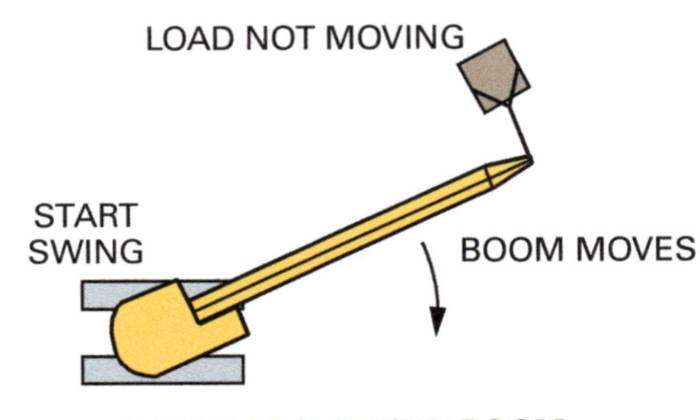

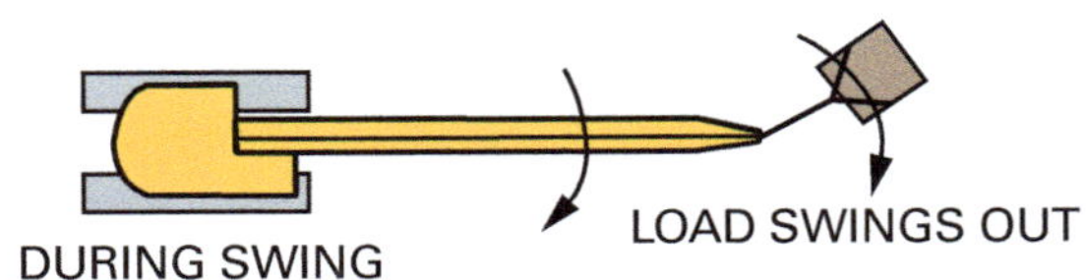

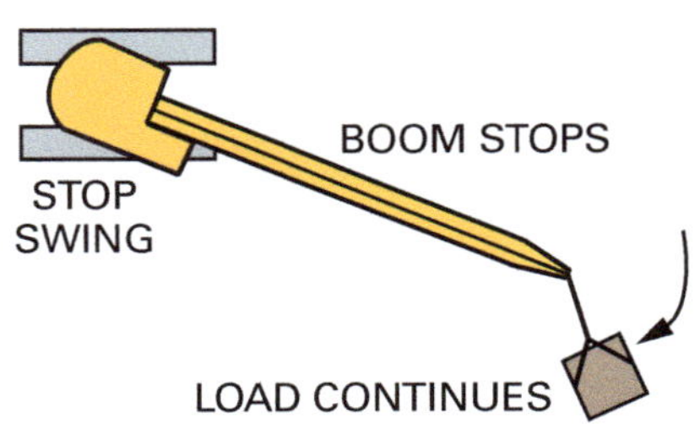

Figure 18 Managing a load while swinging the boom.

If the crane experiences a power failure while in operation, *ASME Standard B30.5* requires the operator to:

- Set all brakes and locking devices
- Move all clutches or other power controls to the Off or Neutral position
- Land any load presently on the hook, under brake control if practical

These actions in response to a power failure and how they are conducted in the crane being used for training should be committed to memory as you study the operator's manual. Operator trainees may be asked to simulate a power failure.

Remember that anyone can display a Stop or Emergency Stop signal to the operator at any time, and the operator is required to obey the signal regardless of its source.

Emergency Training

When a new pilot is being taught to fly a small aircraft, one of the first things that must be taught is how to recover from a stall. Stalls can be scary events, especially for the faint of heart. Stalls caused by pilots often occur as a result of insufficient power and airspeed while the nose of the aircraft is pointed up. The aircraft reaches a point where it literally stops flying, takes a fast turn, and heads toward the ground in a hurry. At low altitudes, the results can be devastating. Stall recovery is therefore taught very early in the process.

Crane operators may also receive emergency training very early in the training process. An emergency can occur as quickly in practice as it can in any other lifting task. Do not be surprised if you are trained immediately in what actions to take in the cab during an emergency before you have a great deal of practice lifting a simple load.

2.3.3 Shutdown and Post-Operation Inspection

When the crane's work is done, it must be shut down properly for safety reasons. If the crane is on outriggers, leave them set as is. If the crane is on rubber, move it to level ground and/or block the tires. The following actions are required for most cranes:

- Set the crane's air brakes and place the transmission in neutral. To combat the possibility of leakage and loss of air pressure over extended periods, set the parking brake.
- Engage the swing brake and lock. If it is a telescopic-boom crane, lower the boom and park the unit somewhere so that if the boom continues to lower or settle overnight, no damage or injury will result. If using a lattice-boom crane, you may want to lower the boom. As an alternative, you can orient the crane to take the prevailing wind at the back of the boom (boom pointed in the direction the wind is blowing) or orient the crane so that the boom points directly into the wind. In either case, set the boom to an angle no higher than 60 degrees. Keep in mind that wind direction often shifts from its prevailing direction.
- Any hanging block must be secured and all **hoist pawls** must be engaged.
- Disengage any clutches and shut off the engine. Lock the control panel and/or door to prevent access as you exit with the key.

Post-operation inspections are not mandated by OSHA or ASME standards. However, if the crane has developed a problem during operation, it is best to be aware of it before the next shift or use begins. In addition, daily inspections are often done when the crane is cold and its components are not at their normal operating temperature. Inspecting the crane immediately following shutdown allows the operator to see the crane in a different condition. Some crane owners and employers require post-operation inspections, regardless of the standards. A thorough post-operation inspection includes the following procedures:

- Check the wire rope to make sure that no damage occurred during operation.
- Check all fluid levels to make sure they are adequate. Remember to ensure that a given system is not pressurized before opening any tanks, and that most fluids are likely hot.
- Lubricate components if necessary.
- Check for any fluid leaks.
- Make sure the control levers are in the neutral position.
- Drain any accumulated water and debris from the air reservoirs to prevent rust and deterioration.

> **WARNING!**
>
> Hydraulic reservoirs may contain hot oil and be pressurized. Engine coolant systems may also be extremely hot and under pressure. Wear appropriate protective gear and carefully depressurize the systems before checking the oil and coolant levels.

Be sure to also follow any post-operation guidance offered in the manufacturer's operating manual. Following these procedures will reduce the likelihood of discovering problems during the next pre-start or daily inspection.

Additional Resources

ASME Standard B30.5, Mobile and Locomotive Cranes. Current edition. New York, NY: American Society of Mechanical Engineers.

29 *CFR* 1926, Subpart CC, *Cranes and Derricks in Construction.* Current edition. **www.ecfr.gov**

Mobile Crane Safety Manual. 2014. Milwaukee, WI: Association of Equipment Manufacturers.

Cranes: Design, Practice, and Maintenance. Second Edition. 2002. Ing J. Verschoof. Hoboken, NJ: John Wiley and Sons, Inc.

North American Crane Bureau, Inc. **www.cranesafe.com**.

2.0.0 Section Review

1. The travel distance from the neutral position for a two-way hand lever may not be greater than _____.

   a. 10 inches (25.4 cm)
   b. 14 inches (35.6 cm)
   c. 20 inches (50.8 cm)
   d. 24 inches (61.0 cm)

2. While performing a shift inspection on a crane, you notice that there is a pool of oil under your crane. You should _____.

   a. make a note of it for later inspection and proceed with the day's lifting tasks
   b. check all the fluid levels again and then proceed with the day's lifting tasks
   c. note it on the inspection checklist and report it to your instructor or supervisor before operating the crane
   d. report the leakage to an OSHA representative so that they can ensure that repairs are made in the allotted time frame

3. If the weight of a load is not known before a lift, the crane operator is allowed to initiate the lift while monitoring the weight. The lift must be stopped and the weight verified by an industry-recognized source if it exceeds _____.

   a. 65 percent of the rated load capacity
   b. 70 percent of the rated load capacity
   c. 75 percent of the rated load capacity
   d. 85 percent of the rated load capacity

SUMMARY

Operating a crane safely and successfully depends on a variety of factors. Although a thorough site assessment and consistent inspections help to ensure the safety of the crane, the primary responsibility for its safe operation falls squarely on the crane operator. The ability to adequately prepare the crane, use the controls effectively and smoothly, and to comprehend and respond to the various hand signals are all key skills that an operator must develop and practice regularly to become proficient.

1. The methodical process that attempts to identify local hazards and factors that may interfere with a lift and then devise plans to overcome or eliminate those issues is referred to as a ______.
   a. blind lift
   b. periodic inspection
   c. lift plan
   d. site assessment

2. What is the recommended clearance between any part of the rotating superstructure and any nearby obstruction?
   a. 2 feet (0.6 meters)
   b. 4 feet (1.2 meters)
   c. 6 feet (1.8 meters)
   d. 8 feet (2.4 meters)

3. If a crane must be operated from a public street or highway, it may be necessary to ______.
   a. seek a crane operator's license from the manufacturer
   b. obtain a permit from the proper authorities
   c. avoid setting up crane barricades
   d. use only voice communication instead of hand signals

4. A load moment indicator must be installed on any crane manufactured after March 29, 2003 with a rated load capacity that exceeds ______.
   a. 3,000 pounds (1,361 kg)
   b. 5,000 pounds (2,268 kg)
   c. 6,000 pounds (2,722 kg)
   d. 10,000 pounds (4,536 kg)

5. Which of the following specific operational aids are required when hoisting personnel (except in pile-driving operations), regardless of when the crane was manufactured?
   a. An active anti-two-blocking system
   b. An outrigger control panel outside the cab
   c. A luffing jib
   d. Basic crane controls outside the cab

6. Per *ASME Standard B30.5*, the force required to move hand levers may not be more than ______.
   a. 15 pounds (67 Newtons)
   b. 25 pounds (111 Newtons)
   c. 35 pounds (156 Newtons)
   d. 50 pounds (222 Newtons)

7. Per 29 *CFR* 1926.1412, which of the following inspections must be documented?
   a. All inspections—shift, monthly, and annual
   b. Only the monthly and annual inspections
   c. Only the shift inspections
   d. Only the annual inspections

8. Before using a crane that has been idle for three months or more, what level of inspection is required by OSHA?
   a. A daily inspection
   b. A frequent inspection
   c. A monthly inspection
   d. An annual inspection

9. While a load is suspended, the crane operator ______.
   a. is not allowed to leave the controls for any reason
   b. may leave the controls only in an emergency if certain precautions are taken
   c. may leave the controls only if relieved by another operator
   d. may leave the controls only under special conditions defined in 29CFR1926.1417

10. When shutting down a crane for the night at the jobsite, all of the following should be done *except* ______.
    a. retracting the outriggers
    b. setting the parking brake
    c. locking the control panel
    d. placing the transmission in neutral

Trade Terms Quiz

Fill in the blank with the correct term that you learned from your study of this module.

1. An inspection type outlined by *ASME Standard B30.5* that is conducted at monthly-to-annual intervals is called a(n) __________.

2. When the crane operator cannot see the load during any part of its movement, it is referred to as a(n) __________.

3. Frequent and periodic inspections, as outlined by ASME standards, are both considered to be a(n) __________.

4. An area where a suspended load could possibly drop into is considered to be a __________.

5. A device that reads pressure and transmits a signal proportional to the reading is called a(n) __________.

6. A(n) __________ is a watertight enclosure designed to be pumped out and used to create a dry area for work below the water line.

7. A hinged device intended to interlock with a mechanism and prevent a hoist from moving is called a(n) __________.

8. All new and significantly altered cranes must undergo a(n) __________.

9. A(n) __________, as outlined in *ASME Standard B30.5*, is conducted at daily-to-monthly intervals.

Trade Terms

Blind lift
Cofferdam
Fall zone

Frequent inspection
Hoist pawl
Initial inspection

Periodic inspection
Pressure transducer
Regular inspection

Larry "Cowboy" Proemsey

President **www.Cowboyscranes.com**

Please give a brief synopsis of your construction career and your current position.

I am the owner and operator of a full-service hoisting, rigging and crane operations company. We do work in all 48 states in the continental US. I started in the construction industry by hanging iron, working on structural steel buildings, communications towers, and a variety of projects. That got me started working with cranes, then I became a crane owner, and things sprung up from there.

How did you get started in the construction industry?

You can't really make a living as a cowboy. I was involved in rodeo and it became obvious very early on that I would have to make a living somehow. When I started in construction, it was in the days of the Wild West before OSHA was so prevalent, so we were working in a high-risk environment. There weren't a lot of people who wanted to do it because of the risk. There weren't many requirements other than you couldn't be afraid of heights. It was an acknowledged risk to do that kind of work, so it was very exciting and it took a while to figure out that it wasn't supposed to be that way.

How has training in construction impacted your life and career? What types of training have you completed?

Through the years, I have done almost everything on the training side. When I started in the trade, there was a high potential for danger. Now with the safety culture in the industry, that's no longer acceptable. We strive now to have zero accidents. People who were around in the 1960s and 70s will remember that speed and production were everything. Then we realized how dangerous those practices were, and that's really when the industry started training.

Training is a way to perfect somebody's skills, but it also teaches somebody how to do things safely. One thing begets the other—the more efficient the work is, the safer it usually is. There are much higher quality and safety standards these days. The complexity of cranes also has taken huge strides just in the last few years. So I really support training as a path to becoming a good crane operator.

Why do you think credentials are important in the construction industry?

More than ever, the industry is fluid. People used to go to work for a company and grow up with that company as they advanced through their career. It gave the company a chance to know what a person is capable of doing. Today's market is different, and people move from company to company. When you deal with people in a transient situation like that, without credentialing and verification of training, you never know who you are getting. It's one thing for a person to claim they have skills, but another for them to show you. When they show you credentials, you know they are headed down the right path. Even if you don't know if they have every skill, at least you know they've had training.

What do you enjoy most about your career?

There are really two things that I enjoy about my career. First, the more challenging projects which are more difficult and complicated or require an in-depth understanding of the job really excite me. I don't like to do the same thing every day, I like to be challenged. The complexity of the jobs is getting more intense and we are always striving to meet those demands, so I enjoy those challenging projects.

The other thing that I really enjoy is taking on young people to mentor. Everybody looks at me as the old man, but in a respectful way. It's nice to see people who are in the beginning stages of their careers who look up to you and listen to you. You can pass on your skills and knowledge and watch them grow up in the trade. As they go on into their careers, it is very rewarding to see that transformation.

Would you recommend construction as a career to others? Why?

I think our country has become more driven to being college-bound and highly educated. The way I see it, the real problem is that construction careers are challenging and they also require education. But there are entry-level opportunities for people who are not college-bound, either because they can't afford it or are just not college-driven individuals. There are opportunities for advancement for those who work hard and it is another pathway to success in life. In many respects, people in the construction industry earn more money than those with college degrees. Our education system tries to push everyone to college careers. More and more people lack the hands-on skills required in construction. There's a huge skills gap, and not many skilled people left to fill it. For that reason, the opportunity in the construction industry is greater than it's ever been and the sky is the limit. It provides long-term job growth and stability for those who want it.

What advice would you give to someone who is new to the construction industry?

I would advise them to take the time to look at different aspects of the industry and get an idea of what exactly they want to do. Shadow somebody in that field so that you get a clear picture of what things are like after spending a day in the life of that job. Get a feel for exactly what a trade requires from a physical, mental, and training standpoint. And get a feel for the satisfaction that comes with building something. Specifically for cranes, if somebody doesn't want to be a crane operator, I would never suggest it or expect them to do it. You only want the people who are driven to become crane operators because it's what they want to do. The ones who want to and strive to be the best will do well in the industry.

How do you define craftsmanship?

Craftsmanship is a combination of skill, quality, production, and safety all rolled together. It is truly the finished product, but also the road it takes to get there. People refer to the craftsmanship in a building, but it's not just the end product. It's what it took to get there. The coordination of the trades and everything that goes into making a finished product as a consortium of individuals is collectively the craftsmanship that represents everybody's efforts. Working together cohesively, and demanding the best for the finished product is how I define craftsmanship.

Blind lift: A lift in which the crane operator cannot see the load during any portion of the lift.

Cofferdam: A watertight and generally temporary enclosure built in a body of water or wet area, designed so that water can be pumped out to create a dry area for work below the water line.

Fall zone: Defined in 29 CFR 1926.1401 as the area beneath a load in which it is reasonably foreseeable that partially or completely suspended materials could fall in the event of an accident.

Frequent inspection: Required recurring crane inspections conducted at daily-to-monthly intervals, directed by ASME Standard B30.5. Frequent inspections are limited in scope and focused primarily on safety.

Hoist pawl: A hinged or pivoting device designed to interlock in the hoist mechanism and physically prevent its movement.

Initial inspection: An inspection required by ASME Standard B30.5 on new or significantly altered cranes, conducted by a qualified person, to ensure the crane's compliance with ASME standards.

Periodic inspection: Required recurring inspections conducted at monthly to annual intervals, directed by ASME Standard B30.5. The scope of these inspections is more complex than the scope of frequent inspections.

Pressure transducer: A device that measures pressure and then converts the pressure value to a proportional analog electrical signal; also referred to as a pressure transmitter.

Regular inspection: Inspections required by ASME Standard B30.5 on all mobile cranes in service. Regular inspections include both frequent and periodic inspections.

Appendix

GROVE

CRANE INSPECTION CHECKLIST EXAMPLE – SHIFT/FREQUENT INSPECTIONS SERVICE CODES

Task to Perform	Code	Task to Perform	Code
		Daily or 10 Hour Check List	
Is item properly installed on machine	1	Check for cracked or broken structures and welds	16
Check for cracked or broken glass	2	Service as required by engine manufacturer	17
Check for proper working order	3	Lubricate	18
Check for proper service/level	4	Check connectors and wiring for proper alignment and insulation	19
Check for proper reading while operating engine	5	Check wire for damaged, frayed, or broken strands	20
Check for excess dirt, grease, or foreign matter	6	Check for proper inflation/pressure	21
Check for proper operation while operating machine	7	Check for loose or damaged wiring	22
Check for proper adjustment	8	Check that clamps are tight and not corroded	23
Check for cracks or leaks	9	Check complete path for any condition that could damage rope	24
Check for excessive wear	10	Replace	25
Check torque	11	Check for leaks at connection points	26
Clean or replace	12	Drain and fill	27
Drain moisture	13		28
Change filter	14		29
Check for deterioration, damage or excessive wear	15		30

Daily or 10 Hour Check List

Published 5-05-2009, Control # 242-01

2

The Service Codes shown here are applied to the inspection/ lubrication items shown on the inspection checklist on the next page.

GROVE

CRANE INSPECTION CHECKLIST EXAMPLE – SHIFT/FREQUENT INSPECTIONS INSPECTION LOG

Line Item	Items to be Inspected/Lubricated	Inspection Code	January — Day of Inspection — Daily or 10 Hour Check List																													
			1	2	3	4	5	6	7	8	9	10	11	12	13	14	15	16	17	18	19	20	21	22	23	24	25	26	27	28	29	30	31
1	Operator's Manual	1																																
2	Signal and Running Lights & Horn	2, 3																																
3	Back-up Alarm	7																																
4	Gauges and Instruments	3, 5, 7																																
5	Brakes (Swing, Foot & Parking)	3, 7																																
6	Boom Angle Indicator	7																																
7	Load Moment Indicator	7, 19																																
8	Antitwo-Block (A2B) Elec. & Hyd.	7, 19																																
9	Brakes and Air System (if equipped)	3, 7, 13																																
10	Tires	21, 10, 6																																
11	Hourmeter	3, 5, 19																																
12	Hydraulic Tank	4, 6																																
13	Hydraulic Oil Filter	4																																
14	Wire Rope	20																																
15	Hoists	1, 6, 7																																
16	Boom and Attachments	1, 6, 7																																
17	Crankcase	4, 6																																
18	Transmission/Torque Converter	4																																
19	Cooling System	4																																
20	Fuel/Water Separator	13																																
21	Air Cleaner	4, 9																																
22	Hook Block & Headache Ball Safety Latch	3, 9, 10																																
23	Lock-out Cylinder & Slew Potentiometer	7, 8																																
24																																		
25																																		
26																																		
27																																		
29																																		
	Inspector's Initials:																																	

Daily or 10 Hour Check List — **January**

Published 5-05-2009, Control # 242-01

GROVE

CRANE INSPECTION CHECKLIST EXAMPLE – MONTHLY/ANNUAL/PERIODIC INSPECTIONS SERVICE CODES

3 Months, 6 Months, 12 Months, or 500 Hour Check List

Task to Perform	Code	Task to Perform	Code
Is item properly installed on machine	1	Check for cracked or broken structures and welds	16
Check for cracked or broken glass	2	Service as required by engine manufacturer	17
Check for proper working order	3	Lubricate	18
Check for proper service/level	4	Check connectors and wiring for proper alignment and insulation	19
Check for proper reading while operating engine	5	Check wire for damaged, frayed, or broken strands	20
Check for excess dirt, grease, or foreign matter	6	Check for proper inflation/pressure	21
Check for proper operation while operating machine	7	Check for loose or damaged wiring	22
Check for proper adjustment	8	Check that clamps are tight and not corroded	23
Check for cracks or leaks	9	Check complete path for any condition that could damage rope	24
Check for excessive wear	10	Replace	25
Check torque	11	Check for leaks at connection points	26
Clean or replace	12	Drain and fill	27
Drain moisture	13		28
Change filter	14		29
Check for deterioration, damage or excessive wear	15		30

3 Months, 6 Months, 12 Months, or 500 Hour Check List

Manitowoc
Crane Care

Published 5-05-2009, Control # 242-01

34

The Service Codes shown here are applied to the inspection/ lubrication items shown on the inspection checklist on the next page.

CRANE INSPECTION CHECKLIST EXAMPLE – MONTHLY/ANNUAL/PERIODIC INSPECTIONS INSPECTION LOG

Line Item	Items to be Inspected/Lubricated	Inspection Code	Interval Months (Or 500 hours, whichever comes first)	3 Months, 6 Months, 12 Months, or 500 Hour Check List											
				Jan	Feb	Mar	Apr	May	June	July	Aug	Sept	Oct	Nov	Dec
1	Steering & Suspension Pivot & King Pins	18	Three (3) Months												
2	Axle	4	Three (3) Months												
3	Crankcase (RT600E, RT9130E)	14, 27	Three (3) Months												
4	Fuel Filter (RT600E)	14	Three (3) Months												
5	Drive Train Slip Joints	18	Three (3) Months												
6	Transmission Filter	14	Three (3) Months												
7	Cooling System (RT9130E)	14	Three (3) Months												
8	Lift Cylinder Pins (Not on RT880E & RT890E)	18	Three (3) Months												
9	Cab Tilt Cylinder (RT880E, RT890E & RT9130E)	18	Three (3) Months												
10	Cab Pillow Block (RT880E, RT890E & RT9130E)	18	Three (3) Months												
11	Mounting Bolts (Trans, Eng, Hoist, Gearbox, Axle)	11	Three (3) Months												
12	Turntable Bolts	11	Three (3) Months												
13	Muffler	9	Three (3) Months												
14	Boom	8	Three (3) Months												
15	Crane Structure	16	Three (3) Months												
16	Differential Breather	3	Three (3) Months												
17	Crankcase (RT530E-2, RT540E & RT700E)	14, 27	Six (6) Months												
18	Fuel Filter (Not RT600E)	14	Six (6) Months												
19	Strainer (RT880E, RT890E, & RT9130E)	25	Six (6) Months												
20	Turntable Bearing & Drive Pinion	18	Six (6) Months												
21	Turntable Gear Box	27	Twelve (12) Months												
22	Boom Mast Sheave (Not on RT9130E)	18	Twelve (12) Months												
23	Hose Reel	18	Six (6) Months												

Inspector's Initials:

Hourmeter Reading:

3 Months, 6 Months, 12 Months, or 500 Hour Check List

Published 5-05-2009, Control # 242-01

Manitowoc
Crane Care

This module is intended as a thorough resource for task training. The following reference works are suggested for further study.

ASME Standard B30.5, Mobile and Locomotive Cranes. Current edition. New York, NY: American Society of Mechanical Engineers.

29 *CFR* 1926, Subpart CC, *Cranes and Derricks in Construction.* **www.ecfr.gov**

Mobile Crane Safety Manual. 2014. Milwaukee, WI: Association of Equipment Manufacturers.

Cranes: Design, Practice, and Maintenance. Second Edition; 2002. Ing J. Verschoof. Hoboken, NJ: John Wiley and Sons, Inc.

North American Crane Bureau, Inc. **www.cranesafe.com**

Figure Credits

© iStockphoto.com/Su san Chiang, Module opener

DICA Outrigger Pads, Figure 2

Quality Mat Company, Figure 4

Topaz Publications, Inc., Figures 5, 6, 9, 11–14

© iStockphoto.com/Tryaging, Figure 7

Robert I. Carr, Ph.D., P.E., Figure 8

Rayco-Wylie Systems, Figure 10

Trimble, Figure 15

Link-Belt Construction Equipment Company, Figure 16

Heavy Equipment Services Company — **www.hescocrane.com**, Figure 17

Netarus, LLC, SA01

The Manitowoc Company, Inc., Appendix

Section Review Answer Key

Answer	Section Reference	Objective
Section One		
1. c	1.1.3	1a
2. d	1.2.2	1b
Section Two		
1. b	2.1.7	2a
2. c	2.2.2	2b
3. c	2.3.2	2c

NCCER CURRICULA — USER UPDATE

NCCER makes every effort to keep its textbooks up-to-date and free of technical errors. We appreciate your help in this process. If you find an error, a typographical mistake, or an inaccuracy in NCCER's curricula, please fill out this form (or a photocopy), or complete the online form at **www.nccer.org/olf**. Be sure to include the exact module ID number, page number, a detailed description, and your recommended correction. Your input will be brought to the attention of the Authoring Team. Thank you for your assistance.

Instructors – If you have an idea for improving this textbook, or have found that additional materials were necessary to teach this module effectively, please let us know so that we may present your suggestions to the Authoring Team.

NCCER Product Development and Revision
13614 Progress Blvd., Alachua, FL 32615

Email: curriculum@nccer.org
Online: www.nccer.org/olf

❏ Trainee Guide   ❏ Lesson Plans   ❏ Exam   ❏ PowerPoints   Other _______________

Craft / Level: ____   Copyright Date: _______________

Module ID Number / Title: ____

Section Number(s): ____

Description: ____

____

____

____

____

Recommended Correction: ____

____

____

____

____

Your Name: ____

Address: ____

____

Email: ____   Phone: _______________

Glossary

Abstraction: Any form of verbal, graphical, or written communication representing a generalized and nonspecific idea or quality of a thing, action, or event.

Accreditation: In the context of crane-operator certification, the act of granting credit or recognition to an organization or institution that maintains suitable standards of testing for the certification process.

American National Standards Institute (ANSI): An organization that oversees the creation, promulgation, and use of thousands of norms and guidelines that directly impact businesses. ANSI also provides third-party accreditation to products and personnel certification programs, recognizing the competence of organizations to carry out these activities in accordance with requirements defined in international standards.

American Society of Mechanical Engineers (ASME): ASME promotes the art, science, and practice of multidisciplinary engineering and allied sciences around the globe through continuing education, training and professional development, codes and standards, research, conferences and publications, government relations, and other forms of outreach. The organization is therefore an engineering society as well as a standards-developing organization.

Anti-two-blocking device: Two-blocking refers to a condition in which the lower load block or hook assembly comes in contact with the boom tip, boom tip sheave assembly or any other component above it as it is being raised. If this occurs, continuing to apply lifting power to the cable can result is serious equipment damage and/or failure of the hoist line. An anti-two-blocking device, therefore, prevents this condition from occurring.

Apprenticeship: As defined by the US Department of Labor, a combination of on-the-job training and related technical instruction in which workers learn the practical and theoretical aspects of a highly skilled occupation. Apprenticeship programs can be sponsored by individual employers, joint employer and labor groups, and/or employer associations.

Avoidance zone: An area both above and below one or more power lines that is defined by the outer perimeter of the prohibited zone. As the name implies, any part of the crane should avoid this area whenever possible, and may not enter the area except under special circumstances.

Backfill: Soil and rock used to level an area or fill voids, such as the perimeter of building foundations or trenches. Areas with fresh backfill may not be stable enough to support a crane.

Base mounting: A crawler crane assembly consisting primarily of the carbody, ring gear drive, crawler frames, and tracks.

Base section: The lowest portion of a telescopic boom that houses the other telescopic sections but does not extend.

Basket hitch: A common hitch made by passing a sling around a load or through a connection and attaching both sling eyes to the hoist line.

Bird caging: A deformation of wire rope that causes the strands or lays to separate and balloon outward like the vertical bars of a bird cage.

Blind hole: A hole that does not penetrate the material completely, leaving a hole with a bottom.

Blind lift: A lift in which the crane operator cannot see the load during any portion of the lift; any lift involving a load that is out of the direct view of the operator. Blind lifts are generally always categorized as critical lifts.

Block and tackle: A system of two or more pulleys, which form a block, with a rope or cable threaded between them, reducing the force needed to lift or pull heavy loads.

Blocking: Wood or a similar material used under outrigger floats to support and distribute loads to the ground. Also referred to as *cribbing*.

Boom torque: A twisting force applied to the crane boom, typically resulting from imbalanced reeving of the boom tip sheave assembly ropes.

Bridge: In relation to overhead cranes, the part of an overhead crane consisting of one or more girders or beams and the supporting trucks. The bridge is the overhead, weight-bearing structure along which the trolley(s) and load block assembly travels.

Bridle hitch: A type of hitch comprised of 2 or more single-leg hitches, used for lifting objects equipped with lifting lugs or other points of connection.

Carbody: The part of a crawler-crane base mounting that carries the rotating upperworks.

Carrier: The base of a wheeled crane that provides crane movement and supports the upperworks.

Center of gravity (CG): The point at which the entire weight of an object is considered to be concentrated, such that supporting the object at this specific point would result in its remaining balanced in position.

Certification: Refers to a candidate's successful completion of a written test of knowledge as well as a practical examination. Certification testing must be based on the equipment capacity and type that will be operated, and can be administered by an accredited third-party organization (such as NCCER) or through an audited testing program operated by the employer.

Check valve: A valve designed to allow flow in one direction but closes as necessary to prevent flow reversal.

Choker hitch: A hitch made by passing a sling around the load, and then passing one eye of the sling through the other. The one eye is then connected to the hoist line, creating a choke-hold on the load.

Cofferdam: A water-tight and generally temporary enclosure built in a body of water or wet area, designed so that water can be pumped out to create a dry area for work below the water line.

Competent person: As defined by OSHA, an individual who is capable of identifying existing and predictable hazards in the surroundings or working conditions which are unsanitary, hazardous, or dangerous to employees, and who has the authorization to take prompt corrective measures to eliminate such hazards.

Consensus standard: A set of proprietary guidelines published and agreed to by a consensus (representative majority) of members of a given industry. While not legally binding, they are often cited in governmental regulations, such as OSHA standards.

Counterweights: Weights added to the crane, usually on the end opposite the boom, to help counter the weight of the load and improve stability.

Crane mat: A portable platform, typically made of large wooden timbers bolted together, used to support and spread the weight of a crane over a larger ground area.

Crawler frames: Crane assemblies comprised of the crawler tracks, track idlers, and track power sources of a crawler crane. Also called *tread members* or *track assemblies*.

Critical lift: Defined in *ASME Standard B30.5* as a hoisting or lifting operation that has been determined to present an increased level of risk beyond normal lifting activities. For example, increased risk may relate to personnel injury, damage to property, interruption of plant production, delays in schedule, release of hazards to the environment, or other significant factors.

Dedicated spotter: An individual qualified as a signal person who is charged with monitoring the separation between power lines and the equipment, load line, and load, so that the minimum approach distance is not compromised per OSHA standards.

Diver tender: One or more individuals assigned to attend to a diver's needs, including providing assistance in equipment preparation and managing the diver's cables and hoses.

Duty cycle: An expression of equipment use over time. In the case of mobile cranes, an 8-, 16-, or 24-hour rating expressed as a percentage.

Dynamic loads: A load on a structure (in this case, a crane) that is not constant, but consistently changing as the result of one or more changes in various factors. Also referred to as shock loading, significant dynamic loads can be applied to a crane through abrupt motions and lifting a load from its support too quickly.

Effective weight: The weight of an accessory such as a boom extension or jib that reflects the effect of its weight on the lift, usually based on its position, rather than its actual weight. For example, a jib folded and stored on the main boom will have different effective weight than when it is installed on the main boom tip.

Equalizer beams: Beams used to distribute the load weight on multi-crane lifts. The beam attaches to the load below, with two or more cranes attached to lifting eyes on the top.

Equalizer plates: A type of rigging plate that has three or more holes, used to level loads when sling lengths are unequal.

Fall zone: Defined in OSHA Standard 1926.1401 as the area beneath a load in which it is reasonable foreseeable that partially or completely suspended materials could fall in the event of an accident.

Floats: The portion of outriggers that touches the ground; the feet of the outriggers.

Frequent inspection: Required recurring crane inspections conducted at daily to monthly intervals, directed by *ASME Standard B30.5*. Frequent inspections are limited in scope and focused primarily on safety.

Gantry: A framed overhead structure supported by legs on each end, used to cross over obstructions. Gantries can be portable or permanent, providing support for hoisting equipment or raising and supporting lighting, cameras, and similar equipment.

Grapples: Devices used to pick up bulk items, containers, rocks, trees and tree limbs, etc. Grapples typically have several jaws that operate like fingers to pick up material, using mechanical or hydraulic power.

Gross capacity: The total amount a crane can safely lift under a given set of conditions. The gross capacity includes but is not limited to the load block, ropes, and rigging as well the primary load.

Hardpan: A hard, compacted layer of subsoil, usually with a major clay component.

Hauling line: The portion of a rope or chain on hoisting equipment that the operator uses to raise or lower the load. Also known as a *hauling part*.

Headache ball: A heavy round weight often attached to a load line to provide sufficient weight to allow the load line to unspool from the drum when there is no live load. Larger versions of headache balls are used to swing into structures to demolish them.

High-voltage proximity warning device: An early-warning device that senses the electric fields created by high-voltage power lines and alerts the crane operator and/or the lift team to the hazard.

Hoist drum: A drum is a cylindrical component around which a rope is wound. The hoist drum is used to wind or unwind the rope for hoisting or lowering the load; the part of a crane that spools and unspools the lifting line.

Hoist pawl: A hinged or pivoting device designed to interlock in the hoist mechanism and physically prevent its movement.

Hoist reeving: The reeving pattern applied to the hoist sheaves. Single- or multiple-line hoist reeving is used for whip, boom, and jib lines.

Hydraulic motors: Motors powered by hydraulic pressure provided by an external pump. Hydraulic motors are often used to power the tracks of crawler cranes, instead of complex drive systems connected directly to the diesel engine.

Idlers: Pulleys, wheels, or rollers that do not transmit power, but guide or place tension on a belt or crawler-crane track.

Impact loads: The dynamic effects on a stationary or mobile body as imparted by the forcible contact of another moving body or the sudden stop of a fall.

Independent wire rope core (IWRC): Wire rope with a core consisting of wire rope, as opposed to a fiber or single-stranded core; considered to be the most durable for rigging applications.

Initial inspection: An inspection required by *ASME Standard B30.5* on new or significantly altered cranes, conducted by a qualified person, to ensure the crane's compliance with ASME standards.

Insulating link: An electrical insulating device used on the crane hook to protect workers in contact with the load from the danger of electrocution in the event the crane contacts a powerline. The link can also provide some level of protection for the crane if the load alone contacts a power line.

Interpolation: The process of estimating or calculating unknown values between two known values.

Jib backstay: A piece of standing rigging that is routed from the jib mast back to the main boom to help support the jib.

Jib forestay: A piece of standing rigging that is routed from the far tip of the jib back to the jib mast, holding the tip of the jib up.

Jib mast: A structure mounted on the main boom that provides a fixed distance for the point of connection of the jib forestay and jib backstay. Also referred to as a *jib strut*.

Jibs: Extensions attached to the boom point to provide added boom length for reaching and lifting loads. Jibs may be in line with the boom, offset to another angle, or adjustable to a variety of angles. A jib is sometimes referred to as a *fly*.

Lattice boom: A boom constructed of steel angles or tubing to create a relatively light-weight but strong, rigid structure.

Leads: Steel structures that provide support for a pile hammer and help to align and position the hammer with the pile to be driven. The hammer can travel up or down in the leads as necessary.

Leverage: The mechanical advantage in power gained by using a lever.

Lift director: As defined in ASME Standard B30.5, the lift director directly oversees the work being performed by a crane and the associated rigging and signaling crew. The lift director is responsible for addressing all hazards related to the lift as well as those brought forward by the crane operator, signal persons, riggers, or other members of the crew.

Line of sight: The straight-line path between an observer's eyes and the thing being observed.

Load moment: The force applied to the crane by the load; the leverage of the load, opposing the leverage of the crane. The load moment is calculated by multiplying the gross load weight by the horizontal distance from the tipping fulcrum to the center of gravity of the suspended load. The load moment is usually reported to the operator as a percentage of the crane's capacity at the present set of conditions. As those conditions change, such as the boom angle, the load moment changes as well.

Load rating chart: One or more charts associated with each individual crane model that specify the crane's capacity at various distances and angles, and in various configurations. Per *ASME Standard B30.5*, Section 5.1.1.3, load charts should include a "full and complete range of manufacturer's crane load ratings at all stated operating radii, boom angles, work areas, boom lengths and configurations, jib lengths and angles (or offset), as well as alternate ratings for the use and nonuse of optional equipment on the crane, such as outriggers and extra counterweights, which affect ratings."

Lowboy: A trailer with a low frame for transporting very tall or heavy loads. A typical lowboy has two drops in deck height: one right after the gooseneck connecting it to the tractor, and one right before the wheels. This allows the trailer deck to be extremely low compared with common trailers.

Luffing: Changing a boom angle by varying the length of the suspension ropes.

Luffing jib: A jib mounted on the end of a boom that can be positioned at different angles relative to the main boom.

Master Trainer: An individual certified in accordance with NCCER's Master Trainer Instructor Certification Training Program (MT-ICTP) and authorized to train NCCER Craft Instructors, Performance Evaluators, and Proctors.

Minimum breaking strength (MBS): The amount of stress required to bring a rigging component to its breaking point. The MBS is a factor in determining a components' rated load capacity.

Minimum clearance distance: The OSHA-required distance that cranes, load lines, and loads must maintain from energized power lines. This OSHA term is synonymous with the ASME term prohibited zone.

National consensus standard: As defined in 29 *CFR* 1910.2(g), "any standard or modification thereof which (1) has been adopted and promulgated by a nationally recognized standards-producing organization under procedures whereby it can be determined by the Secretary of Labor or by the Assistant Secretary of Labor that persons interested and affected by the scope or provisions of the standard have reached substantial agreement on its adoption, (2) was formulated in a manner which afforded an opportunity for diverse views to be considered, and (3) has been designated as such a standard by the Secretary or the Assistant Secretary, after consultation with other appropriate Federal agencies."

Net capacity: The weight of the item(s) that can be lifted by the crane; the gross capacity of a crane minus all noted capacity deductions.

Non-ferrous: Having no iron. Ferrous metals, such as steel, contain iron and are magnetic as a result.

Nonverbal communication: All communication that does not use words. This includes appearance, personal environment, use of time, and body language.

North American Crane Bureau, Inc. (NACB): NACB's responsibility is to provide the most comprehensive services available in crane training and inspection services, while maintaining current state-of-the-art safety information applicable to crane operation. NACB is committed to providing the industry with a standard of excellence unparalleled in the field of safety. A division of the NACB, NACB Interactive Educational Systems, Inc., provides and distributes crane training simulators. NCCER and NACB work as partners in support of NCCER's Mobile Crane Operator Certification Program.

Occupational Safety and Health Administration (OSHA): A division of the US Department of Labor that is responsible for establishing and enforcing safety and health standards for all places of employment in the United States.

Open mike: In electronic communications, the condition where a radio's Transmit button or switch is held continuously without releasing it, even during pauses in speaking.

Open-throat boom: A lattice boom with an opening in the boom structure near the far end, allowing the hoist lines to drop through the boom rather than over the end of the boom.

Operating radius: The distance from the center of the boom's mounting point (usually the ring gear drive) to the center of gravity of the load.

Outriggers: Extendable or fixed members attached to a crane base that rest on ground supports at the outer end to stabilize and support the crane.

Paraphrasing: Expressing the perceived meaning of something read or heard in one's own words, generally to ensure clarity. Paraphrasing is an important component of active listening.

Parts of line: When a line is reeved more than once, the resulting number of lines that are supporting the load block.

Pendants: Ropes or strands of a specified length with fixed end connections, used to support a lattice boom or boom components. According to 29 *CFR* 1926.1401, a pendant may also consist of a solid bar.

Periodic inspection: Required recurring inspections conducted at monthly to annual intervals, directed by ASME Standard B30.5. The scope of these inspections is more complex than the scope of frequent inspections.

Practical examiner: An NCCER-certified individual with the appropriate qualifications and training to administer practical examinations for NCCER certification programs.

Pressure transducer: A device that measures pressure and then converts the pressure value to a proportional analog electrical signal; also referred to as a pressure transmitter.

Prohibited zone: An area of specific dimensions, based on the voltage of a power line(s) that no part of the crane is allowed to enter during normal operations. Special considerations and preparations are required if the crane's task must place any part of it within the prohibited zone. The prohibited zone is a term used by ASME that is synonymous with the term minimum clearance distance used by OSHA.

Quadrant of operation: The direction of the boom relative to the base mounting or carrier body.

Qualified person: As defined in 29 *CFR* 1926.1401, a person who, by possession of a recognized degree, certificate, or professional standing, or who by extensive knowledge, training, or experience, has successfully demonstrated the ability to solve problems relating to the subject matter, the work, or the project.

Rated load: The maximum working load permitted by a component manufacturer under a specific set of conditions. Alternate names for rated load include *working load limit (WLL)*, *rated capacity*, and *safe working load (SWL)*.

Reach: The combined operating height and radius of a boom, or the combination of boom and jib.

Recloser: A device that functions much like a circuit breaker, or in conjunction with a circuit breaker, in power distribution and transmission systems that automatically recloses the circuit after a fault has been detected and the circuit has been opened. Reclosers allow the power system to be re-energized quickly after a transient (temporary) condition, such as a tree limb falling across power lines and then falling to the ground, has occurred. If the fault reoccurs upon closure, the circuit will typically remain open until the situation has been addressed by power line workers or operators.

Reeving: A method often used to multiply the pulling or lifting capability by using wire rope routed through multiple pulleys or sheaves a number of times.

Regular inspection: Inspections required by *ASME Standard B30.5* on all mobile cranes in service. Regular inspections include both frequent and periodic inspections.

Rigger: An individual who specializes in the preparation and movement of often large and heavy objects. Riggers attach all types of loads to cranes and other equipment using cables, slings, hooks, and other specialized hardware, and also specialize in techniques used to move and position objects without the use of cranes and similar heavy equipment.

Rigging links: Links or plates with two holes used as termination hardware to appropriate lifting points.

Ring gear drive: Sometimes referred to as the swing circle. An assembly that provides the point of attachment and pivot point for the upperworks of a crane. The ring gear is typically driven by hydraulic pressure, allowing the upperworks to rotate on a set of bearings that reduce friction and transfer the weight of the upperworks (and any load) to the carbody.

Saddle: The portion of a hook directly below the center of the lifting eye.

Sheaves: Wheels that have a groove for a belt, rope, or cable to run in. The terms *sheave* and *pulley* are often used interchangeably.

Shock loading: A sudden, dramatically increased load imposed on a crane and rigging, usually as the result of momentum from the load that occurs due to swinging side-to-side, dropping the load and then stopping it suddenly, and similar actions that create momentum.

Site supervisor: As defined in *ASME Standard B30.5*, a site supervisor exercises supervisory control over the work site on which a crane is being used and over the work that is being performed on that site. Responsibilities include ensuring that a qualified person is designated as a lift director and being fully aware of any site conditions that may affect the lift, such as buried utilities.

Sling angle: The angle formed by the legs of a sling with respect to the horizontal plane when tension is placed on the rigging.

Spreader beams: Beams or bars used to distribute the load of a lift across more than one point to increase stability. Spreader beams are often used when the object being lifted is too long or large to be lifted from a single point, or when the use of slings around the load may crush the sides.

Spur track: A relatively short branch leading from a primary railroad track to a destination for loading or unloading. A spur is typically connected to the main at its origin only (a dead end).

Standard lift: A lift that can be accomplished through standard procedures, allowing load-handling and lift team personnel to execute it using common methods, materials, and equipment.

Standards: As defined by OSHA, statements that require conditions, or the adoption or use of one or more practices, means, methods, operations, or processes, that are reasonably necessary or appropriate to provide safe or healthful employment and places of employment. Standards developed by some organizations are voluntary in nature, while OSHA standards and those they incorporate by reference are enforceable by law.

Swallow: The space between the sheave and the frame of a block, through which the rope is passed.

Tagline: A rope attached to a lifted load for the purpose of controlling load spinning and swinging, or used to stabilize and control suspended attachments.

Telescopic boom: A crane boom that extends and retracts in sections that slide in and out, powered by hydraulic pressure.

Tilt-up construction: A building technique involving concrete walls. The walls are formed and poured at the construction site, creating large concrete panels. Each panel is then raised into position to form the exterior walls as the panels are joined.

Tipping fulcrum: The point of crane contact with the ground where it would pivot if it were to tip over; the fulcrum of the leverage applied by the load. Depending on the attitude and type of crane, the tipping fulcrum may be the edge of one crawler assembly, one or more outriggers, or similar locations.

Training credentials: Recognition provided by documents and records such as transcripts, certificates, and wallet cards that provide proof of a trainee's successful completion of specific training curricula through an NCCER-accredited organization.

Trucks: In relation to overhead cranes, a mechanical assembly consisting of a frame, wheels, bearings, and axles that support the bridge of an overhead crane and provide the ability for it to move along a set of parallel tracks.

Upperworks: A term that refers to the assembly of components above the ring gear drive; the rotating collection of components on top of the base mounting or carrier; may also be referred to as the *house,* or as the *superstructure* as defined in 29 *CFR* 1926.1401.

Vertical hitch: A simple hitch that uses one end of a sling to connect to a point on the load and the opposite end to connect to the hoist line. Also known as a *straight-line hitch.*

Wheelbase: The distance between the front and rear axles of a vehicle.

Whip line: A secondary hoisting rope usually of lower capacity than that provided by the main hoisting system. When a whip line exists, it is typically out at the tip of a jib, while the main hoist line is closer to the crane and operated from the tip of the main boom.

Index

Boom torque, (21102):1, 19, 50
Bottle jack, (38102):47
Bowline knot, (38102):33–34
Bridge, (53101):6, 21–22, 29
Bridge Travel, (53101):23
Bridle hitch, (38102):1, 5, 24–25, 54
    multi-leg, (38102):25
Buckets
    clamshell buckets, (21102):10–11, 14
    concrete buckets, (21102):11, 14
    drag buckets, (21102):11
    self-dumping bins, (21102):11, 14
Burke, Harold "Ed," (38102):52–53

C

Cameras, (21105):17
Capelli, Robert, (53101):27–28
Carbody, (21102):1, 3–4, 50
Carbon steel slings, (38102):28
Carriers
    defined, (21102):1, 3, 50
    rough-terrain cranes, (21102):6
    wheeled truck cranes, (21102):5
Carry-deck cranes, (21102):6
Center of gravity (CG)
    calculating, (38102):27–28
    defined, (21102):23, 50, (38102):1, 54
    lift capacity and, (21102):24
    suspending a load near, (38102):7
Certification, (21101):1, 4, 7–10, 21
CG. *See* Center of gravity (CG)
Chain
    applications, (38102):17
    characteristics, (38102):19
    common configurations, (38102):19
    inspection, (38102):20
    storage, (38102):20
Chain fall, (38102):41
Chain hoists
    characteristics, (38102):41–42
    electric, (38102):43, 44
    hook-style, (38102):43
    inspection and use of, (38102):43–44
    load capacity, (38102):43
    manual, (38102):43
    selection of, (38102):43–44
    spur-geared, (38102):42
    trolly-style, (38102):42–43
Check valve, (21102):1, 15, 50
Cherry picker, (21102):6
Choker hardware, synthetic web slings, (38102):16
Choker hitch, (38102):11, 12, 25–26, 54
    double-wrap, (38102):26
Clamps
    beam clamps, (38102):5
    function, (21102):12
    highway barrier, (21102):16
    plate clamps, (38102):5, 7
    serrated-jaw clamps, (38102):7
Clamshell buckets, (21102):10–11, 14
Clove hitch, (38102):34–36
Cofferdam, (21105):9, 17, 25
Cold-related hazards, (21106):18, (38102):38
Come-alongs, (38102):44–45
Communication
    barriers to
        abstractions, (53101):1, 3, 29
        environmental factors, (53101):4
        fear, (53101):3–4
        lack of common experience, (53101):2–3
        language, (53101):2–3
    devices
        cell phones, (53101):11
        DECT-compatible devices, (53101):11, 13
        hardwired systems, (53101):11–12
        portable radios, (53101):10–11
        walkie-talkies, (53101):11
    electronic, (53101):8
    filters, (53101):2
    flags, (21106):14
    nonverbal
        audible signals, (53101):12
        defined, (53101):1, 29
        flags, (21106):21, (53101):13
        horn signals, (53101):22
        warning devices, (21106):14–15
    paraphrasing, (53101):1, 2, 29
    receiver's role, (53101):1–2
    sender's role, (53101):1–2
    standards
        ASME standards, (53101):7
        OSHA standards, (53101):6–7
    verbal, (53101):10–12
    voice signals, (21106):7–8, (53101):8
Communication process
    feedback, (53101):1–2
    receiving the message, (53101):1–2
    sending the message, (53101):1
Competent person, (21106):1, 4, 33, (53101):10
Compound block and tackle, (38102):41, 42
Concrete buckets, (21102):11, 14
Consensus standard, (53101):6, 13, 29
Cooperation, success and, (21101):15
Counterweights, (21102):1, 3, 15, 50
Craft professionals, characteristics of successful
    being present and on time, (21101):15–16
    cooperative, (21101):15
    honesty, (21101):14–15
    loyalty, (21101):15
    professionalism, (21101):14
    responsible, (21101):15
    willingness to learn, (21101):15
Crane attachments
    buckets
        clamshell buckets, (21102):10–11, 14
        concrete buckets, (21102):11, 14
        drag buckets, (21102):11
        self-dumping bins, (21102):11, 14
    clamps, (21102):12, 16
    counterweights, (21102):1, 3, 15
    grapples, (21102):1, 12, 14, 50
    magnets, (21102):12, 15
    personnel platforms, (21102):14, 17
    pile drivers
        pile hammers, (21102):14–15
        vibratory, (21102):14–15, 18
    specialty lifting devices, (21102):12, 16
Crane mats, (21102):1, 26–27, 50, (21105):2, 4
Crane operations. *See also* Mobile cranes
    general practices
        load handling, (21105):17–18
        pre-start, (21105):16
        shutdown, (21105):19
    lift operations
        blind lift, (21105):1, 7, 25
        protecting the public, (21105):6, 7

idle equipment requirement, (21105):16
initial, (21105):9, 14, 25
manufacturer's recommendations, (21105):16
periodic, (21105):9, 15, 25
post-operation, (21105):19
regular, (21105):9, 15, 25
severe service requirement, (21105):16
shift requirements, (21105):15–16
standards, (21105):14
Insulating links, (21106):10, 14–15, 33, (38102):31
Interpolation, (21102):23, 51
ISO standards, *Hand Signals Used with Cranes*, (53101):3
IWRC. *See* Independent wire rope core (IWRC)

J

Jacks
bottle jack, (38102):47
hydraulic, (38102):45, 47
ratchet jacks, (38102):45, 46
screw jacks, (38102):45, 46
toe jack, (38102):47
use and maintenance, (38102):46
Jib, (21102):1, 12, 51
Jib backstay, (21102):2, 8, 51
Jib forestay, (21102):2, 8, 51
Jib mast, (21102):2, 9, 51

K

King of knots, (38102):33
Knots
bowline, (38102):33–34
center of gravity calculations, (38102):27–28
hitch
applications, (38102):23
basket hitch, (38102):11, 12, 26–27, 54
bridle hitch, (38102):1, 5, 24–25, 54
choker hitch, (38102):11, 12, 25–26, 54
clove hitch, (38102):34–36
common, (38102):23–24
double-wrap basket hitch, (38102):27
double-wrap choker hitch, (38102):26
half hitch, (38102):34, 35
multi-leg bridle hitch, (38102):25
single vertical hitch, (38102):24
vertical hitch, (38102):11, 15, 54
king of, (38102):33

L

Labels
RFID chips, (38102):14
wire-rope slings, (38102):12, 15
Laird, Richard, (21102):48–49
Language barriers, (53101):2–3
Lattice boom
configurations, (21102):8–10, 12
controls, (21102):7–8
defined, (21102):2, 51
sag, (21102):34
Lattice-boom cranes
anti-two-blocking devices, (21105):12
assembly/disassembly clearance, (21105):3–4
computer-controlled, (21102):7–8
crawler cranes, (21102):3, 6, 8, 12, 18
railcar-mounted cranes, (21102):6
setup requirements, (21102):8
truck-mounted, (21102):8, 11
wheeled truck cranes, (21102):5
Leads, (21102):2, 14, 51
Leverage, (21102):23, 51

LICCON System, (21102):17
Liebherr LTM 11200-9.1, (21101):5
Lift crane, (21102):10
Lift director, (21101):1, 5, 21, (21105):1, (53101):10
Lifting lugs, (38102):7
Lifts
operations
blind lift, (21105):1, 7, 25
protecting the public, (21105):6
proximity to other equipment, (21105):5–6
safety, (21105):7
preparations
assembly/disassembly clearance, (21105):3–4
ground conditions, assessing, (21105):2–3
site assessment, (21105):1–3
site conditions suitability, responsibility for, (21105):1
transit routes, (21105):5
Lift teams, (21105):7
Lightning, (21106):18, 20, (38102):38
Line of sight, (53101):6, 7, 29
Links, (38102):1, 8, 54
Load block, (21102):8
Load charts, (21102):36–40
Load-handling safety
ASME standards, (21106):3, 10–11
basics, (21105):17–18
mobile cranes, (21106):7
pre-lift meetings, (21106):10–11
sudden descent stops, (21106):7
Load moment, (21102):2, 8, 51
Load moment indicators, (21105):11, 12
Load rating chart, (21101):1, 21
Loads
dynamic, (21102):23, 36, 50
impact, (21102):23, 24, 51
normal working load (NWL), (38102):2
static, (21102):36
working load limit (WLL), (38102):1, 2
Lockout systems, (21105):12
Lowboy, (21102):2, 4, 6, 51
Lower, (53101):13
Lower Boom, (53101):15
Lower Boom and Raise the Load, (53101):15
Loyalty, success and, (21101):15
Luffing, (21102):2, 8, 51
Luffing jib, (21102):2, 8, 51
Luffing jib limiting devices, (21105):11

M

Magnet Disconnected, (53101):23
Magnets, (21102):12, 15
Master Trainer, (21101):1, 9, 21
Materials
delivery, (38102):28
labeling
RFID chips, (38102):14
slings, (38102):12, 14, 15
storage
chain, (38102):20
rebar, (38102):28
slings, (38102):12–14, 17, 20
Mats, (21102):1, 26–27, 50
MBS. *See* Minimum breaking strength (MBS)
Metal-mesh slings
advantages, (38102):20
ASME standards, (38102):20
characteristics, (38102):20
inspection, (38102):20
storage, (38102):20

Nonverbal communication, (21106):14, 21, (53101):1, 29
Normal working load (NWL), (38102):2
North American Crane Bureau, Inc. (NACB), (21101):1, 9, 22
NWL. *See* Normal working load (NWL)

O

Occupational Safety and Health Administration (OSHA), (21101):2, 22
Occupational Safety and Health Administration (OSHA) standards
  assembly/disassembly (A/D) director, (21105):3
  clearance, (21105):3
  cranes
    communications, (53101):6–7
    Cranes and Derricks in Construction, (21101):3, (21106):3
    inspections, (21105):9, 14, 15
    safety, (21102):15, 17, (21106):1–3
    supporting divers in the water, (53101):9
  Crawler Locomotive and Truck Cranes, (21101):2
  dedicated spotters, (53101):9–10
  electronic communications, (53101):8
  fire extinguishers, (21106):24
  hand signal charts, (53101):8
  hand signals, (53101):3, 7, 13, 21–22
  Hoisting Personnel, (53101):9
  lift directors, (53101):10
  mobile crane operators, qualification and certification, (21101):8
  operational aids, (21105):10
  personnel lifts, (21106):21–22
  power lines, working around, (21106):12–13, 14, 15, (53101):7
  Power Line Safety, (53101):7
  safety devices, (21105):10
  signal flags, (53101):13
  signal persons, (53101):7, 9
  site conditions suitability, responsibility for, (21105):1
  voice signals, (53101):8
Office of Apprenticeship (DOL), (21101):12
On rubber, (21102):26
On-the-job learning, (21101):12, 13
Open mike, (53101):6, 29
Open-throat boom, (21102):2, 9, 51
Operating radius, (21102):23, 24, 51
Operator cab
  crawler cranes, (21102):3
  fixed cab, (21102):6
  swing cab, (21102):6
OSHA. *See* Occupational Safety and Health Administration (OSHA)
Outrigger floats, blocking, (21105):2, 3, 4
Outriggers
  bearing surfaces, (21102):26, 28
  crossbeam, (21102):8
  defined, (21102):2, 51
  function, (21106):7
  pick-and-carry cranes, (21102):6
  wheeled truck cranes, (21102):6
Overhead cranes
  bridge, (53101):21–22
  hand signals
    Bridge Travel, (53101):23
    Magnet Disconnected, (53101):23
    Multiple Trolleys, (53101):23
    Trolley Travel, (53101):21
  trucks, (53101):6, 22, 29
  types of, (21101):2

P

Paraphrasing, (53101):1, 2, 29
Parts of line, (21102):2, 20, 51, (38102):41, 54
Pendants, (21102):2, 8, 52
Periodic inspection, (21105):9, 15, 25
Personal protective equipment (PPE)
  crane safety, (21106):4
  gloves, (21106):4, 5, (53101):22
  signal persons, (53101):22
Personnel lifts/platforms, (21102):14, (21106):21–23
Pick-and-carry cranes, (21102):6, 26
Pile drivers, vibratory, (21102):14–15, 18
Pile hammers, (21102):14–15
Pipe
  delivery, (38102):28
  hooks, (21102):16
  rigging and handling, (38102):28–30
  storage, (38102):28
  weight of, determining, (38102):29
Pipe clamp, (21102):16
Plate clamps, (38102):5, 7
Plates
  equalizer plates, (38102):8
  rigging plates, (38102):8
Power lines, working around
  avoidance zone, (21106):12–13
  de-energized, grounded power lines, (21106):13–14
  energized lines, (21106):14, 15–16
  high-voltage proximity warning device, (21106):10, 14–15, 33
  insulating links, (21106):10, 14–15, 33, (38102):31
  line contact emergency procedures, (21106):16–18
  minimum clearance distance, (21106):12, 13, 16
  owner involvement, (21106):13, 16
  prohibited zones, (21106):12–13, 14, 15–16
  spotters, (21106):14, 15, (53101):7, 9–10
  standards
    ASME standards, (21106):16
    OSHA standards, (21106):12–13, 14, 15, (53101):7
  in transit with no load and boom lowered, (21106):13, 16
  visual aids, (21106):14
  warning lines with flags, (21106):14
PPE. *See* Personal protective equipment (PPE)
Practical examiner, (21101):2, 9, 22
Pressure transducer, (21105):9, 13, 25
Proemsey, Larry "Cowboy," (21105):23–24
Professionalism, success and, (21101):14
Prohibited zone, (21106):10, 12, 33

Q

Quadrant of operation, (21102):23, 24, 52
Qualified person, (21101):2, 4, 22

R

Railcar-mounted cranes
  booms, (21102):6–7
  self-propelled, (21102):6–7
Railroad jack, (38102):45
Rain, (21106):18
Raise Boom, (53101):14
Raise Boom and Lower the Load, (53101):14
Ratchet jacks, (38102):45, 46
Ratchet-lever hoists, (38102):44–45
Rated load, (38102):1, 54
Reach, (21102):2, 3, 52